Advances in Experimental Medicine and Biology

More information about this series at http://www.springer.com/series/5584

John D. Lambris • Kristina N. Ekdahl
Daniel Ricklin • Bo Nilsson
Editors

Immune Responses to Biosurfaces

Mechanisms and Therapeutic Interventions

Springer

Editors
John D. Lambris
Perelman School of Medicine
University of Pennsylvania
Philadelphia, PA, USA

Daniel Ricklin
Perelman School of Medicine
University of Pennsylvania
Philadelphia, PA, USA

Kristina N. Ekdahl
Immunology, Genetics and Pathology
University of Uppsala
Uppsala, Sweden

Bo Nilsson
Immunology, Genetics and Pathology
University of Uppsala
Uppsala, Sweden

ISSN 0065-2598 ISSN 2214-8019 (electronic)
Advances in Experimental Medicine and Biology
ISBN 978-3-319-18602-3 ISBN 978-3-319-18603-0 (eBook)
DOI 10.1007/978-3-319-18603-0

Library of Congress Control Number: 2015943674

Springer Cham Heidelberg New York Dordrecht London

Printed on acid-free paper

Springer International Publishing AG Switzerland is part of Springer Science+Business Media
(www.springer.com)

Preface

This book highlights contemporary and emerging aspects of improving medical treatment modalities employing biomaterials or transplantation. During such treatments, different biosurfaces (i.e., the surfaces of the medical devices or transplanted cells or organs) inevitably come in contact and interact with human blood and tissues. Such interactions frequently trigger activation of multiple defense systems such as the complement, contact, and coagulation cascades and contribute to anaphylactoid reactions, ischemia-reperfusion injury, thrombo-inflammation, and immune responses that negatively affect the clinical outcome. Classical examples of high clinical importance are biomaterial implants, extracorporeal circuits, bioengineered devices (e.g., drug delivery vehicles), soft and hard tissue implants, as well as transplantation of cells (e.g., mesenchymal stromal cells or hepatocytes), cell clusters (primarily islets of Langerhans), or whole vascularized organs. Optimal tissue integration and modulation of foreign body reactions are therefore essential for preserving anticipated functions and avoiding adverse effects. Modification of biosurfaces or pharmaceutical interventions are viable strategies that already produced successful results in some cases. However, biosurface-induced complications such as rejection, local and systemic inflammation, and thrombosis remain major problems in the clinic, thereby fueling a demand for novel surface-modification strategies and therapeutic modalities.

This volume compiles data on this rapidly growing field as presented by prominent scientists at the First International Conference on Immune Response to Biosurfaces (September 27–October 2, 2014) in Chania, Greece. Topics covered in this book include mechanistic and applied research within the fields of extracorporeal devices, soft and hard tissue implants, tissue and biomaterial targeting, therapeutic modulation of foreign body reactions, cell encapsulations, as well as cell and whole organ transplantation. We would like to express our sincerest thanks to all the authors for contributing timely and highly informative chapters on this fascinating and emerging topic of modern medicine. We also thank Dimitrios Lambris and Aegean Conferences, Inc., for all the help provided in organizing this exceptional meeting, and Brian Halm of Springer for his supervision of this book's production.

Philadelphia, PA, USA — John D. Lambris
Uppsala, Sweden — Kristina N. Ekdahl
Philadelphia — Daniel Ricklin
Uppsala, Sweden — Bo Nilsson

Contents

Contributors

Toshiharu Abe Department of Microbiology, Penn Dental Medicine, University of Pennsylvania, Philadelphia, PA, USA

Jan-Michael Abicht Department of Anaesthesiology, LMU, Munich, Germany

Ida Arvidsson Department of Pediatrics, Clinical Sciences Lund, Lund University, Lund, Sweden

Sana Asif Department of Immunology, Genetics and Pathology (IGP), Rudbeck Laboratory C5:3, Uppsala University, Uppsala, Sweden

Zivile Békássy Department of Pediatrics, Clinical Sciences Lund, Lund University, Lund, Sweden

Ali-Reza Biglarnia Surgical Science, Department of Transplantation, Uppsala University Hospital, Uppsala University, Uppsala, Sweden

Paolo Brenner Department of Cardiac Surgery, LMU, Munich, Germany

Rolf E. Brenner Division for Biochemistry of Joint and Connective Tissue Diseases, Department of Orthopedics, University of Ulm, Ulm, Germany

Diane J. Burgess Department of Pharmaceutical Sciences, University of Connecticut, Storrs, CT, USA

Marilia Cascalho Transplantation Biology, Department of Surgery, University of Michigan, Ann Arbor, MI, USA

Department of Microbiology & Immunology University of Michigan, Ann Arbor, MI, USA

Triantafyllos Chavakis Department of Clinical Pathobiochemistry, Medical Faculty, Technische Universität Dresden, Dresden, Germany

Kristina N. Ekdahl Immunology, Genetics and Pathology, University of Uppsala, Uppsala, Sweden

Peter Garred Laboratory of Molecular Medicine, Department of Clinical Immunology, Section 7631, Rigshospitalet, Faculty of Health and Medical Sciences, University Hospital of Copenhagen, Copenhagen O, Denmark

Pierre Gianello Université catholique de Louvain, SSS/IREC/CHEX, Brussels, Belgium

Sonja Guethoff Walter Brendel Centre of Experimental Medicine, Ludwig-Maximilians-Universität (LMU), Munich, Germany

George Hajishengallis Department of Microbiology, Penn Dental Medicine, University of Pennsylvania, Philadelphia, PA, USA

Evlambia Hajishengallis Department of Restorative and Preventive Sciences, Penn Dental Medicine, University of Pennsylvania, Philadelphia, PA, USA

Estrid Hein Laboratory of Molecular Medicine, Department of Clinical Immunology, Section 7631, Rigshospitalet, Faculty of Health and Medical Sciences, University Hospital of Copenhagen, Copenhagen O, Denmark

Markus Huber-Lang Department of Orthopedic Trauma, Hand-, Plastic-, and Reconstructive Surgery, University of Ulm, Ulm, Germany

Anita Ignatius Institute of Orthopedic Research and Biomechanics, University of Ulm, Ulm, Germany

Karl Johansson Department of Pediatrics, Clinical Sciences Lund, Lund University, Lund, Sweden

Geoffrey B. Johnson Department of Radiology, Mayo Clinic, Rochester, MN, USA

Department of Immunology, Mayo Clinic, Rochester, MN, USA

Nina Jonsson Department of Immunology, Genetics and Pathology (IGP), Rudbeck Laboratory C5:3, Uppsala University, Uppsala, Sweden

Linnæus Center of Biomaterials Chemistry, Linnæus University, Kalmar, Sweden

Robin Kahn Department of Pediatrics, Clinical Sciences Lund, Lund University, Lund, Sweden

Diana Karpman Department of Pediatrics, Clinical Sciences Lund, Lund University, Lund, Sweden

Michail Kastellorizios Department of Pharmaceutical Sciences, University of Connecticut, Storrs, CT, USA

Alexander Kind Livestock Biotechnology, Technical University Munich, Munich, Germany

Klara Kotlabova Department of Clinical Pathobiochemistry, Medical Faculty, Technische Universität Dresden, Dresden, Germany

Ioannis Kourtzelis Department of Clinical Pathobiochemistry, Medical Faculty, Technische Universität Dresden, Dresden, Germany

Ann-Charlotte Kristoffersson Department of Pediatrics, Clinical Sciences Lund, Lund University, Lund, Sweden

Themis R. Kyriakides Vascular Biology and Therapeutics Program, Yale University, New Haven, CT, USA

Biomedical Engineering, Yale University, New Haven, CT, USA

Department of Pathology, Yale University School of Medicine, New Haven, CT, USA

John D. Lambris Perelman School of Medicine, University of Pennsylvania, Philadelphia, PA, USA

Sebastian Loos Department of Pediatrics, Clinical Sciences Lund, Lund University, Lund, Sweden

Barbara Ludwig Department of Medicine III, Technical University Dresden, Dresden, Germany

Tomoki Maekawa Department of Microbiology, Penn Dental Medicine, University of Pennsylvania, Philadelphia, PA, USA

Peetra U. Magnusson Department of Immunology, Genetics and Pathology (IGP), Rudbeck Laboratory C5:3, Uppsala University, Uppsala, Sweden

Tanja Mayr Walter Brendel Centre of Experimental Medicine, Ludwig-Maximilians University (LMU), Munich, Germany

Laura Beth Moore Department of Genetics, Yale University, New Haven, CT, USA

Vascular Biology and Therapeutics Program, Yale University, New Haven, CT, USA

Maria Mossberg Department of Pediatrics, Clinical Sciences Lund, Lund University, Lund, Sweden

Bo Nilsson Immunology, Genetics and Pathology, University of Uppsala, Uppsala, Sweden

Jeffrey L. Platt Transplantation Biology, Department of Surgery, University of Michigan, Ann Arbor, MI, USA

Department of Microbiology & Immunology, University of Michigan, Ann Arbor, MI, USA

Thomas Poettinger Department of Cardiac Surgery, LMU, Munich, Germany

Bruno Reichart Walter Brendel Centre of Experimental Medicine, Ludwig-Maximilians University (LMU), Munich, Germany

Daniel Ricklin Perelman School of Medicine, University of Pennsylvania, Philadelphia, PA, USA

Anne-lie Ståhl Department of Pediatrics, Clinical Sciences Lund, Lund University, Lund, Sweden

Ramesh Tati Department of Pediatrics, Clinical Sciences Lund, Lund University, Lund, Sweden

Yuji Teramura Department of Bioengineering, The University of Tokyo, Bunkyo-ku, Tokyo, Japan

Namita Tipnis Department of Pharmaceutical Sciences, University of Connecticut, Storrs, CT, USA

Eckhard Wolf Chair for Molecular Animal Breeding and Biotechnology, Gene Center, LMU, Munich, Germany

Lucile E. Wrenshall Department of Surgery, Wright State University, Dayton, OH, USA

Department of Neuroscience, Cell Biology and Physiology, Wright State University, Dayton, OH, USA

Part I
Mechanisms

Chapter 1
Thromboinflammation in Therapeutic Medicine

Kristina N. Ekdahl, Yuji Teramura, Sana Asif, Nina Jonsson, Peetra U. Magnusson, and Bo Nilsson

Abstract Thromboinflammation is primarily triggered by the humoral innate immune system, which mainly consists of the cascade systems of the blood, i.e., the complement, contact/coagulation and fibrinolytic systems. Activation of these systems subsequently induces activation of endothelial cells, leukocytes and platelets, finally resulting in thrombotic and inflammatory reactions. Such reactions are triggered by a number of medical procedures, e.g., treatment with biomaterials or drug delivery devices as well as in transplantation with cells, cell clusters or whole vascularized organs. Here, we (1) describe basic mechanisms for thromboinflammation; (2) review thromboinflammatory reactions in therapeutic medicine; and (3) discuss emerging strategies to dampen thromboinflammation.

Keywords Coagulation • Complement • Extracorporeal treatment • Platelets • Therapeutic medicine • Thromboinflammation

---

K.N. Ekdahl (✉) • B. Nilsson
Immunology, Genetics and Pathology, University of Uppsala, Uppsala, Sweden
e-mail: Kristina.Nilsson_Ekdahl@igp.uu.se

Y. Teramura
Department of Bioengineering, The University of Tokyo
7-3-1 Hongo, Bunkyo-Ku, Tokyo 113-8656, Japan

S. Asif • P.U. Magnusson
Department of Immunology, Genetics and Pathology (IGP), Rudbeck Laboratory C5:3, Uppsala University, 751 85 Uppsala, Sweden

N. Jonsson
Department of Immunology, Genetics and Pathology (IGP), Rudbeck Laboratory C5:3, Uppsala University, 751 85 Uppsala, Sweden

Linnæus Center of Biomaterials Chemistry, Linnæus University, 391 82 Kalmar, Sweden

J.D. Lambris et al. (eds.), *Immune Responses to Biosurfaces*,
Advances in Experimental Medicine and Biology 865,
DOI 10.1007/978-3-319-18603-0_1

1.1 Introduction

Innate immunity is fundamental to our defense against microorganisms and foreign substances and controls the discrimination between self and non-self structures in the human body. As a consequence of its properties and actions, it is responsible for many of the incompatibility reactions that occur when foreign substances, materials, cells, and organs are introduced into the body. These reactions pose a major problem when modern biotechnological treatment modalities are used, including biomaterial devices, drug delivery systems, various bioengineered implants, cell therapies, and transplantation. Furthermore, the intravascular innate immune system can cause severe side effects in patients, as well as rejection and dysfunction of implanted devices and tissues, as reviewed in [1].

1.2 Innate Immune Recognition in Thromboinflammation

1.2.1 Recognition Mechanisms and Activation of the Cascade Systems of the Blood

The intravascular innate immune system acts as a purging system: It identifies and removes foreign substances, including microorganisms, apoptotic cell debris, and foreign bodies/materials and orchestrates subsequent immune/thromboinflammatory responses. It consists of the cascade systems of the blood (complement, contact, coagulation, and fibrinolysis systems), blood cells (polymorphonuclear cells [2] monocytes, platelets), and endothelial cells [3]. When innate immune reactions occur on the endothelial cell surface, they also involve overcoming the actions of endogenous anti-inflammatory and thrombotic agents, such as endothelial-derived developmental endothelial locus-1 (Del-1) and the NTPDase CD39, which antagonize, respectively, leukocyte adhesion to and platelet aggregation on the endothelium [4, 5]. Because of this interplay, innate immunity is a key contributor to the adverse effects that have been observed in many therapies using biomaterials and therapeutic cells/organs [1, 6, 7].

1.2.2 The Complement System

The unique capacity of complement to sense, differentiate between, and distinctively react to healthy but injured to apoptotic self-cells and foreign intruders relies on an intricate interplay between pattern recognition molecules, protein scaffolds, enzymes, regulators, and cell-surface receptors. The hub-like organization of complement and its cell surface-directed action, involving ~50 constituents, is essential for adjusting the complement response to various triggers. When the body is faced with foreign intruders, recognition of molecular surface patterns triggers distinct initiation pathways: In the classical pathway (CP), C1q binds to antibody patches, C-reactive protein (CRP), and other cell markers; in the lectin pathway (LP), microbial carbohydrates are recognized by mannose-binding lectin (MBL) or ficolins in complex with MBL-associated serine proteases (MASP). Both pathways lead to the assembly of C3 convertases, which cleave component C3 into the anaphylatoxin C3a and the opsonin C3b. Once deposited on target surfaces, C3b fuels an amplification loop via the alternative pathway (AP), forming additional C3 convertases with factors B and D (FB, FD). Continuous deposition of C3b favors the generation of C5 convertases that convert component C5 into C5b, which then forms membrane attack complexes (MAC) that lyse susceptible cells. Cleavage of C5 also releases the anaphylatoxin C5a, which, together with C3a, attracts immune cells to sites of activation via binding to the anaphylatoxin receptors C5aR1 and C3aR. Phagocytic cells recognize opsonized surfaces via complement receptors (CR1, CR3, CR4, CRIg) that bind to C3b and its degradation products iC3b and C3dg. iC3b and C3dg also interact with CR2, which is part of the B-cell co-receptor complex, and reduces the threshold for B-cell and dendritic cell activation. Host cells are protected from autologous complement attack by expressing membrane-bound regulators that either destabilize convertases or act as cofactors for the factor I-mediated degradation of C3b to iC3b and C3dg. In addition, the soluble regulators of complement activation (RCAs) C4b-binding protein (C4BP) and factor H (FH) recognize host cell-surface patterns and contribute to the regulation of the CP/LP and AP convertases. Finally, the regulators CD59, vitronectin, and clusterin prevent the formation of the lytic MAC or C5b-9 complex on host cells. Apoptotic cells induce yet another response that lies in between the one observed for foreign cells and that for host cells: While the detection of surface modifiers on apoptotic cells by pattern recognition molecules induces opsonization, the presence of regulators prevents excessive amplification and the generation of C5a or the MAC. Thus, complement facilitates the elimination of apoptotic cells, immune complexes, and cellular debris without inducing inflammatory triggers [8] (Figs. 1.1 and 1.2).

1.2.3 The Contact/Kallikrein System

Binding of factor (F) XII to a material surface induces a conformational change in the protein, ultimately leading to its transformation into an active-enzyme form, FXIIa, through a process known as autoactivation. FXIIa generated at the surface can, in turn, cleave prekallikrein to kallikrein in the form of a complex with high molecular weight kininogen (HMWK). This mutual activation of prekallikrein and FXII creates an amplification loop. Ultimately, FXIIa activates surface-bound FXI as a complex with HMWK to generate FXIa, leading to the propagation of subsequent coagulation cascade reactions and thrombin formation. In addition, the fibrin clots formed are able to promote further activation of FXII, thereby creating another amplification loop. Most of the generated proteases are subsequently regulated by the C1 inhibitor (C1-INH) or antithrombin (AT) [9–11].

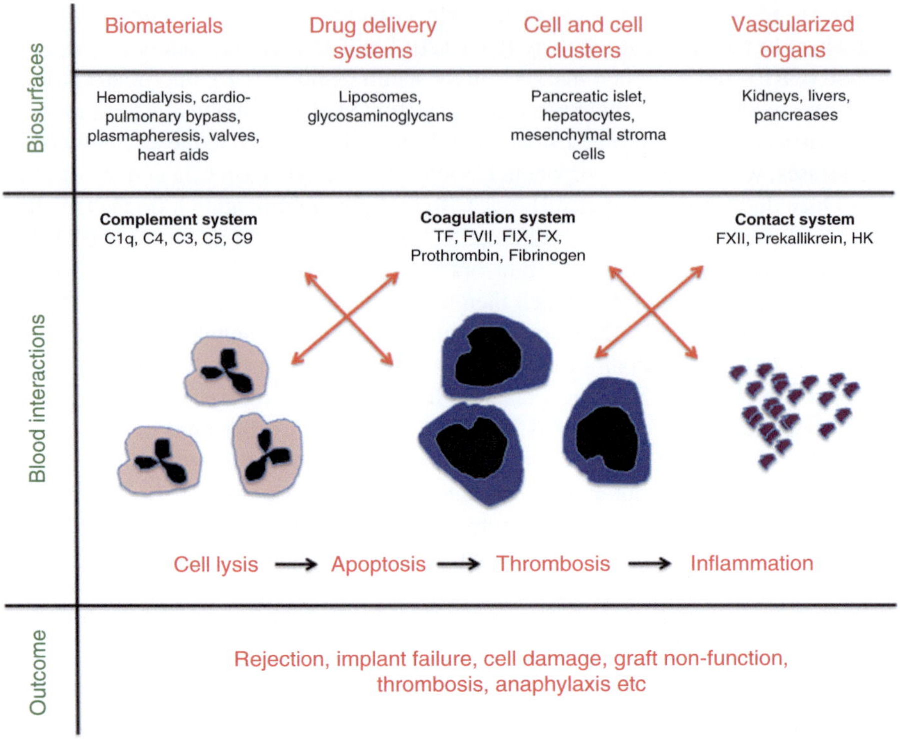

Fig. 1.1 Biosurfaces in therapeutic medicine. Incompatibility reactions triggered by innate immune responses to altered-self and non-self structures on biomaterials, drug delivery devices, cells, or cell clusters, and vascularized organs for therapeutic use. Upon exposure to blood, recognition molecules belonging to different cascade systems target altered-self and non-self structures on biomaterials and cells. C1q, mannose-binding lectin (MBL), and properdin are "recognition/trigger" molecules in the complement system, tissue factor (TF), factor (F) VII, and fibrinogen in the TF pathway of the coagulation system, and FXII in the contact system. The activation of each cascade system triggers amplification reactions. In the complement cascade, there is a powerful amplification of C3 that initiates the generation of the anaphylatoxins C3a and C5a as well as the lytic C5b-9 complex. The generated activation products in turn trigger the activation of platelets, PMNs, and monocytes/macrophages, resulting in thrombotic and inflammatory reactions. Activation of the coagulation cascade leads to the generation of thrombin from prothrombin. Further activation of the contact system elicits the generation of the potent vasoactive peptide bradykinin from high molecular weight kininogen (HMWK). These adverse events, together with complement-mediated cell lysis and coagulation-mediated sequestration, may lead to rejection or serious damage to the biomaterial or transplanted graft

1.2.4 Interaction Between Platelets, Complement, and Contact System Proteins

Apart from their role in primary hemostasis, platelets also have a vital role as innate immune cells, bringing together constituents of the complement and contact/kallikrein systems and ultimately producing inflammation. One important player in this

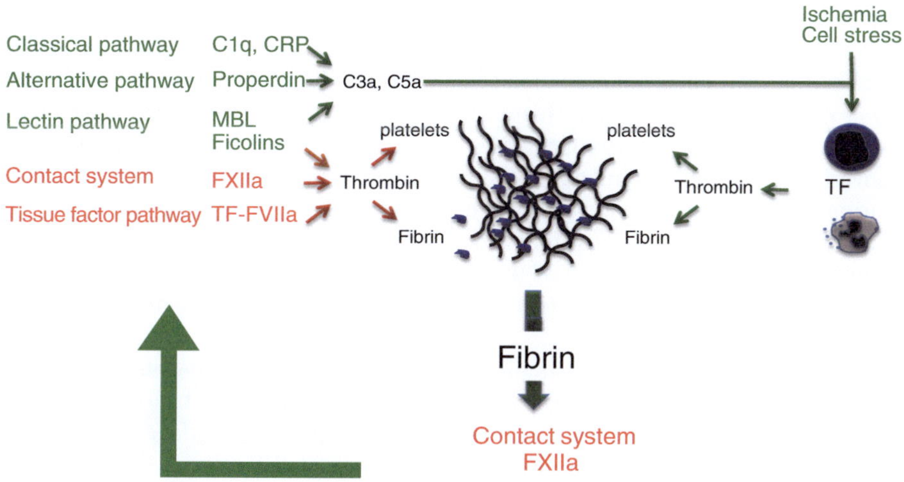

Fig. 1.2 The concept of thromboinflammation. Recognition molecules C1q and C-reactive protein (CRP); properdin, mannan-binding lectin (MBL) and the ficolins of the three activation pathways of the complement system; and factor (F) XII, FVII, and tissue factor (TF) of the two activation pathways of the coagulation system initiate the activation of the various cascade systems of the blood, resulting in the generation of the anaphylatoxins C3a and C5a (which activates PMNs and monocytes) and thrombin (which elicits platelet activation and fibrin formation). In addition, ischemia or cell stress induces the expression of TF on monocytes and other cells of various tissues, which leads to further thrombin generation. The fibrin formed induces the activation of FXII of the contact system, thereby establishing an amplification loop. The ultimate result is a thrombus, in which activated leukocytes and platelets are trapped in a fibrin network

context is chondroitin sulfate A (CS-A), which is released from the α-granules during platelet activation. CS-A is a potent activator of complement in the fluid phase [12] and also mediates the binding of multiple complement proteins to the platelet surface [13]. Of particular interest is the fact that C3 binds to the surface of activated platelets in a conformationally altered conformation, $C3H_2O$, which is not a result of activation by either of the C3 convertases or other proteolytic enzymes [14]. In addition, platelet activation induces the activation of FXI and FXII within the contact activation system; when activated by platelets, these activated proteases are preferentially inactivated by AT, rather than by C1-INH [9, 10, 15].

1.3 Thromboinflammation: The Crosstalk of Complement, Coagulation, and Platelets with Leukocytes and the Endothelium

The intravascular innate immune response is largely mediated by thromboinflammatory reactions that take place on biomaterial or endothelial cell surfaces. The endothelial surface is constitutively anti-thrombotic and has anti-adhesive and anti-inflammatory properties, mainly because of its surface composition of proteoglycans and phospholipids found within the glycocalyx [16]. However, in the course of intravascular inflammation (e.g., whole-body inflammation after hemodialysis (HD) or in ischemia-reperfusion injury or transplant vasculopathy), the endothelium is activated by cytokines, C5a, or the insertion of the MAC and thereby converted to a pro-coagulatory and pro-adhesive/pro-inflammatory state [17]. During ischemia-reperfusion injury, metabolites produced during the ischemic phase by the vasculature as well as reactive oxygen species (ROS) from recruited neutrophils cause the shedding of the protective glycocalyx [18], leading to exposure of endothelial adhesion proteins, followed by additional innate immune activation and cellular infiltration [19].

Also, endothelial thrombomodulin is downregulated, reducing its endothelial anti-thrombotic properties [7]. The exocytosis of von Willebrand factor from endothelial granulae promotes platelet adhesion on the endothelial surface, whereas the upregulation of a multitude of adhesion molecules of the selectin and immunoglobulin superfamilies confers on the endothelium a pro-adhesive attraction for neutrophils and other leukocytes [20, 21]. The leukocyte-endothelial adhesion cascade is initiated by the selectin-mediated rolling of leukocytes along the endothelial cell surface, which brings the leukocytes into contact with chemokines deposited on the apical endothelial surface. These chemokines rapidly activate integrins on the leukocytes, which mediate the firm adhesion of the leukocytes to the endothelium. The most important integrin adhesion receptors on neutrophils are LFA-1 (CD11a/CD18) and CR3 (CD11b/CD18 or Mac-1), which interact with endothelial ICAM-1 and -2 and other adhesion molecules, thereby mediating firm leukocyte adhesion to the endothelium. These interactions are also involved in the subsequent transmigration of neutrophils through the endothelium [21]. Neutrophil adhesion to the endothelium is mainly mediated by endothelial-adherent platelets. This platelet-mediated "bridge" between leukocytes and the endothelium is a major component of intravascular thrombo-inflammatory injury. Neutrophils interact with endothelial-adherent platelets via the binding of P- selectin to P-selectin glycoprotein ligand-1 (PSGL1) [22] and the binding between Mac-1 on neutrophils and platelet glycoprotein Ib (GPIb) as well as platelet junctional adhesion molecule-3 [23–25]. Blocking neutrophil-platelet interactions, e.g., by blocking Mac-1 binding to platelet GPIb, can prevent a major part of the acute and subacute leukocyte recruitment in inflammatory and autoimmune pathologies [26–28].

We have recently studied the protein adsorption from plasma and the complement activation and cytokine generation in whole blood that are induced by a number of well-characterized artificial polymers. Our main finding was a series of strong positive correlations between the ratio of the complement activating protein C4 to its inhibitor C4BP (i.e., the C4/C4BP ratio) and the generation of 10 (mainly proinflammatory) cytokines, including IL-17, IFN-γ, and IL-6. The levels of generated C3a (reflecting complement activation) correlated weakly with four of these cytokines, but no correlations were found with either C5a or sC5b-9, confirming their poor predictive value [29, 30].

1.4 Thromboinflammatory Reactions in Therapeutic Medicine

1.4.1 Biomaterials

Treatments in which extracorporeal circulation devices are used (i.e., treatments in which blood is deviated from the circulation of the patient to a device outside the body) include hemodialysis (HD), hemofiltration, cardiopulmonary bypass (CPB), extracorporeal membrane oxygenation (ECMO), plasmapheresis, leukapheresis, and thrombapheresis. In most cases, whole blood is led through the device in the form of unseparated blood or plasma. The plasma is produced by either separating whole blood using a membrane/filter or by centrifugation, procedures that trigger incompatibility reactions to varying degrees.

Complement-mediated tissue injury triggered by extracorporeal treatments has to a great extent been found to be a consequence of bioincompatibility reactions [6, 31]. Historically, HD was associated with severe anaphylactic reactions, and even fatal events were reported. Today these problems are less frequent, although adverse reactions still occur. Of increasing concern in uremic patients undergoing maintenance HD is whole-body inflammation and accelerating arteriosclerosis. The risk of myocardial infarction in these patients is five to ten times higher than in healthy individuals. It is becoming more and more obvious that the chronic whole-body inflammation triggered by HD is a major contributor to arteriosclerosis in uremic patients, leading to a life expectancy of only 4 years after the installation of this treatment modality [32, 33]. This inflammation is to a great extent driven by complement activation and is clearly undesirable. It was recently shown that extracorporeal circulation of blood though polysulfone membranes results in C5a generation and increased expression of functional tissue factor (TF), the primary initiator of coagulation, by blood neutrophils, potentially contributing to the elevated risk of thrombosis in HD patients [34]. Importantly, complement inhibition with the peptidic C3 inhibitor compstatin significantly reduces TF expression [35], supporting a link between complement and coagulation activation and suggesting wide-ranging beneficial effects for therapeutic complement inhibition in HD.

CPB and ECMO procedures have also increased over the past decade as a result of increasing vascular bypass surgery; respiratory or cardiac failure in newborns, children, and adults; and long-lasting infections affecting the lungs, such as swine influenza and SARS (acute respiratory distress syndrome). The CPB/ECMO procedures are, like HD, associated with a number of side effects related to contact between the blood and the material surface. The resulting cellular and humoral defense reactions are termed the systemic inflammatory response syndrome (SIRS). During long-term treatments, there is often an increased incidence of bleeding (a side-effect of extensive anticoagulant treatment) and thrombotic events, including postoperative myocardial infarction and stroke, which may lead to serious neurological symptoms and inflammation-induced pulmonary fibrosis. The risk of bleeding as a result of anticoagulant treatment clearly demonstrates that the biocompatibility in such devices is poor.

Other applications of complement/thromboinflammation-regulatory surfaces include small implants within in the blood circulation, which also react with the blood cascade systems, including complement, although no systemic reactions occur. Instead, other effects of the interactions are manifested: Vascular stents elicit fibrosis, re-stenosis, and thrombosis at the implantation site in the vessel, and cardiac aids and pumps can trigger thrombotic reactions, leading to emboli.

1.4.2 Cell Transplantation

Thrombo-inflammatory injury is also a major component of the immediate blood-mediated inflammatory reaction (IBMIR) against transplanted pancreatic islets or other cell clusters or cells of non-hematologic origin, such as hepatocytes [36] and mesenchymal stromal cells (MSCs) that attenuate graft-versus-host disease (GVHD) [37, 38]. The IBMIR involves the detrimental effects of the concomitant activation of innate immune reactions, including the accumulation of neutrophils and platelets, and activation of the coagulation cascade, which result in complement-mediated injury [39–42]. Exposure of collagen types I and IV on the islets and activation of prothrombin to thrombin activate platelets. The intravascular complement activation further perpetuates the vicious cycle of inflammation, since complement anaphylatoxins C3a and C5a can act as neutrophil chemoattractants in concert with various chemokines (e.g., MCP-1, IL-8, and MIF) that are expressed by the islets. Furthermore, the anaphylatoxins also upregulate pro-coagulant TF expression on the leukocytes, further propagating fibrin formation, which eventually forms a capsule around the islets that entraps activated leukocytes (PMNs and monocytes) together with platelets. In clinical islet transplantation, this reaction results in a rapid and severe tissue loss [1, 43, 44].

1.4.3 Ischemia-Reperfusion

Ischemia-reperfusion injury is a complex condition triggered by cells or tissues exposed to ischemia and/or hypoxia [45] that results in tissue injury and a profound inflammatory response. Ischemia occurs during the obstruction of blood flow in situations such as cardiac arrest or vascular occlusion, or after a complete disruption of blood flow during situations such as organ transplantation. The former hypoxic state is a an example of warm ischemia, whereas the latter is defined as cold ischemia, in which the organ is perfused with perfusion solution upon organ procurement [45]. A loss of vascular protective barriers such as the vascular glycocalyx, a decrease in adenylate cyclase activity, and an increase in vascular permeability are all common in ischemia-reperfusion injury [19]. The hypoxic endothelial and stromal cells are converted into an inflammatory phenotype that not only expresses TF and proinflammatory cytokines and chemokines but also deposits complement on the endothelial surface, leading to the triggering of a local inflammation, with the binding of platelets and infiltration of leukocytes (particularly PMNs). This response leads to a further loss of integrity of the endothelial cells, ultimately causing vascular damage. After exposure to circulating blood, the hypoxic cells are also attacked by the innate immune system, which recognizes the cells as "altered self." This attack further aggravates the condition and finally leads to cell death and apoptosis. Recognition molecules in this process are FXII and MBL, but CRP and natural antibodies of the IgM isotype have also been implicated in these reactions.

1.5 Strategies to Dampen Thromboinflammation

1.5.1 State-of-the-Art Procedures

Biomaterial and cell surfaces that come into contact with blood are routinely used untreated or are treated with nonspecific coatings (e.g., polyethylene glycol [46], and heparin), which do not suffice for long-term treatment. More suitable biomaterials and therapeutic tissues (cells or organs) intended for contact with blood, such as those used in HD and kidney transplantation, have been developed in stages with regard to their reactivity with the intravascular immune system. Initially, materials and cells were used without any further processing; only a systemic anticoagulant treatment was used in some instances. Trial and error led to a selection of materials that were superior with regard to innate immune activation after contact with blood. Certain polymers, *eng*., polystyrene and polyvinyl chloride, caused a relatively low level of activation, while titanium had the opposite effect. Similarly, cells and organs were originally used without being given any protection. In a second stage of this evolution, the material or cell surfaces were coated with nonspecific coatings such as PEG, heparin, and recently, betaine [47–49]. A strategy to protect the vasculature from the effects of devastating ischemia-reperfusion injury is to immobilize a heparin conjugate on the vascular surface. This immobilization creates a local deposit of heparin and inhibition of thromboinflammation, as shown in models utilizing human whole blood and primary endothelial cells (Nordling et al., *in press*). Coatings like these will lower the activation of innate immunity to varying degrees, but not completely; they are not universal inhibitors of all aspects of innate immunity. In particular, complement activation is less inhibited by both PEG and heparin coatings. They have also side effects that lower their efficiency and usability. PEG surfaces, for instance, can be recognized by natural antibodies and thereby activate complement [50]. The coatings are also difficult to apply to surfaces, necessitating several steps of processing before the surface is covered.

1.5.2 Autoregulation by Regulator of Complement Activation (RCA)-Capturing Peptides

A more specifically targeted approach for increasing hemocompatibility is to make the surface auto-regulatory, either by directly immobilizing regulatory molecules or specifically targeting them to the biomaterial surface. Immobilized RCAs, FH and decay accelerating factor (DAF), reduce surface-induced complement activation, as we [51, 52] and others [53] have demonstrated.

Encouraged by these results as well as the impact of adsorbed C4BP mentioned above [30], we further developed this approach by using synthetic peptides with an affinity for the RCAs FH and C4BP; we succeeded in recruiting these regulators from the plasma (i.e., in the fluid phase) onto the material surface, with a resulting attenuation in AP and CP complement activation [54, 55].

1.5.3 ADP Depletion

The endothelial cells play an active role not only in propagating inflammatory/thrombotic events [3] but also providing platelet-inhibitory compounds, including nitric oxide, prosta-cyclin, and ecto-nucleotidases of the CD39/NTPDase (EC 3.6.1.5) family [56]. In addition, CD39 has been shown to have anti-inflammatory and immuno-suppressive properties [57, 58] and to contribute to the regulation of vascular integrity by regulating thromboinflammation [59]. Studies with $CD39^{-/-}$ knockout mice have highlighted the important role of this enzyme in the control of acute immune responses [60]. The strategy of blocking platelet recruitment and aggregation by hydrolyzing released ADP and ATP has produced satisfactory results in various *in vivo* settings [61, 62]. Overexpression of CD39 has led to an increased anti-inflammatory capacity [63] and protection in experimental transplantation models of ischemia reperfusion injury [64]. In addition, we have recently successfully inhibited platelet-dependent activation of the coagulation system as well as platelet activation by immobilizing an ADP-degrading apyrase (from potato) onto the surface of model biomaterials [65].

1.5.4 Auto-Protection by Combined Inhibition

Our next step was to co-immobilize potato apyrase and peptide 5C6 onto model biomaterial surfaces, again using the biotin-avidin system, and in this way we succeeded in creating a surface that is auto-protective against the activation of both the complement system (by recruiting FH), and platelets (by degrading ADP), thereby suppressing not only platelet activation but also subsequent coagulation activation. We then made an analogous modification of cellular surfaces, demonstrating this modus operandi to be efficient not only for biomaterial surfaces but also for cells intended for therapeutic transplantation [66]. The next step in refining this combined auto-protective surface is to replace the apyrase with CD39, a human counterpart of apyrase, in order to increase the efficacy and, more important, to minimize the risk of immunization against proteins of non-human origin.

1.5.5 PEG-Lipid Coatings

The amphiphilic polymer PEG-conjugated phospholipid (PEG-lipid) is spontaneously incorporated into the lipid bilayer membrane, where hydrophobic interactions take place. In the case of PEG-lipid derivatives, the cell surface can be coated with PEG and various functional substances for further functionalization. For example, PEG-lipid derivatives have been used to immobilize various bioactive substances on the surface of cells or islets, including urokinase, argatroban-loaded liposomes, the soluble domain of complement receptor 1 (sCR1), apyrase, and factor H-binding peptide (5C6). When these anticoagulants and complement regulators were attached to the surfaces, the cells were protected from host immune responses, and graft survival was improved. In addition, this approach has been extended to fabricating an ultra-thin immune-isolation membrane composed of PEG and to microencapsulate islets with living cells. This new approach of creating an immune-isolation capsule holds potential for islet transplantation into the liver, because it does not appreciably increase the graft volume [67–70].

The approaches mentioned in Sects. 1.5.4 and 1.5.5 are discussed in greater detail by Dr. Yuji Teramura in Chap. 12.

Acknowledgment We thank Dr. Deborah McClellan for excellent editorial assistance. This work was supported by grant 2013-65X-05647-34-4 from the Swedish Research Council (VR), the European Community's Seventh Framework Programme under the grant agreement no. 602699 (DIREKT), the Carl Tryggers Foundation, the Swedish Foundation for International Cooperation in Research and Higher Education (STINT), the Bilateral Joint Research Projects (Japan-Sweden) of the Japan Society for the Promotion of Science (JSPS), and from the AFA Insurance. AFA is a nonprofit insurance company that is jointly owned by the Swedish Labor Organization and the Swedish Employers Organization, and faculty grants from the Linnæus University.

References

1. Nilsson B, Korsgren O, Lambris JD, Ekdahl KN. Can cells and biomaterials in therapeutic medicine be shielded from innate immune recognition? Trends Immunol. 2010;31:32–8.
2. Nilsson B, Larsson R, Hong J, Elgue G, Ekdahl KN, Sahu A, et al. Compstatin inhibits complement and cellular activation in whole blood in two models of extracorporeal circulation. Blood. 1998;92:1661–7.
3. Markiewski MM, Nilsson B, Ekdahl KN, Mollnes TE, Lambris JD. Complement and coagulation: strangers or partners in crime? Trends Immunol. 2007;28:184–92.

4. Choi EY, Chavakis E, Czabanka MA, Langer HF, Fraemohs L, Economopoulou M, et al. Del-1, an endogenous leukocyte-endothelial adhesion inhibitor, limits inflammatory cell recruitment. Science. 2008;322:1101–4.
5. Croix DA, Ahearn JM, Rosengard AM, Han S, Kelsoe G, Ma M, et al. Antibody response to a T-dependent antigen requires B cell expression of complement receptors. J Exp Med. 1996;183:1857–64.
6. Nilsson B, Ekdahl KN, Mollnes TE, Lambris JD. The role of complement in biomaterial-induced inflammation. Mol Immunol. 2007;44:82–94.
7. Isermann B, Vinnikov IA, Madhusudhan T, Herzog S, Kashif M, Blautzik J, et al. Activated protein C protects against diabetic nephropathy by inhibiting endothelial and podocyte apoptosis. Nat Med. 2007;13:1349–58.
8. Ricklin D, Hajishengallis G, Yang K, Lambris JD. Complement: a key system for immune surveillance and homeostasis. Nat Immunol. 2010;11:785–97.
9. Bäck J, Huber-Lang M, Elgue G, Kalbitz M, Sanchez J, Ekdahl KN, et al. Distinctive regulation of contact activation by antithrombin and C1-inhibitor on activated platelets and material surfaces. Biomaterials. 2009;30:6573–80.
10. Bäck J, Sanchez J, Elgue G, Ekdahl KN, Nilsson B. Activated human platelets induce factor XIIa-mediated contact activation. Biochem Biophys Res Commun. 2010;391:11–7.
11. Bäck J, Lood C, Bengtsson AA, Ekdahl KN, Nilsson B. Contact activation products are new potential biomarkers to evaluate the risk of thrombotic events in systemic lupus erythematosus. Arthritis Res Ther. 2013;15:R206.
12. Hamad OA, Ekdahl KN, Nilsson PH, Andersson J, Magotti P, Lambris JD, et al. Complement activation triggered by chondroitin sulfate released by thrombin receptor-activated platelets. J Thromb Haemost. 2008;6:1413–21.
13. Hamad OA, Nilsson PH, Lasaosa M, Ricklin D, Lambris JD, Nilsson B, et al. Contribution of chondroitin sulfate A to the binding of complement proteins to activated platelets. PLoS One. 2010;5:12889.
14. Hamad OA, Nilsson PH, Wouters D, Lambris JD, Ekdahl KN, Nilsson B. Complement component C3 binds to activated normal platelets without preceding proteolytic activation and promotes binding to complement receptor 1. J Immunol. 2010;84:2686–92.
15. Hamad OA, Bäck J, Nilsson PH, Nilsson B, Ekdahl KN. Platelets, complement, and contact activation: partners in inflammation and thrombosis. Adv Exp Med Biol. 2012;946:185–205.
16. Kolářová H, Ambrůzová B, Svihálková Šindlerová L, Klinke A, Kubala L. Modulation of endothelial glycocalyx structure under inflammatory conditions. Mediators Inflamm. 2014;2014:694312.
17. Yang G, Lucas R, Caldwell R, Yao L, Romero MJ, Caldwell RW. Novel mechanisms of endothelial dysfunction in diabetes. J Cardiovasc Dis Res. 2010;1:59–63.
18. van Golen RF, van Gulik TM, Heger M. Mechanistic overview of reactive species-induced degradation of the endothelial glycocalyx during hepatic ischemia/reperfusion injury. Free Radic Biol Med. 2012;52:1382–402.
19. Singh A, Ramnath RD, Foster RR, Wylie EC, Friden V, Dasgupta I, et al. Reactive oxygen species modulate the barrier function of the human glomerular endothelial glycocalyx. PLoS One. 2013;8, e55852.
20. Hajishengallis G, Chavakis T. Endogenous modulators of inflammatory cell recruitment. Trends Immunol. 2013;34:1–6.
21. Langer HF, Chavakis T. Leukocyte-endothelial interactions in inflammation. J Cell Mol Med. 2009;13:1211–20.
22. Huo Y, Schober A, Forlow SB, Smith DF, Hyman MC, Jung S, et al. Circulating activated platelets exacerbate atherosclerosis in mice deficient in apolipoprotein E. Nat Med. 2003;9:61–7.
23. Simon DI, Chen Z, Xu H, Li CQ, Dong J, McIntire LV, et al. Platelet glycoprotein ibalpha is a counterreceptor for the leukocyte integrin Mac-1 (CD11b/CD18). J Exp Med. 2000;192: 193–204.

24. Santoso S, Sachs UJ, Kroll H, Linder M, Ruf A, Preissner KT, et al. The junctional adhesion molecule 3 (JAM-3) on human platelets is a counterreceptor for the leukocyte integrin Mac-1. J Exp Med. 2002;196:679–91.
25. Chavakis T, Santoso S, Clemetson KJ, Sachs UJ, Isordia-Salas I, Pixley RA, et al. High molecular weight kininogen regulates platelet-leukocyte interactions by bridging Mac-1 and glycoprotein Ib. J Biol Chem. 2003;278:45375–81.
26. Langer HF, Choi EY, Zhou H, Schleicher R, Chung KJ, Tang Z, et al. Platelets contribute to the pathogenesis of experimental autoimmune encephalomyelitis. Circ Res. 2012;110:1202–10.
27. Wang Y, Sakuma M, Chen Z, Ustinov V, Shi C, Croce K, et al. Leukocyte engagement of platelet glycoprotein Ibalpha via the integrin Mac-1 is critical for the biological response to vascular injury. Circulation. 2005;112:2993–3000.
28. Simon DI. Inflammation and vascular injury: basic discovery to drug development. Circ J. 2012;76:1811–8.
29. Engberg AE, Rosengren-Holmberg JP, Chen H, Nilsson B, Lambris JD, Nicholls IA, et al. Blood protein-polymer adsorption: implications for understanding complement-mediated hemoincompatibility. J Biomed Mater Res A. 2011;97:74–84.
30. Engberg AE, Nilsson PH, Huang S, Fromell K, Hamad OA, Mollnes TE, et al. Prediction of inflammatory responses induced by biomaterials in contact with human blood using protein fingerprint from plasma. Biomaterials. 2015;36:55–65.
31. Ekdahl KN, Lambris JD, Elwing H, Ricklin D, Nilsson PH, Teramura Y, et al. Innate immunity activation on biomaterial surfaces: a mechanistic model and coping strategies. Adv Drug Deliv Rev. 2011;63:1042–50.
32. Vazquez MA. Vascular access for dialysis: recent lessons and new insights. Curr Opin Nephrol Hypertens. 2009;18:116–21.
33. Cavalli A, Del Vecchio L, Manzoni C, Locatelli F. Hemodialysis: yesterday, today and tomorrow. Minerva Urol Nefrol. 2010;62:1–11.
34. Kourtzelis I, Markiewski MM, Doumas M, Rafail S, Kambas K, Mitroulis I, et al. Complement anaphylatoxin C5a contributes to hemodialysis-associated thrombosis. Blood. 2010;116:631–9.
35. Qu H, Magotti P, Ricklin D, Wu EL, Kourtzelis I, Wu YQ, et al. Novel analogues of the therapeutic complement inhibitor compstatin with significantly improved affinity and potency. Mol Immunol. 2011;48:481–9.
36. Gustafson EK, Elgue G, Hughes RD, Mitry RR, Sanchez J, Haglund U, et al. The instant blood-mediated inflammatory reaction characterized in hepatocyte transplantation. Transplantation. 2011;91:632–8.
37. Moll G, Jitschin R, von Bahr L, Rasmusson-Duprez I, Sundberg B, Lonnies L, et al. Mesenchymal stromal cells engage complement and complement receptor bearing innate effector cells to modulate immune responses. PLoS One. 2011;6, e21703.
38. Moll G, Rasmusson-Duprez I, von Bahr L, Connolly-Andersen AM, Elgue G, Funke L, et al. Are therapeutic human mesenchymal stromal cells compatible with human blood? Stem Cells. 2012;30:1565–74.
39. Moberg L, Johansson H, Lukinius A, Berne C, Foss A, Kallen R, et al. Production of tissue factor by pancreatic islet cells as a trigger of detrimental thrombotic reactions in clinical islet transplantation. Lancet. 2002;360:2039–45.
40. Johansson H, Lukinius A, Moberg L, Lundgren T, Berne C, Foss A, et al. Tissue factor produced by the endocrine cells of the islets of Langerhans is associated with a negative outcome of clinical islet transplantation. Diabetes. 2005;54:1755–62.
41. Goto M, Tjernberg J, Dufrane D, Elgue G, Brandhorst D, Ekdahl KN, et al. Dissecting the instant blood-mediated inflammatory reaction in islet xenotransplantation. Xenotransplantation. 2008;15:225–34.
42. van der Windt DJ, Bottino R, Casu A, Campanile N, Cooper DK. Rapid loss of intraportally transplanted islets: an overview of pathophysiology and preventive strategies. Xenotransplantation. 2007;14:288–97.

43. Nilsson B, Ekdahl KN, Korsgren O. Control of instant blood-mediated inflammatory reaction to improve islets of Langerhans engraftment. Curr Opin Organ Transplant. 2011;16:620–6.
44. Nilsson B, Teramura Y, Ekdahl KN. The role and regulation of complement activation as part of the thromboinflammation elicited in cell therapies. Mol Immunol. 2014;61:185–90.
45. Eltzschig HK, Eckle T. Ischemia and reperfusion–from mechanism to translation. Nat Med. 2011;17:1391–401.
46. Lappegard KT, Bergseth G, Riesenfeld J, Pharo A, Magotti P, Lambris JD, et al. The artificial surface-induced whole blood inflammatory reaction revealed by increases in a series of chemokines and growth factors is largely complement dependent. J Biomed Mater Res A. 2008;87:129–35.
47. Yang W, Bai T, Carr LR, Keefe AJ, Xu J, Xue H, et al. The effect of lightly crosslinked poly(carboxybetaine) hydrogel coating on the performance of sensors in whole blood. Biomaterials. 2012;33:7945–51.
48. Chien HW, Tsai WB, Jiang S. Direct cell encapsulation in biodegradable and functionalizable carboxybetaine hydrogels. Biomaterials. 2012;33:5706–12.
49. Leijon J, Carlsson F, Brännström J, Sanchez J, Larsson R, Nilsson B, et al. Attachment of flexible heparin chains to gelatin scaffolds improves endothelial cell infiltration. Tissue Eng Part A. 2013;19:1336–48.
50. Richter AW, Akerblom E. Polyethylene glycol reactive antibodies in man: titer distribution in allergic patients treated with monomethoxy polyethylene glycol modified allergens or placebo, and in healthy blood donors. Int Arch Allergy Appl Immunol. 1984;74:36–9.
51. Andersson J, Larsson R, Richter R, Ekdahl KN, Nilsson B. Binding of a model regulator of complement activation (RCA) to a biomaterial surface: surface-bound factor H inhibits complement activation. Biomaterials. 2001;22:2435–43.
52. Andersson J, Bexborn F, Klinth J, Nilsson B, Ekdahl KN. Surface-attached PEO in the form of activated Pluronic with immobilized factor H reduces both coagulation and complement activation in a whole-blood model. J Biomed Mater Res A. 2006;76:25–34.
53. Watkins NJ, Braidley P, Bray CJ, Savill CM, White DJ. Coating of human decay accelerating factor (hDAF) onto medical devices to improve biocompatibility. Immunopharmacology. 1997;38:111–8.
54. Wu YQ, Chen H, Ricklin D, Nilsson B, Ekdahl KN, Lambris JD. Protection of non-self surfaces from complement attack by factor H-binding peptides: implications for therapeutic medicine. J Immunol. 2011;186:4269–77.
55. Engberg AE, Sandholm K, Bexborn F, Persson J, Nilsson B, Lindahl G, et al. Inhibition of complement activation on a model biomaterial surface by streptococcal M protein-derived peptides. Biomaterials. 2009;30:2653–9.
56. Jin RC, Voetsch B, Loscalzo J. Endogenous mechanisms of inhibition of platelet function. Microcirculation. 2005;12:247–58.
57. Eltzschig HK, Thompson LF, Karhausen J, Cotta RJ, Ibla JC, Robson SC, et al. Endogenous adenosine produced during hypoxia attenuates neutrophil accumulation: coordination by extracellular nucleotide metabolism. Blood. 2004;104:3986–92.
58. Reutershan J, Cagnina RE, Chang D, Linden J, Ley K. Therapeutic anti-inflammatory effects of myeloid cell adenosine receptor A2a stimulation in lipopolysaccharide-induced lung injury. J Immunol. 2007;179:1254–63.
59. Roberts V, Lu B, Rajakumar S, Cowan PJ, Dwyer KM. The CD39-adenosinergic axis in the pathogenesis of renal ischemia-reperfusion injury. Purinergic Signal. 2013;9:135–43.
60. Eltzschig HK, Ibla JC, Furuta GT, Leonard MO, Jacobson KA, Enjyoji K, et al. Coordinated adenine nucleotide phosphohydrolysis and nucleoside signaling in posthypoxic endothelium: role of ectonucleotidases and adenosine A2B receptors. J Exp Med. 2003;189:783–96.
61. Buergler JM, Maliszewski CR, Broekman MJ, Kaluza GL, Schulz DG, Marcus AJ, et al. Effects of SolCD39, a novel inhibitor of Platelet Aggregation, on Platelet Deposition and Aggregation after PTCA in a Porcine Model. J Thromb Thrombolysis. 2005;19:115–22.

62. Cai M, Huttinger ZM, He H, Zhang W, Li F, Goodman LA, et al. Transgenic over expression of ectonucleotide triphosphate diphosphohydrolase-1 protects against murine myocardial ischemic injury. J Mol Cell Cardiol. 2011;51:927–35.
63. Dwyer KM, Robson SC, Nandurkar HH, Campbell DJ, Gock H, Murray-Segal LJ, et al. Thromboregulatory manifestations in human CD39 transgenic mice and the implications for thrombotic disease and transplantation. J Clin Invest. 2004;113:1440–6.
64. Crikis S, Lu B, Murray-Segal LM, Selan C, Robson SC, D'Apice AJ, et al. Transgenic overexpression of CD39 protects against renal ischemia-reperfusion and transplant vascular injury. Am J Transplant. 2010;10:2586–95.
65. Nilsson PH, Engberg AE, Bäck J, Faxälv L, Lindahl TL, Nilsson B, et al. The creation of an antithrombotic surface by apyrase immobilization. Biomaterials. 2010;31:4484–91.
66. Nilsson PH, Ekdahl KN, Magnusson PU, Qu H, Iwata H, Ricklin D, et al. Autoregulation of thromboinflammation on biomaterial surfaces by a multicomponent therapeutic coating. Biomaterials. 2012;34:985–94.
67. Teramura Y, Iwata H. Surface modification of islets with PEG-lipid for improvement of graft survival in intraportal transplantation. Transplantation. 2009;88:624–30.
68. Teramura Y, Iwata H. Improvement of graft survival by surface modification with poly(ethylene glycol)-lipid and urokinase in intraportal islet transplantation. Transplantation. 2011;91:271–8.
69. Teramura Y, Minh LN, Kawamoto T, Iwata H. Microencapsulation of islets with living cells using PolyDNA-PEG-lipid conjugate. Bioconjugate Chem. 2010;21:792–6.
70. Teramura Y, Oommen OP, Olerud J, Hilborn J, Nilsson B. Microencapsulation of cells, including islets, within stable ultra-thin membranes of maleimide-conjugated PEG-lipid with multifunctional crosslinkers. Biomaterials. 2013;34:2683–93.

Chapter 2
Complement Interactions with Blood Cells, Endothelial Cells and Microvesicles in Thrombotic and Inflammatory Conditions

Diana Karpman, Anne-lie Ståhl, Ida Arvidsson, Karl Johansson, Sebastian Loos, Ramesh Tati, Zivile Békássy, Ann-Charlotte Kristoffersson, Maria Mossberg, and Robin Kahn

Abstract The complement system is activated in the vasculature during thrombotic and inflammatory conditions. Activation may be associated with chronic inflammation on the endothelial surface leading to complement deposition. Complement mutations allow uninhibited complement activation to occur on platelets, neutrophils, monocytes, and aggregates thereof, as well as on red blood cells and endothelial cells. Furthermore, complement activation on the cells leads to the shedding of cell derived-microvesicles that may express complement and tissue factor thus promoting inflammation and thrombosis. Complement deposition on red blood cells triggers hemolysis and the release of red blood cell-derived microvesicles that are prothrombotic. Microvesicles are small membrane vesicles ranging from 0.1 to 1 μm, shed by cells during activation, injury and/or apoptosis that express components of the parent cell. Microvesicles are released during inflammatory and vascular conditions. The repertoire of inflammatory markers on endothelial cell-derived microvesicles shed during inflammation is large and includes complement. These circulating microvesicles may reflect the ongoing inflammatory process but may also contribute to its propagation. This overview will describe complement activation on blood and endothelial cells and the release of microvesicles from these cells during hemolytic uremic syndrome, thrombotic thrombocytopenic purpura and vasculitis, clinical conditions associated with enhanced thrombosis and inflammation.

Keywords Complement • Microvesicles • Platelets • Neutrophils • Monocytes • Red blood cells • Endothelial cells • Hemolytic uremic syndrome • Thrombotic thrombocytopenic purpura • Vasculitis

---

D. Karpman (✉) • A.-l. Ståhl • I. Arvidsson • K. Johansson • S. Loos
R. Tati • Z. Békássy • A.-C. Kristoffersson • M. Mossberg • R. Kahn
Department of Pediatrics, Clinical Sciences Lund, Lund University, Lund, Sweden
e-mail: diana.karpman@med.lu.se

J.D. Lambris et al. (eds.), *Immune Responses to Biosurfaces*,
Advances in Experimental Medicine and Biology 865,
DOI 10.1007/978-3-319-18603-0_2

2.1 Complement Activation in Thrombotic and Inflammatory Disease

The complement system is an assembly of proteins activated via three pathways, the classical, lectin and alternative pathways [1]. All ultimately result in activation of the common terminal complement pathway. Complement activation results in disposal of foreign cells, such as bacterial pathogens, unwanted host cells, such as apoptotic cells, or immune complexes, by opsonization or cytolysis. Certain complement components have anaphylatoxic, anti-microbial and chemotactic properties [1]. Due to its potent effects the complement system needs to be strictly regulated to prevent uncontrolled activation on host cells. Even so, complement may become activated, with secondary harmful effects, during infectious, hereditary and autoimmune diseases.

Complement activation may be so extensive that regulators are overwhelmed thus allowing continued activity. Alternatively, if the regulators are dysfunctional or the complement components are hyperfunctional, complement activation may proceed in an uninhibited manner. Regardless of the cause, inappropriate activation will be deleterious to host cells in the vasculature at the interface of blood cells and the endothelium. Activation on blood cells and endothelial cells will predispose the host to thrombosis, inflammation and hemolysis by rendering the endothelium prothrombotic while activating platelets, leukocytes and red blood cells. Activated blood and endothelial cells may shed microvesicles [2].

2.2 Complement Interaction with the Endothelium

Complement activation occurs on endothelial cells during inflammatory and thrombotic processes and enhances cell injury as well as vascular permeability and leukocyte recruitment to the vascular wall and underlying tissues. In infectious and inflammatory conditions the endothelium may be markedly affected, by bacterial virulence factors, platelet microthrombi, as well by neutrophil proteases and migration. Complement activation is either a primary or secondary phenomenon. Endothelial cells express complement factors, regulators and their receptors [3, 4]. Endothelial complement deposition will lead to cell activation, expression of adhesion molecules [5, 6], release of cytokines and chemokines [7–10], membrane attack complex (MAC) formation [11] and ultimately cytolysis. C5b-9 formation on the endothelial cell induces von Willebrand factor (VWF) secretion and expression of the catalytic surface for the prothrombinase enzyme complex as well as shedding of microvesicles with assembled C5b-9 on their surfaces [12]. C3a and C5a are released activation products that induce cytokine release [13, 14]. C5a and IL-8 (also secreted by the endothelium) are chemotactic for polymorphonuclear leukocytes. Vascular permeability is increased by complement activation [15] in part due to C3a and C5a [16] and the terminal complement complex, C5b-9, which also induces increased endothelial permeability [17] followed by leukocyte migration into the extravascular space [18]. Tissue factor release by endothelial cells is mediated by IL-1α and regulated by complement [19] promoting a procoagulant phenotype.

2.3 Microvesicles

Microvesicles are extracellular organelles that have an important role in inflammatory and thrombotic processes. These plasma membrane vesicles are shed from cells during activation or apoptosis [20]. Complement deposition on cells may trigger microvesicle shedding [21–23]. Microvesicles are larger (100–1000 nm) than exosomes and are not preformed within the cell's multivesicular bodies but rather shed directly from the cell membrane by the process of ectocytosis [21]. The phospholipid bilayer plasma membrane of these vesicles exposes receptors from the parent cell, and the cytosolic content is similarly derived from the cell of origin. Microvesicles can communicate with cells in their immediate vicinity or at remote sites from the parent cell by receptor binding of the entire microvesicle, by detachment of a protein from the microvesicle that serves as a ligand binding to the target cell, by fusion of the microvesicle to the cell or by endocytosis [24]. This interaction may lead to a signal cascade affecting essential cell functions such as proliferation, differentiation and apoptosis [25]. Increased numbers of circulating microvesicles occur in thrombotic and inflammatory diseases [20].

Most microvesicles in the systemic circulation originate from platelets [26] but they may originate from a variety of cells, including leukocytes [27], red blood cells [28] and endothelial cells [29]. As they transport substances derived from the cell of origin they may contain a broad spectrum of receptors, DNA, mRNA, micro RNA, histones and other proteins, as well as lipids capable of interacting with cells.

Of specific importance are the proinflammatory and prothrombotic properties of circulating microvesicles. Leukocyte-derived microvesicles harbor complement, chemokines, cytokines, proteases, HLA antigens, and adhesion molecules [27]. Upon contact these microvesicles can activate resting platelets and endothelial cells. Endothelial cells respond by expression of adhesion molecules [30]. Platelet-derived microvesicles bind to endothelial cells and induce the synthesis of pro-inflammatory cytokines and arachidonic acid pathway products [31]. In the context of inflammation, they also promote binding of monocytes to endothelial cells [32]. Furthermore, platelet microvesicles modulate adaptive immunity by affecting immunoglobulin production by B cells [33]. Endothelial-derived microvesicles have been shown to express markers involved in cellular adhesion and inflammation such as CD62E (E-selectin), CD62P (P-selectin), ICAM-1 (intercellular adhesion molecule-1), integrin $\alpha_v\beta_3$, CD31 (PECAM-1: platelet endothelial cell adhesion molecule-1), CD105 (endoglin), CD144 (VE-cadherin), VWF and CD146 (S-endo-1) [34]. Microvesicles released from cytokine-stimulated endothelial cells induced secretion of soluble ICAM-1 from targeted endothelial cells [35]. Endothelial microvesicles may be considered biomarkers of sustained endothelial activation [23].

The prothrombotic potential of microvesicles is mostly related to their exposure of phosphatidylserine and tissue factor. The release of microvesicles by cell activation is a calcium-dependent enzymatic process, involving scramblase, floppase and translocase/flippase activities [36, 37], leading to disruption of phospholipid membrane asymmetry during which phosphatidylserine is translocated to the outer membrane of the released microvesicles [20, 37, 38]. Not all microvesicles expose phosphatidylserine on the outer leaflet [38]. Phosphatidylserine can bind coagulation factors VII, IX, X, and prothrombin thus initiating clotting. Microvesicles from platelets were shown to have a higher procoagulant potential than the parent cell [39]. The prothrombotic effect was also demonstrated in red blood cell-derived microvesicles [40].

Tissue factor is the receptor for factor VIIa and initiates the extrinsic coagulation cascade and thrombin generation. Tissue factor has been demonstrated on microvesicles derived from monocytes and endothelial cells [38] and also on platelet-derived microvesicles released from platelet-leukocyte aggregates [41]. Furthermore, tissue factor may be transferred to activated platelets on the surface of endothelial and monocytic microvesicles [34].

2.3.1 Detection of Microvesicles

Microvesicle detection requires specific techniques. The vesicles are isolated from biological fluids by a series of centrifugations and detected using flow cytometry and cell-specific markers. The use of multiple markers enables higher specificity and exclusion of contaminating subpopulations. Most microvesicles expose phosphatidylserine and will therefore bind annexin V. Their size is determined by comparison with fluorescently-labeled particles/beads of a defined size and the quantity/mL is calculated using a specified quantity of blank calibration particles [42].

In addition to flow cytometry-based methodology microvesicles can be studied using proteomics, electron microscopy, ELISA (using immobilized annexin V) as well as by nanoparticle tracking.

2.3.2 Complement Activation and Microvesicle Release in Clinical Conditions

Elevated levels of blood cell- and endothelial cell-derived microvesicles have been demonstrated in prothrombotic states such as acute coronary syndrome [43], atherosclerotic disease [44], end stage renal disease [23], hemolytic uremic syndrome (HUS) [41, 42, 45, 46], thrombotic thrombocytopenic purpura (TTP) [47, 48], autoimmune thrombocytopenia [49] and vasculitis [50]. Our current understanding of the contribution of complement activation and microvesicles to the pathogenesis of the thrombotic and inflammatory entities comprised in thrombotic microangiopathies and vasculitides will be summarized below.

2.4 Hemolytic Uremic Syndrome

HUS is defined by the simultaneous occurrence of non-immune hemolytic anemia, thrombocytopenia and acute renal failure [51]. The pathological kidney lesion, known as thrombotic microangiopathy (TMA), is characterized by vessel wall thickening, detachment of the endothelium from the basement membrane with accumulation of amorphous material in the subendothelium and formation of microthrombi in glomerular capillaries leading to vessel occlusion. This lesion is typical of both HUS and TTP although the underlying pathogenetic mechanisms differ.

HUS is subdivided based on etiology into cases associated with specific infections, or not, the latter including atypical HUS and secondary causes (Table 2.1) [52, 53]. The most common subtype of HUS is associated with enterohemorrhagic *Escherichia coli* (EHEC) infection. The second most common cause is atypical HUS, aHUS, associated with mutations in complement regulators or factors, as well as auto-antibodies to the complement regulator factor H.

Table 2.1 Classification and subtypes of hemolytic uremic syndrome (HUS)

	Cause
Infection-associated	
EHEC-HUS[a]	Shiga toxin-producing non-invasive strains that can cause hemorrhagic colitis: enterohemorrhagic *E. coli* (EHEC) or *Shigella dysenteriae*
Streptococcus pneumoniae	Neuraminidase-producing invasive strains
HIV[b]	
Not associated with specific infection	
aHUS[c]	Mutations in complement factors and regulators and auto-antibodies to factor H
DGKE[d] deficiency	Recessive DGKE mutations
Cobalamin metabolism defect	Cobalamin type C MMACHC mutations, methylmalonic aciduria and homocystinuria
Drugs	Quinine, mitomycin, anti-VEGF
Malignancies	Tumors and anti-cancer treatment, drugs and irradiation
Solid organ transplantation	Including use of calcineurin inhibitors
SLE[e] and antiphospholipid syndrome	
Pregnancy HELLP[f] syndrome	

[a]HUS caused by bacterial strains that produce Shiga toxin and induce hemorrhagic colitis, also known as D+ HUS, diarrhea-associated
[b]HIV: human immunodeficiency virus
[c]aHUS: atypical HUS
[d]DGKE: diacylglycerol kinase ε
[e]SLE: systemic lupus erythematosus
[f]HELLP: HEmolytic anemia, elevated Liver enzymes, and Low Platelets

2.4.1 EHEC-Associated HUS: Toxin Interaction with Blood Cells, the Endothelium and Complement Activation

EHEC bacteria produce a unique virulence factor known as Shiga toxin (Stx). The strains themselves are non-invasive, thus remaining in the intestinal tract after colonization [54]. EHEC infection is associated with enteritis followed, in approximately 15 % of cases by the development of HUS [55]. The remote target organ damage occurring in the kidneys and other organs, including the central nervous system, during HUS, is secondary to systemic spread of the toxin and other bacterial virulence factors together with a potent intestinal and systemic host response [56–58]. Stx has been demonstrated in the circulation of HUS patients bound to platelets, neutrophils and monocytes [41, 59] but not in free form [60]. It has also been demonstrated in the kidneys of affected HUS patients [61, 62].

Stx binds to the glycosphingolipid receptor globotriaosylceramide (Gb3/CD77), also known as the P^k antigen on red blood cells. The Gb3 receptor is present on a variety of cells including platelets, monocytes and endothelial cells [63–65]. The toxin receptor on neutrophils has been proposed to be toll-like receptor 4 (TLR4) [66]. After binding, the holotoxin undergoes endocytosis. Intracellular toxin is transported via a retrograde route from early endosomes to the endoplasmic reticulum. From there the A subunit is translocated to rRNA where it inhibits protein synthesis leading to cell death. The cytotoxic effect has been demonstrated in renal glomerular endothelial cells [67] and tubular cells [68, 69]. However, not all cells are affected in this way. Certain cells, primarily blood cells, are activated by the toxin to secrete inflammatory mediators [58, 64, 70–72]. Platelets, in particular, when partially activated, can be further activated by Stx [41, 73, 74]. The combination of platelet activation and endothelial cell injury will promote a prothrombotic interaction and enhance inflammation.

Thrombocytopenia occurring during HUS is secondary to platelet consumption on the endothelium in the microangiopathic lesion. Both Stx and lipopolysaccharide circulate bound to platelets as well as to other blood cells [73, 75] and both virulence factors activate platelets, either directly and/or in the presence of inflammatory mediators, particularly under conditions of high shear stress [41]. Our group has shown that co-stimulation with both factors induced formation of platelet-leukocyte complexes that expressed tissue factor [41].

During HUS, platelets are activated [76] and degranulated [77, 78], as indicated by elevated platelet-derived factors such as β-thromboglobulin, platelet factor-4 and VWF [79]. The latter is also secreted from endothelial cells [80, 81] and mediates Stx-induced platelet adhesion to activated endothelial cells involving the platelet receptor GPIIb/IIIa and P-selectin [82]. Platelets may also be activated by chemokines and cytokines released by Stx-stimulated monocytes [71] or endothelial cells [72].

Platelet activation and thrombus formation during HUS are thus caused by a direct interaction of bacterial virulence factors, especially Stx, with platelets and endothelial cells, as well as an indirect inflammatory effect, reflected by cytokine release, which further enhances the platelet-endothelial interaction [74]. Our group showed that Stx induced the formation of platelet-leukocyte aggregates coated with C3 and C9, an effect enhanced by co-stimulation with *E. coli* O157 lipopolysaccharide [83]. Platelets are capable of binding C3, C3b, C3d and C4, as well as expressing P selectin and CR2 that bind C3b and C3d, respectively [74, 84–87]. Complement activation progresses to the terminal pathway and the cytolytic membrane attack complex may be assembled on the platelet surface [22, 83]. Complement deposition on platelets has an activating prothrombotic effect on the cells [88–91]. The reverse was also shown, that the complement system can become sequentially activated on activated platelets [84]. Thus, in order to prevent excessive complement activation platelets express and bind complement regulators, including factor H [92–98]. In vitro studies have, however, shown that Stx2, at rather high concentrations, could bind to factor H at its host surface recognition domain, thus inhibiting its regulatory effect and allowing activation of the alternative pathway of complement in serum [99].

Components of the coagulation cascade are not consumed as a consequence of the thrombotic process occurring during HUS, but prothrombin fragment 1+2 [100–102], tissue plasminogen activator, tissue plasminogen activator inhibitor-1 [103], and D dimers are elevated suggesting prothrombotic activity and impaired fibrinolysis.

In addition to platelets, neutrophils and monocytes, Stx can bind to red blood cells via the P^k antigen (an antigen within the P1PK blood group system) present on all red blood cells except for the rare p phenotype [42, 104]. Binding was shown, by our group, to induce hemolysis as indicated by release of lactic dehydrogenase and hemoglobin [42]. The hemolytic effect of Stx was complement-mediated, via the alternative pathway, as it was reduced in the absence of plasma and when plasma was heat-inactivated as well as in the presence of EDTA or eculizumab, a humanized anti-C5 antibody.

Stx has been shown to have a direct cytotoxic effect on endothelial cells [105] including glomerular endothelial cells. This effect may be enhanced in the presence of proinflammatory cytokines such as interleukin 1 and tumor necrosis factor [65, 105]. The toxin induces the expression of P selectin on microvascular endothelial cells. P selectin binds C3, thus activating the alternative pathway and reducing thrombomodulin. This process was found to promote thrombus formation under perfusion in vitro, and in a mouse model in vivo [106].

The induction of endothelial cell injury and microthrombi formation in glomeruli alongside severe tubular damage [68] would be expected to induce secondary complement activation during EHEC-mediated HUS. This is indeed the case as patients have been shown to have low levels of serum C3 [107, 108] and elevated levels of complement products C3a, factor Bb and soluble C5b-9 [83, 109], as well as complement C3 and C9 on platelet-leukocyte aggregates suggesting that bacterial virulence factors and the host response induce complement activation on these blood cells. C3 has also been demonstrated on patient red blood cells [42] implicating a role in the induction of hemolysis.

Taken together, the findings in patients, animal models and in vitro studies indicate that complement is activated on blood cells and the endothelium during

EHEC-associated HUS and that Stx plays a major role in this process. Complement activation could aggravate the procoagulant interaction in the vasculature and thus contribute to formation of the pathological lesion manifest in thrombotic microangiopathy.

2.4.2 Microvesicles in EHEC-HUS

Patients with EHEC-associated HUS have elevated levels of circulating microvesicles derived from platelets, monocytes, neutrophils [41, 83, 110] and red blood cells [42]. Plasma microvesicle concentrations were higher during the acute phase of disease, decreasing to normal levels after recovery. The microvesicles exposed phosphatidylserine and a large proportion bore tissue factor [41], which could trigger thrombosis. In addition, blood cell-derived microvesicles had deposits of C3 and C9 [42, 83], suggesting that complement was either activated on the parent cells before the vesicle bud off, or occurred directly on the microvesicle. Complement deposition contributed to opsonization of labeled microvesicles [83]. A recent finding was that microvesicles circulating in EHEC-HUS patients bore Stx which could thus be transferred to kidney cells and evade immune detection [46].

In vitro studies showed that Stx and lipopolysaccharide from *E. coli* O157:H7 induced the release of blood cell-derived microvesicles, mostly from platelets but also from monocytes. These microvesicles exhibit deposits of tissue factor [41], C3 and C9 [83] as well as Stx [46]. Stx also triggered the release of C5b-9 coated red blood cell-derived microvesicles, an effect mediated by complement activation via the alternative pathway and inhibited by purinergic receptor antagonists [42]. Thus the findings in patients could be reproduced in vitro by stimulation with Stx.

2.4.3 Atypical HUS: Complement Interactions with Blood Cells and the Endothelium

aHUS is in many cases familial and recurrent leading to renal failure and vascular complications [111, 112]. It has been associated with mutations in complement regulators factor H, factor I, membrane cofactor protein (MCP/CD46), thrombomodulin, clusterin, or complement factors C3 or factor B [111, 113, 114]. Mutations are mostly heterozygous. Their impact on protein function is described in Table 2.2. Less than 10 % of patients have mutations in more than one complement factor. A subset of patients have auto-antibodies to factor H [115], which are, in some cases, associated with deletions or rearrangements in factor H and factor H-related proteins (CFHRs) resulting in hybrid genes but usually not associated with other complement mutations.

Table 2.2 The effect of complement mutations in aHUS on protein function

Complement factor	Pathway	Soluble or membrane-bound	Complement factor or regulator	Function	Consequence of mutation in aHUS
Factor H	Alternative	Soluble	Regulator	• Cofactor for factor I in C3b cleavage • Accelerates decay of the C3 convertase • Host cell recognition	Enhanced complement activation on host cells
Factor H related protein-1	Terminal	Soluble	Regulator	• Inhibits the C5 convertase	Increased complement activation via the terminal pathway
Factor I	Alternative and classical	Soluble	Regulator	• Cleaves C3b to iC3b (inactive form) in the presence of cofactors: factor H, C4 binding protein, MCP or complement receptor 1	Enhanced complement activation
MCP (CD46)	Alternative, classical or terminal	Membrane-bound	Regulator	• Cofactor for factor I-mediated C3b cleavage	Enhanced complement activation
Thrombo-modulin	Alternative, classical or terminal	Membrane-bound	Regulator	• Enhanced factor I C3b cleavage with cofactor factor H • Generates TAFI that inactivates C3a and C5a	Enhanced complement activation and more circulating C3a and C5a
Clusterin	Terminal	Soluble	Regulator	• Inhibits MAC formation	Enhanced MAC formation
C3	All	Soluble	Factor	• C3 cleavage to C3a and C3b having anaphylactic, chemotactic and anti-microbial properties • C3b forms the C3 convertase with factor B and further binding forms the C5 convertase • C3b and its inactive form iC3b are opsonins	Hyperfunctional C3 forms a stable convertase which may be resistant to degradation
Factor B	Alternative	Soluble	Factor	• Binds to C3 and is cleaved by factor D to form the C3 convertase C3bBb	Hyperfunctional C3 convertase

MCP: membrane co-factor protein. TAFI: thrombin-activatable fibrinolysis inhibitor. MAC: membrane attack complex

Mutations in complement regulators lead to loss-of-function of the regulators thus allowing uninhibited complement activation to occur on endothelial cells [116] and platelets [45] in vitro. Mutations in, and antibodies to, factor H mainly affect the C terminus [117], which is involved in host recognition, differentiating host cells from foreign surfaces. Neutralization of host recognition results in undesirable complement activation on host cells. Mutations in complement factors C3 and factor B may lead to gain-of-function resulting in a hyperfunctional C3 convertase [118–120]. Thus mutations trigger complement activation, which, in patients, results in complement deposition on renal endothelial cells [121] and platelets [45, 114, 120]. Complement activation occurs via the alternative pathway followed by activation of the terminal complement cascade with deposition of C3 and C9 [45]. Complement activation occurs both systemically and locally in the renal vasculature.

The effect of aHUS mutations on interactions with blood cells and the endothelium has been studied in vitro and in animal models. Factor H was shown to bind to platelets mostly via its C terminal domain [96]. An aHUS mutant form of factor H, FH-E1198Stop, mutated in the C terminus, exhibited reduced binding [45] and promoted C3 deposition on normal washed platelets. Likewise, patient sera, containing mutated forms of factor H or C3, allowed deposition of both C3 and C9 on normal washed platelets leading to their activation [45, 120].

In addition to direct complement activation on platelet surfaces, platelet activation during aHUS may occur due to complement-mediated endothelial damage [116]. Uninhibited complement activation on endothelial cells has been shown to occur in the presence of mutated factor H [122], C3 and factor B mutations [118, 119] due to decreased regulatory capacity of factor H or hyperfunctional C3 or factor B leading to excess formation of the C3 convertase which may also be resistant to decay [123]. Glomerular endothelial cells stimulated with pro-inflammatory mediators (TNF and IFNγ) and exposed to aHUS serum with hyperfunctional C3 expressed tissue factor and a prothrombotic phenotype [118]. Similarly, cytokine-stimulated human umbilical vein endothelial cells exposed to certain factor B mutations bound more C3 [112, 123]. These in vitro studies exemplify how mutated over-functional complement mediated excessive complement deposition on the endothelium. An in vivo model of aHUS was developed in mice transgenically expressing a factor H protein lacking the C terminal region. These mice exhibited C3 and C9 staining in glomeruli. Thrombotic microangiopathy did not develop in mice that, in addition to mutant factor H, lacked C5, indicating that the terminal complement pathway was involved in formation of the specific glomerular lesion [124].

It would seem likely that complement should deposit on red blood cells during aHUS, as complement deposition has been demonstrated on platelets and endothelial cells. This aspect, and if complement mediates hemolysis during aHUS, has, however, not been studied. Nonetheless, once hemolysis has occurred and hemoglobin is released, heme is liberated and induces complement activation [125]. Heme was shown to activate complement via the alternative pathway in serum as well as on endothelial cells in vitro. The interaction between heme and C3 resulted in an overactive C3 convertase [125].

2.4.4 *Microvesicles in aHUS*

Sera from aHUS patients with defined factor H mutations located at the C terminal domain induced the release of tissue factor-bearing and phosphatidylserine exposing microvesicles from normal washed platelets [45]. This effect was markedly decreased when the normal platelets were preincubated with normal factor H before addition of the patient sera indicating that the mutated complement regulator contributed to shedding of prothrombotic microvesicles. Although complement deposition, secondary to excessive complement activation, has been demonstrated on blood and endothelial cells in aHUS, this has yet to be demonstrated on patient microvesicles.

2.5 Thrombotic Thrombocytopenic Purpura

TTP is defined as the simultaneous occurrence of hemolytic anemia, thrombocytopenia, renal and neurological manifestations and fever. It may be recurrent. TTP is associated with dysfunctional or deficient ADAMTS13 (a disintegrin and metalloproteinase with a thrombospondin type 1 motif, member 13) known as the VWF cleaving protease. Deficiency of the protease occurs in congenital TTP due to compound heterozygous or homozygous mutations in the protein whereas acquired dysfunction of the protease occurs due to the presence of circulating auto-antibodies to the protease [126]. Deficiency or dysfunction of ADAMTS13 results in uncleaved ultra-large VWF multimers with high biological potency to bind platelets and form thrombi, thus predisposing patients to a prothrombotic condition. Lack of ADAMTS13 promotes the formation of platelet-VWF strings on endothelial cells in vitro [127].

The primary pathogenetic event during TTP is the formation of microthrombi and the typical microangiopathic lesion. However, even during TTP secondary complement activation occurs, which may enhance the detrimental interaction between endothelial cells and platelets. Patients with TTP, both congenital and acquired, exhibit complement activation as shown by low C3 [128], elevated C3a, C5a and C5b-9 in the circulation during acute episodes [129, 130]. Renal, skin and cardiac tissue from TTP patients exhibited complement deposition [48, 131–133]. Sera from TTP patients induced C3 deposition and MAC formation on microvascular endothelial cells and promoted neutrophil-induced endothelial cytotoxicity, effects abrogated by complement inhibition [134].

The ultra-large VWF strings formed on the endothelial surface in vitro were found to activate complement via the alternative pathway. Complement activation occurred on platelets, endothelial cells and on VWF itself under perfusion [48, 135] and the presence of ADAMTS13 abrogated this effect. Our group has shown that exposure of histamine-stimulated glomerular endothelial cells to patient plasma, but not normal plasma, induced C3 deposition on platelet-VWF strings and the endothelium [48].

Interestingly, factor H was found to bind to VWF via its C terminal region [136] and thereby affect ADAMTS13-mediated cleavage, although conflicting data have been reported suggesting it may either enhance [136], or inhibit [137], this interaction. Factor H was shown to reduce the size of VWF multimers [138] which would affect their capability to bind platelets. Moreover, VWF binding to factor H enhanced cofactor activity in conjunction with factor I [137] thus inactivating complement progression. Both factor H and VWF are secreted by endothelial cells and may thus cooperate in reducing the inflammatory and thrombotic reaction on the endothelial cell surface.

2.5.1 Microvesicles in TTP

In TTP platelet and endothelial microvesicles are released into the circulation [47, 139]. Calcium-dependent proteolytic activity (Calpain) was associated with platelet-derived microvesicles [47]. Patient endothelial microvesicles co-expressed CD62E (E-selectin) and VWF [140]. Expression of VWF may enhance the platelet aggregative potential of the microvesicles [141]. Our group has shown that endothelial microvesicles in TTP patient plasma bore deposits of C3 and C9 [48] reflecting complement activation on the vascular lining.

TTP plasma induced the release of procoagulant endothelial cell-derived microvesicles from brain and renal microvascular endothelial cells [139]. The endothelial microvesicles expressed CD62E, ICAM-1, platelet endothelial cell adhesion molecule (PECAM-1), CD105 (endoglin) and VWF [140]. In addition, TTP plasma combined with normal platelets and perfused over histamine-treated glomerular endothelial cells induced significant C3 and C9 deposition on microvesicles released from the endothelial cells, in comparison to control plasma [48].

2.6 Vasculitis

Vasculitides are a group of autoimmune diseases characterized by inflammation in and around vessel walls. They are predominantly classified according to the size of the affected vessels [142]. A subset of patients with vasculitides have anti-neutrophil-cytoplasmic-antibodies (ANCAs) commonly directed to proteinase 3 and/or myeloperoxidase. Multiple organs may be affected during vasculitis, such as the kidneys, respiratory tract, gastrointestinal (GI) tract, joints and skin and thus patients may exhibit a variety of symptoms such as renal dysfunction, respiratory symptoms, sinusitis, GI-bleeding, arthritis and purpura. The clinical severity of vasculitides varies from mild and transient to life-threatening conditions, which may recur. Although the pathogenesis is not fully elucidated ANCAs, complement, neutrophil- and T cell-mediated inflammation are believed to be essential in the development of disease [143–147].

The innate immune system, and especially the complement system, have been shown to play a part in the inflammatory process in vasculitis [3]. In both immune-complex mediated vasculitides, such as in Systemic Lupus Erythematosus (SLE), and in pauci-immune vasculitides, such as ANCA-associated vasculitides, the complement system is activated [147, 148]. In SLE both immunoglobulins and complement components including C1q, C3 and C4 are deposited in the renal glomerulus, thus indicating activation of the classical pathway of complement. In addition, the importance of the classical pathway in SLE is illustrated by the association with C1q and C2 deficiency [149, 150]. On the other hand, animal models of pauci-immune vasculitides have demonstrated activation of the alternative pathway of complement [151, 152]. In these animal models C4 deficiency (classical pathway) did not affect the course of disease whereas C5 and factor B deficiency (alternative pathway) were protective suggesting that the alternative pathway mediated inflammatory injury during ANCA-related disease [153].

Neutrophil migration appears to be an essential component of tissue injury and particularly endothelial damage during ANCA-associated vasculitis. The importance of C5a, acting as a chemotactic agent and an anaphylatoxin, was demonstrated as patient neutrophils activated normal serum to produce C5a. The pivotal role of C5a was demonstrated by attenuation of the disease in a C5a receptor-deficient mouse model of necrotizing crescentic glomerulonephritis [151].

2.6.1 The Role of Microvesicles in Vasculitis

Both adults and children with vasculitis have elevated levels of microvesicles in the circulation that have been shown to correlate to disease activity [154, 155]. Endothelial-, neutrophil- and platelet-derived microvesicles have been demonstrated in vasculitis patients [50, 156]. As identification of circulating microvesicles, especially endothelial cell-derived, may reflect endothelial damage seen in vasculitis, it has been suggested that circulating microvesicles could be used as a biomarker for vasculitis activity [154, 155, 157].

Neutrophil-derived microvesicles have been shown to induce endothelial cell activation and subsequent release of cytokines [30, 158] as well as endothelial cell damage [159]. ANCA, which are present in a subset of vasculitides, have been shown to induce release of microvesicles from neutrophils in vitro, followed by microvesicle-dependent activation of endothelial cells, suggesting a pathogenic role of microvesicles in vasculitis [157].

2.7 Clinical and Therapeutic Implications of Complement Activation and Microvesicle Release

In the conditions described, HUS (both EHEC-associated and aHUS), TTP and vasculitis, complement activation on blood- and endothelial cells activates platelets, monocytes, neutrophils, red blood cells as well as endothelial cells promoting thrombotic and inflammatory damage. Activated or apoptotic cells will release microvesicles, and the presence of complement on these microvesicles reflects the detrimental in vivo process. Once complement activation and microvesicle release manifest the vascular damage may already be, to a certain degree, irreversible. However, an ongoing continuous process may be amenable to treatment.

During EHEC infection and prior to the development of HUS antibiotic treatment may expose the patient to an increased risk of developing HUS [160]. Nonetheless, antibiotic therapy during the course of HUS may be beneficial [161]. The infection, and the renal and neurological complications that may develop, are best managed by supportive care [161, 162] as no specific treatment has, to-date, been shown to be effective. Patients exhibit excessive complement activation [42, 83, 109] and a few patients were suggested to respond to eculizumab, humanized anti-C5 antibody (Alexion, Cheshire, CT) [163], but, by the time patients present with HUS, treatment aimed at blocking complement activation may be ineffective, as reported during the large *E. coli* O104:H4 outbreak in 2011 [161, 164, 165].

In contrast, the disease process occurring during aHUS leads to continuous complement activation due to the presence of complement mutations, or antibodies to factor H. In these patients treatment with eculizumab is extremely effective, preventing disease recurrence and renal failure [166] and allowing patients who previously developed kidney failure to undergo renal transplantation [167].

Most patients with TTP are successfully treated with plasma infusions or exchange, and patients with acquired TTP respond to rituximab (anti-CD20) [126]. Complement blockade was described in one TTP patient whose disease activity was refractory to other treatments (plasma exchange, glucocorticoids, rituximab, and vincristine) [133]. Although this patient had antibodies to ADAMTS13 he was later shown to also have antibodies to factor H [168], which may explain the beneficial response to eculizumab.

Patients with vasculitis are treated with non-specific immune suppression, usually achieving remission. Eculizumab was shown to be effective in one severe pediatric case of SLE with vasculitis [169] and in an adult case of SLE complicated by thrombotic microangiopathy [170]. As animal models of necrotizing glomerulonephritis have shown that C5 is involved in the renal damage, using C5a receptor deficient mice [151], further study of complement blockade as a treatment modality in vasculitis and SLE is warranted.

In addition to the commercially available eculizumab, multiple therapeutic agents are being developed in order to block complement activation at different levels and in the different pathways [171]. These agents may prove to be promising therapeutic options for certain thrombotic microangiopathies and vasculitides in the future.

Elevated circulating microvesicles may be biomarkers of an ongoing inflammatory condition. Still, cells may attempt to eliminate unwanted substances by shedding these within microvesicles. Thus future investigations will need to address the contribution of microvesicles to the inflammatory process and if therapeutic interventions directed at blocking microvesicle release are beneficial.

2.8 Conclusions

Whether as a primary event or as a secondary phenomenon, complement activation during thrombotic and inflammatory conditions affecting the vasculature will further perturb injury to the endothelial lining. Microvesicles released from blood and endothelial cells are prothrombotic and via secretion of their contents predispose to inflammation. Conditions such as HUS, TTP and vasculitis differ in their etiologies and pathogenesis but in these conditions complement activation and microvesicle release may exacerbate the course of disease. For this reason therapeutic interventions directed at these processes are worthy of future study.

Acknowledgements This work was supported by grants from The Swedish Research Council (K2013-64X-14008 and K2015-99X-22877-01-6), The Torsten Söderberg Foundation, The Konung Gustaf V:s 80-årsfond, SUS stiftelser och donationer (all to DK). The Swedish Renal foundation (to ZB) and the Fanny Ekdahl Foundation (to ZB and RK). Crown Princess Lovisa's Society for Child Care (to DK and RK), Alfred Österlund's Foundation, Greta and Johan Kock's Foundation, the Samariten Foundation, and the Royal Physiographic Society (all to RK). SL was supported by a research fellowship from the Deutsche Forschungsgemeinschaft (LO 2021/2-1).

References

1. Walport MJ. Complement. First of two parts. N Engl J Med. 2001;344(14):1058–66.
2. Lopez JA, del Conde I, Shrimpton CN. Receptors, rafts, and microvesicles in thrombosis and inflammation. J Thromb Haemost. 2005;3(8):1737–44.
3. Acosta J, Qin X, Halperin J. Complement and complement regulatory proteins as potential molecular targets for vascular diseases. Curr Pharm Des. 2004;10(2):203–11.
4. Langeggen H, Pausa M, Johnson E, Casarsa C, Tedesco F. The endothelium is an extrahepatic site of synthesis of the seventh component of the complement system. Clin Exp Immunol. 2000;121(1):69–76.
5. Lozada C, Levin RI, Huie M, Hirschhorn R, Naime D, Whitlow M, et al. Identification of C1q as the heat-labile serum cofactor required for immune complexes to stimulate endothelial expression of the adhesion molecules E-selectin and intercellular and vascular cell adhesion molecules 1. Proc Natl Acad Sci U S A. 1995;92(18):8378–82.
6. Kilgore KS, Shen JP, Miller BF, Ward PA, Warren JS. Enhancement by the complement membrane attack complex of tumor necrosis factor-alpha-induced endothelial cell expression of E-selectin and ICAM-1. J Immunol. 1995;155(3):1434–41.
7. van den Berg RH, Faber-Krol MC, Sim RB, Daha MR. The first subcomponent of complement, C1q, triggers the production of IL-8, IL-6, and monocyte chemoattractant peptide-1 by human umbilical vein endothelial cells. J Immunol. 1998;161(12):6924–30.
8. Kilgore KS, Flory CM, Miller BF, Evans VM, Warren JS. The membrane attack complex of complement induces interleukin-8 and monocyte chemoattractant protein-1 secretion from human umbilical vein endothelial cells. Am J Pathol. 1996;149(3):953–61.
9. Benzaquen LR, Nicholson-Weller A, Halperin JA. Terminal complement proteins C5b-9 release basic fibroblast growth factor and platelet-derived growth factor from endothelial cells. J Exp Med. 1994;179(3):985–92.
10. Fischetti F, Tedesco F. Cross-talk between the complement system and endothelial cells in physiologic conditions and in vascular diseases. Autoimmunity. 2006;39(5):417–28.

11. Hamilton KK, Ji Z, Rollins S, Stewart BH, Sims PJ. Regulatory control of the terminal complement proteins at the surface of human endothelial cells: neutralization of a C5b-9 inhibitor by antibody to CD59. Blood. 1990;76(12):2572–7.
12. Hamilton KK, Hattori R, Esmon CT, Sims PJ. Complement proteins C5b-9 induce vesiculation of the endothelial plasma membrane and expose catalytic surface for assembly of the prothrombinase enzyme complex. J Biol Chem. 1990;265(7):3809–14.
13. Monsinjon T, Gasque P, Chan P, Ischenko A, Brady JJ, Fontaine MC. Regulation by complement C3a and C5a anaphylatoxins of cytokine production in human umbilical vein endothelial cells. FASEB J. 2003;17(9):1003–14.
14. Kerr H, Richards A. Complement-mediated injury and protection of endothelium: lessons from atypical haemolytic uraemic syndrome. Immunobiology. 2012;217(2):195–203.
15. Karpman D, Kahn R. The contact/kinin and complement systems in vasculitis. APMIS Suppl. 2009;127:48–54.
16. Bossi F, Peerschke EI, Ghebrehiwet B, Tedesco F. Cross-talk between the complement and the kinin system in vascular permeability. Immunol Lett. 2011;140(1-2):7–13.
17. Bossi F, Fischetti F, Pellis V, Bulla R, Ferrero E, Mollnes TE, et al. Platelet-activating factor and kinin-dependent vascular leakage as a novel functional activity of the soluble terminal complement complex. J Immunol. 2004;173(11):6921–7.
18. Dobrina A, Pausa M, Fischetti F, Bulla R, Vecile E, Ferrero E, et al. Cytolytically inactive terminal complement complex causes transendothelial migration of polymorphonuclear leukocytes in vitro and in vivo. Blood. 2002;99(1):185–92.
19. Saadi S, Holzknecht RA, Patte CP, Stern DM, Platt JL. Complement-mediated regulation of tissue factor activity in endothelium. J Exp Med. 1995;182(6):1807–14.
20. Morel O, Jesel L, Freyssinet JM, Toti F. Cellular mechanisms underlying the formation of circulating microparticles. Arterioscler Thromb Vasc Biol. 2011;31(1):15–26.
21. Stein JM, Luzio JP. Ectocytosis caused by sublytic autologous complement attack on human neutrophils. The sorting of endogenous plasma-membrane proteins and lipids into shed vesicles. Biochem J. 1991;274(Pt 2):381–6.
22. Sims PJ, Faioni EM, Wiedmer T, Shattil SJ. Complement proteins C5b-9 cause release of membrane vesicles from the platelet surface that are enriched in the membrane receptor for coagulation factor Va and express prothrombinase activity. J Biol Chem. 1988;263(34): 18205–12.
23. Rabelink TJ, de Boer HC, van Zonneveld AJ. Endothelial activation and circulating markers of endothelial activation in kidney disease. Nat Rev Nephrol. 2010;6(7):404–14.
24. Meckes Jr DG, Raab-Traub N. Microvesicles and viral infection. J Virol. 2011;85(24): 12844–54.
25. Mause SF, Weber C. Microparticles: protagonists of a novel communication network for intercellular information exchange. Circ Res. 2010;107(9):1047–57.
26. Flaumenhaft R, Dilks JR, Richardson J, Alden E, Patel-Hett SR, Battinelli E, et al. Megakaryocyte-derived microparticles: direct visualization and distinction from platelet-derived microparticles. Blood. 2009;113(5):1112–21.
27. Angelillo-Scherrer A. Leukocyte-derived microparticles in vascular homeostasis. Circ Res. 2012;110(2):356–69.
28. Rubin O, Canellini G, Delobel J, Lion N, Tissot JD. Red blood cell microparticles: clinical relevance. Transfus Med Hemother. 2012;39(5):342–7.
29. Chironi GN, Boulanger CM, Simon A, Dignat-George F, Freyssinet JM, Tedgui A. Endothelial microparticles in diseases. Cell Tissue Res. 2009;335(1):143–51.
30. Mesri M, Altieri DC. Leukocyte microparticles stimulate endothelial cell cytokine release and tissue factor induction in a JNK1 signaling pathway. J Biol Chem. 1999;274(33): 23111–8.
31. Pap E, Pallinger E, Pasztoi M, Falus A. Highlights of a new type of intercellular communication: microvesicle-based information transfer. Inflamm Res. 2009;58(1):1–8.

32. Barry OP, Pratico D, Savani RC, FitzGerald GA. Modulation of monocyte-endothelial cell interactions by platelet microparticles. J Clin Invest. 1998;102(1):136–44.
33. Sprague DL, Elzey BD, Crist SA, Waldschmidt TJ, Jensen RJ, Ratliff TL. Platelet-mediated modulation of adaptive immunity: unique delivery of CD154 signal by platelet-derived membrane vesicles. Blood. 2008;111(10):5028–36.
34. Dignat-George F, Boulanger CM. The many faces of endothelial microparticles. Arterioscler Thromb Vasc Biol. 2011;31(1):27–33.
35. Curtis AM, Wilkinson PF, Gui M, Gales TL, Hu E, Edelberg JM. p38 mitogen-activated protein kinase targets the production of proinflammatory endothelial microparticles. J Thromb Haemost. 2009;7(4):701–9.
36. Zwaal RF, Schroit AJ. Pathophysiologic implications of membrane phospholipid asymmetry in blood cells. Blood. 1997;89(4):1121–32.
37. Daleke DL. Regulation of transbilayer plasma membrane phospholipid asymmetry. J Lipid Res. 2003;44(2):233–42.
38. Owens 3rd AP, Mackman N. Microparticles in hemostasis and thrombosis. Circ Res. 2011;108(10):1284–97.
39. Sinauridze EI, Kireev DA, Popenko NY, Pichugin AV, Panteleev MA, Krymskaya OV, et al. Platelet microparticle membranes have 50- to 100-fold higher specific procoagulant activity than activated platelets. Thromb Haemost. 2007;97(3):425–34.
40. Rubin O, Delobel J, Prudent M, Lion N, Kohl K, Tucker EI, et al. Red blood cell-derived microparticles isolated from blood units initiate and propagate thrombin generation. Transfusion. 2013;53(8):1744–54.
41. Ståhl AL, Sartz L, Nelsson A, Békássy ZD, Karpman D. Shiga toxin and lipopolysaccharide induce platelet-leukocyte aggregates and tissue factor release, a thrombotic mechanism in hemolytic uremic syndrome. PLoS One. 2009;4(9), e6990.
42. Arvidsson I, Ståhl AL, Manea Hedström M, Kristoffersson AC, Rylander C, Westman JS, Storry JR, Olsson ML, Karpman D. Shiga toxin–induced complement-mediated hemolysis and release of complement-coated red blood cell–derived microvesicles in hemolytic uremic syndrome. J Immunol. 2015;194:2309–18.
43. Mallat Z, Benamer H, Hugel B, Benessiano J, Steg PG, Freyssinet JM, et al. Elevated levels of shed membrane microparticles with procoagulant potential in the peripheral circulating blood of patients with acute coronary syndromes. Circulation. 2000;101(8):841–3.
44. Leroyer AS, Tedgui A, Boulanger CM. Role of microparticles in atherothrombosis. J Intern Med. 2008;263(5):528–37.
45. Ståhl AL, Vaziri-Sani F, Heinen S, Kristoffersson AC, Gydell KH, Raafat R, et al. Factor H dysfunction in patients with atypical hemolytic uremic syndrome contributes to complement deposition on platelets and their activation. Blood. 2008;111(11):5307–15.
46. Ståhl AL, Arvidsson I, Johansson K, Chromek M, Rebetz J, Loos S, Kristoffersson AC, Békássy ZD, Mörgelin M, Karpman D. A novel mechanism of bacterial toxin transfer within host blood cell-derived microvesicles. PLoS Pathog. 2015;11, e1004619.
47. Kelton JG, Warkentin TE, Hayward CP, Murphy WG, Moore JC. Calpain activity in patients with thrombotic thrombocytopenic purpura is associated with platelet microparticles. Blood. 1992;80(9):2246–51.
48. Tati R, Kristoffersson AC, Ståhl AL, Rebetz J, Wang L, Licht C, et al. Complement activation associated with ADAMTS13 deficiency in human and murine thrombotic microangiopathy. J Immunol. 2013;191(5):2184–93.
49. Khan I, Zucker-Franklin D, Karpatkin S. Microthrombocytosis and platelet fragmentation associated with idiopathic/autoimmune thrombocytopenic purpura. Br J Haematol. 1975;31(4):449–60.
50. Brogan PA, Shah V, Brachet C, Harnden A, Mant D, Klein N, et al. Endothelial and platelet microparticles in vasculitis of the young. Arthritis Rheum. 2004;50(3):927–36.
51. Karpman D. Haemolytic uremic syndrome and thrombotic thrombocytopenic purpura. Curr Paediatr. 2002;12:569–74.

52. George JN, Nester CM. Syndromes of thrombotic microangiopathy. N Engl J Med. 2014;371(19):1847–8.
53. Besbas N, Karpman D, Landau D, Loirat C, Proesmans W, Remuzzi G, et al. A classification of hemolytic uremic syndrome and thrombotic thrombocytopenic purpura and related disorders. Kidney Int. 2006;70(3):423–31.
54. McKee ML, O'Brien AD. Investigation of enterohemorrhagic *Escherichia coli* O157:H7 adherence characteristics and invasion potential reveals a new attachment pattern shared by intestinal *E. coli*. Infect Immun. 1995;63(5):2070–4.
55. Tarr PI, Gordon CA, Chandler WL. Shiga-toxin-producing *Escherichia coli* and haemolytic uraemic syndrome. Lancet. 2005;365(9464):1073–86.
56. Proulx F, Seidman EG, Karpman D. Pathogenesis of Shiga toxin-associated hemolytic uremic syndrome. Pediatr Res. 2001;50(2):163–71.
57. Zoja C, Buelli S, Morigi M. Shiga toxin-associated hemolytic uremic syndrome: pathophysiology of endothelial dysfunction. Pediatr Nephrol. 2010;25(11):2231–40.
58. Karpman D, Ståhl AL. Enterohemorrhagic *Escherichia coli* pathogenesis and the host response. Microbiol Spectrum. 2014;2(5):1–15.
59. Tazzari PL, Ricci F, Carnicelli D, Caprioli A, Tozzi AE, Rizzoni G, et al. Flow cytometry detection of Shiga toxins in the blood from children with hemolytic uremic syndrome. Cytometry B Clin Cytom. 2004;61(1):40–4.
60. Brigotti M, Tazzari PL, Ravanelli E, Carnicelli D, Rocchi L, Arfilli V, et al. Clinical relevance of Shiga toxin concentrations in the blood of patients with hemolytic uremic syndrome. Pediatr Infect Dis J. 2011;30(6):486–90.
61. Uchida H, Kiyokawa N, Horie H, Fujimoto J, Takeda T. The detection of Shiga toxins in the kidney of a patient with hemolytic uremic syndrome. Pediatr Res. 1999;45(1):133–7.
62. Chaisri U, Nagata M, Kurazono H, Horie H, Tongtawe P, Hayashi H, et al. Localization of Shiga toxins of enterohaemorrhagic *Escherichia coli* in kidneys of paediatric and geriatric patients with fatal haemolytic uraemic syndrome. Microb Pathog. 2001;31(2):59–67.
63. Cooling LL, Walker KE, Gille T, Koerner TA. Shiga toxin binds human platelets via globotriaosylceramide (Pk antigen) and a novel platelet glycosphingolipid. Infect Immun. 1998;66(9):4355–66.
64. van Setten PA, Monnens LA, Verstraten RG, van den Heuvel LP, van Hinsbergh VW. Effects of verocytotoxin-1 on nonadherent human monocytes: binding characteristics, protein synthesis, and induction of cytokine release. Blood. 1996;88(1):174–83.
65. van de Kar NC, Monnens LA, Karmali MA, van Hinsbergh VW. Tumor necrosis factor and interleukin-1 induce expression of the verocytotoxin receptor globotriaosylceramide on human endothelial cells: implications for the pathogenesis of the hemolytic uremic syndrome. Blood. 1992;80(11):2755–64.
66. Brigotti M, Carnicelli D, Arfilli V, Tamassia N, Borsetti F, Fabbri E, et al. Identification of TLR4 as the receptor that recognizes Shiga toxins in human neutrophils. J Immunol. 2013;191(9):4748–58.
67. Louise CB, Obrig TG. Specific interaction of *Escherichia coli* O157:H7-derived Shiga-like toxin II with human renal endothelial cells. J Infect Dis. 1995;172(5):1397–401.
68. Karpman D, Hakansson A, Perez MT, Isaksson C, Carlemalm E, Caprioli A, et al. Apoptosis of renal cortical cells in the hemolytic-uremic syndrome: in vivo and in vitro studies. Infect Immun. 1998;66(2):636–44.
69. Hughes AK, Stricklett PK, Kohan DE. Cytotoxic effect of Shiga toxin-1 on human proximal tubule cells. Kidney Int. 1998;54(2):426–37.
70. Tesh VL, Ramegowda B, Samuel JE. Purified Shiga-like toxins induce expression of proinflammatory cytokines from murine peritoneal macrophages. Infect Immun. 1994;62(11):5085–94.
71. Guessous F, Marcinkiewicz M, Polanowska-Grabowska R, Keepers TR, Obrig T, Gear AR. Shiga toxin 2 and lipopolysaccharide cause monocytic THP-1 cells to release factors which activate platelet function. Thromb Haemost. 2005;94(5):1019–27.

72. Guessous F, Marcinkiewicz M, Polanowska-Grabowska R, Kongkhum S, Heatherly D, Obrig T, et al. Shiga toxin 2 and lipopolysaccharide induce human microvascular endothelial cells to release chemokines and factors that stimulate platelet function. Infect Immun. 2005;73(12):8306–16.
73. Karpman D, Papadopoulou D, Nilsson K, Sjögren AC, Mikaelsson C, Lethagen S. Platelet activation by Shiga toxin and circulatory factors as a pathogenetic mechanism in the hemolytic uremic syndrome. Blood. 2001;97(10):3100–8.
74. Karpman D, Manea M, Vaziri-Sani F, Ståhl AL, Kristoffersson AC. Platelet activation in hemolytic uremic syndrome. Semin Thromb Hemost. 2006;32(2):128–45.
75. Ståhl AL, Svensson M, Morgelin M, Svanborg C, Tarr PI, Mooney JC, et al. Lipopolysaccharide from enterohemorrhagic *Escherichia coli* binds to platelets through TLR4 and CD62 and is detected on circulating platelets in patients with hemolytic uremic syndrome. Blood. 2006;108(1):167–76.
76. Bolande RP, Kaplan BS. Experimental studies on the hemolytic-uremic syndrome. Nephron. 1985;39(3):228–36.
77. Fong JS, Kaplan BS. Impairment of platelet aggregation in hemolytic uremic syndrome: evidence for platelet "exhaustion". Blood. 1982;60(3):564–70.
78. Sassetti B, Vizcarguenaga MI, Zanaro NL, Silva MV, Kordich L, Florentini L, et al. Hemolytic uremic syndrome in children: platelet aggregation and membrane glycoproteins. J Pediatr Hematol Oncol. 1999;21(2):123–8.
79. Appiani AC, Edefonti A, Bettinelli A, Cossu MM, Paracchini ML, Rossi E. The relationship between plasma levels of the factor VIII complex and platelet release products (beta-thromboglobulin and platelet factor 4) in children with the hemolytic-uremic syndrome. Clin Nephrol. 1982;17(4):195–9.
80. Tsai HM, Chandler WL, Sarode R, Hoffman R, Jelacic S, Habeeb RL, et al. von Willebrand factor and von Willebrand factor-cleaving metalloprotease activity in *Escherichia coli* O157:H7-associated hemolytic uremic syndrome. Pediatr Res. 2001;49(5):653–9.
81. van de Kar NC, van Hinsbergh VW, Brommer EJ, Monnens LA. The fibrinolytic system in the hemolytic uremic syndrome: in vivo and in vitro studies. Pediatr Res. 1994;36(2):257–64.
82. Morigi M, Galbusera M, Binda E, Imberti B, Gastoldi S, Remuzzi A, et al. Verotoxin-1-induced up-regulation of adhesive molecules renders microvascular endothelial cells thrombogenic at high shear stress. Blood. 2001;98(6):1828–35.
83. Stahl AL, Sartz L, Karpman D. Complement activation on platelet-leukocyte complexes and microparticles in enterohemorrhagic *Escherichia coli*-induced hemolytic uremic syndrome. Blood. 2011;117(20):5503–13.
84. Del Conde I, Cruz MA, Zhang H, Lopez JA, Afshar-Kharghan V. Platelet activation leads to activation and propagation of the complement system. J Exp Med. 2005;201(6):871–9.
85. Endresen GK, Mellbye OJ. Studies on the binding of complement factor C3 to the surface of human blood platelets. Haemostasis. 1984;14(3):269–80.
86. Sandvik T, Endresen GK, Forre O. Studies on the binding of complement factor C4 in human platelets. Complement activation by means of cold agglutinins. Int Arch Allergy Appl Immunol. 1984;74(2):152–7.
87. Nunez D, Charriaut-Marlangue C, Barel M, Benveniste J, Frade R. Activation of human platelets through gp140, the C3d/EBV receptor (CR2). Eur J Immunol. 1987;17(4):515–20.
88. Zimmerman TS, Kolb WP. Human platelet-initiated formation and uptake of the C5-9 complex of human complement. J Clin Invest. 1976;57(1):203–11.
89. Polley MJ, Nachman RL. Human complement in thrombin-mediated platelet function: uptake of the C5b-9 complex. J Exp Med. 1979;150(3):633–45.
90. Polley MJ, Nachman RL. Human platelet activation by C3a and C3a des-arg. J Exp Med. 1983;158(2):603–15.
91. Dixon RH, Rosse WF. Mechanism of complement-mediated activation of human blood platelets in vitro: comparison of normal and paroxysmal nocturnal hemoglobinuria platelets. J Clin Invest. 1977;59(2):360–8.

92. Nicholson-Weller A, March JP, Rosen CE, Spicer DB, Austen KF. Surface membrane expression by human blood leukocytes and platelets of decay-accelerating factor, a regulatory protein of the complement system. Blood. 1985;65(5):1237–44.
93. Tschopp J, Jenne DE, Hertig S, Preissner KT, Morgenstern H, Sapino AP, et al. Human megakaryocytes express clusterin and package it without apolipoprotein A-1 into alpha-granules. Blood. 1993;82(1):118–25.
94. Yu GH, Holers VM, Seya T, Ballard L, Atkinson JP. Identification of a third component of complement-binding glycoprotein of human platelets. J Clin Invest. 1986;78(2):494–501.
95. Morgan BP. Isolation and characterization of the complement-inhibiting protein CD59 antigen from platelet membranes. Biochem J. 1992;282(Pt 2):409–13.
96. Vaziri-Sani F, Hellwage J, Zipfel PF, Sjöholm AG, Iancu R, Karpman D. Factor H binds to washed human platelets. J Thromb Haemost. 2005;3(1):154–62.
97. Licht C, Pluthero FG, Li L, Christensen H, Habbig S, Hoppe B, et al. Platelet-associated complement factor H in healthy persons and patients with atypical HUS. Blood. 2009;114(20):4538–45.
98. Kenney DM, Davis 3rd AE. Association of alternative complement pathway components with human blood platelets: secretion and localization of factor D and beta 1H Globulin. Clin Immunol Immunopathol. 1981;21(3):351–63.
99. Orth D, Khan AB, Naim A, Grif K, Brockmeyer J, Karch H, et al. Shiga toxin activates complement and binds factor H: evidence for an active role of complement in hemolytic uremic syndrome. J Immunol. 2009;182(10):6394–400.
100. Nevard CH, Jurd KM, Lane DA, Philippou H, Haycock GB, Hunt BJ. Activation of coagulation and fibrinolysis in childhood diarrhoea-associated haemolytic uraemic syndrome. Thromb Haemost. 1997;78(6):1450–5.
101. Van Geet C, Proesmans W, Arnout J, Vermylen J, Declerck PJ. Activation of both coagulation and fibrinolysis in childhood hemolytic uremic syndrome. Kidney Int. 1998;54(4):1324–30.
102. Chandler WL, Jelacic S, Boster DR, Ciol MA, Williams GD, Watkins SL, et al. Prothrombotic coagulation abnormalities preceding the hemolytic-uremic syndrome. N Engl J Med. 2002;346(1):23–32.
103. Bergstein JM, Riley M, Bang NU. Role of plasminogen-activator inhibitor type 1 in the pathogenesis and outcome of the hemolytic uremic syndrome. N Engl J Med. 1992; 327(11):755–9.
104. Bitzan M, Richardson S, Huang C, Boyd B, Petric M, Karmali MA. Evidence that verotoxins (Shiga-like toxins) from *Escherichia coli* bind to P blood group antigens of human erythrocytes in vitro. Infect Immun. 1994;62(8):3337–47.
105. Louise CB, Obrig TG. Shiga toxin-associated hemolytic-uremic syndrome: combined cytotoxic effects of Shiga toxin, interleukin-1 beta, and tumor necrosis factor alpha on human vascular endothelial cells in vitro. Infect Immun. 1991;59(11):4173–9.
106. Morigi M, Galbusera M, Gastoldi S, Locatelli M, Buelli S, Pezzotta A, et al. Alternative pathway activation of complement by Shiga toxin promotes exuberant C3a formation that triggers microvascular thrombosis. J Immunol. 2011;187(1):172–80.
107. Monnens L, Molenaar J, Lambert PH, Proesmans W, van Munster P. The complement system in hemolytic-uremic syndrome in childhood. Clin Nephrol. 1980;13(4):168–71.
108. Robson WL, Leung AK, Fick GH, McKenna AI. Hypocomplementemia and leukocytosis in diarrhea-associated hemolytic uremic syndrome. Nephron. 1992;62(3):296–9.
109. Thurman JM, Marians R, Emlen W, Wood S, Smith C, Akana H, et al. Alternative pathway of complement in children with diarrhea-associated hemolytic uremic syndrome. Clin J Am Soc Nephrol. 2009;4(12):1920–4.
110. Ge S, Hertel B, Emden SH, Beneke J, Menne J, Haller H, et al. Microparticle generation and leucocyte death in Shiga toxin-mediated HUS. Nephrol Dial Transplant. 2012;27(7): 2768–75.
111. Loirat C, Noris M, Fremeaux-Bacchi V. Complement and the atypical hemolytic uremic syndrome in children. Pediatr Nephrol. 2008;23(11):1957–72.

112. Békássy ZD, Kristoffersson AC, Cronqvist M, Roumenina LT, Rybkine T, Vergoz L, et al. Eculizumab in an anephric patient with atypical haemolytic uraemic syndrome and advanced vascular lesions. Nephrol Dial Transplant. 2013;28(11):2899–907.
113. Delvaeye M, Noris M, De Vriese A, Esmon CT, Esmon NL, Ferrell G, et al. Thrombomodulin mutations in atypical hemolytic-uremic syndrome. N Engl J Med. 2009;361(4):345–57.
114. Ståhl AL, Kristoffersson A, Olin AI, Olsson ML, Roodhooft AM, Proesmans W, et al. A novel mutation in the complement regulator clusterin in recurrent hemolytic uremic syndrome. Mol Immunol. 2009;46(11-12):2236–43.
115. Dragon-Durey MA, Sethi SK, Bagga A, Blanc C, Blouin J, Ranchin B, et al. Clinical features of anti-factor H autoantibody-associated hemolytic uremic syndrome. J Am Soc Nephrol. 2010;21(12):2180–7.
116. Manuelian T, Hellwage J, Meri S, Caprioli J, Noris M, Heinen S, et al. Mutations in factor H reduce binding affinity to C3b and heparin and surface attachment to endothelial cells in hemolytic uremic syndrome. J Clin Invest. 2003;111(8):1181–90.
117. Skerka C, Zipfel PF. Complement factor H related proteins in immune diseases. Vaccine. 2008;26 Suppl 8:I9–14.
118. Roumenina LT, Frimat M, Miller EC, Provot F, Dragon-Durey MA, Bordereau P, et al. A prevalent C3 mutation in aHUS patients causes a direct C3 convertase gain of function. Blood. 2012;119(18):4182–91.
119. Roumenina LT, Jablonski M, Hue C, Blouin J, Dimitrov JD, Dragon-Durey MA, et al. Hyperfunctional C3 convertase leads to complement deposition on endothelial cells and contributes to atypical hemolytic uremic syndrome. Blood. 2009;114(13):2837–45.
120. Sartz L, Olin AI, Kristoffersson AC, Ståhl AL, Johansson ME, Westman K, et al. A novel C3 mutation causing increased formation of the C3 convertase in familial atypical hemolytic uremic syndrome. J Immunol. 2012;188(4):2030–7.
121. Brackman D, Sartz L, Leh S, Kristoffersson AC, Bjerre A, Tati R, et al. Thrombotic microangiopathy mimicking membranoproliferative glomerulonephritis. Nephrol Dial Transplant. 2011;26(10):3399–403.
122. Vaziri-Sani F, Holmberg L, Sjöholm AG, Kristoffersson AC, Manea M, Fremeaux-Bacchi V, et al. Phenotypic expression of factor H mutations in patients with atypical hemolytic uremic syndrome. Kidney Int. 2006;69(6):981–8.
123. Marinozzi MC, Vergoz L, Rybkine T, Ngo S, Bettoni S, Pashov A, et al. Complement factor B mutations in atypical hemolytic uremic syndrome-disease-relevant or benign? J Am Soc Nephrol. 2014;25(9):2053–65.
124. de Jorge EG, Macor P, Paixao-Cavalcante D, Rose KL, Tedesco F, Cook HT, et al. The development of atypical hemolytic uremic syndrome depends on complement C5. J Am Soc Nephrol. 2011;22(1):137–45.
125. Frimat M, Tabarin F, Dimitrov JD, Poitou C, Halbwachs-Mecarelli L, Fremeaux-Bacchi V, et al. Complement activation by heme as a secondary hit for atypical hemolytic uremic syndrome. Blood. 2013;122(2):282–92.
126. Manea M, Karpman D. Molecular basis of ADAMTS13 dysfunction in thrombotic thrombocytopenic purpura. Pediatr Nephrol. 2009;24(3):447–58.
127. Dong JF, Moake JL, Nolasco L, Bernardo A, Arceneaux W, Shrimpton CN, et al. ADAMTS-13 rapidly cleaves newly secreted ultralarge von Willebrand factor multimers on the endothelial surface under flowing conditions. Blood. 2002;100(12):4033–9.
128. Noris M, Ruggenenti P, Perna A, Orisio S, Caprioli J, Skerka C, et al. Hypocomplementemia discloses genetic predisposition to hemolytic uremic syndrome and thrombotic thrombocytopenic purpura: role of factor H abnormalities. Italian Registry of Familial and Recurrent Hemolytic Uremic Syndrome/Thrombotic Thrombocytopenic Purpura. J Am Soc Nephrol. 1999;10(2):281–93.
129. Reti M, Farkas P, Csuka D, Razso K, Schlammadinger A, Udvardy ML, et al. Complement activation in thrombotic thrombocytopenic purpura. J Thromb Haemost. 2012;10(5):791–8.

130. Westwood JP, Langley K, Heelas E, Machin SJ, Scully M. Complement and cytokine response in acute Thrombotic Thrombocytopenic Purpura. Br J Haematol. 2014;164(6): 858–66.
131. Mant MJ, Cauchi MN, Medley G. Thrombotic thrombocytopenic purpura: report of a case with possible immune etiology. Blood. 1972;40(3):416–21.
132. Weisenburger DD, O'Conner ML, Hart MN. Thrombotic thrombocytopenic purpura with C'3 vascular deposits: report of a case. Am J Clin Pathol. 1977;67(1):61–3.
133. Chapin J, Weksler B, Magro C, Laurence J. Eculizumab in the treatment of refractory idiopathic thrombotic thrombocytopenic purpura. Br J Haematol. 2012;157(6):772–4.
134. Ruiz-Torres MP, Casiraghi F, Galbusera M, Macconi D, Gastoldi S, Todeschini M, et al. Complement activation: the missing link between ADAMTS-13 deficiency and microvascular thrombosis of thrombotic microangiopathies. Thromb Haemost. 2005;93(3):443–52.
135. Turner NA, Moake J. Assembly and activation of alternative complement components on endothelial cell-anchored ultra-large von Willebrand factor links complement and hemostasis-thrombosis. PLoS One. 2013;8(3), e59372.
136. Feng S, Liang X, Cruz MA, Vu H, Zhou Z, Pemmaraju N, et al. The interaction between factor H and Von Willebrand factor. PLoS One. 2013;8(8), e73715.
137. Rayes J, Roumenina LT, Dimitrov JD, Repesse Y, Ing M, Christophe O, et al. The interaction between factor H and VWF increases factor H cofactor activity and regulates VWF prothrombotic status. Blood. 2014;123(1):121–5.
138. Nolasco L, Nolasco J, Feng S, Afshar-Kharghan V, Moake J. Human complement factor H is a reductase for large soluble von Willebrand factor multimers–brief report. Arterioscler Thromb Vasc Biol. 2013;33(11):2524–8.
139. Jimenez JJ, Jy W, Mauro LM, Horstman LL, Ahn YS. Elevated endothelial microparticles in thrombotic thrombocytopenic purpura: findings from brain and renal microvascular cell culture and patients with active disease. Br J Haematol. 2001;112(1):81–90.
140. Jimenez JJ, Jy W, Mauro LM, Horstman LL, Soderland C, Ahn YS. Endothelial microparticles released in thrombotic thrombocytopenic purpura express von Willebrand factor and markers of endothelial activation. Br J Haematol. 2003;123(5):896–902.
141. Jy W, Jimenez JJ, Mauro LM, Horstman LL, Cheng P, Ahn ER, et al. Endothelial microparticles induce formation of platelet aggregates via a von Willebrand factor/ristocetin dependent pathway, rendering them resistant to dissociation. J Thromb Haemost. 2005;3(6): 1301–8.
142. Hunder GG, Arend WP, Bloch DA, Calabrese LH, Fauci AS, Fries JF, et al. The American College of Rheumatology 1990 criteria for the classification of vasculitis. Introduction. Arthritis Rheum. 1990;33(8):1065–7.
143. Xiao H, Heeringa P, Liu Z, Huugen D, Hu P, Maeda N, et al. The role of neutrophils in the induction of glomerulonephritis by anti-myeloperoxidase antibodies. Am J Pathol. 2005;167(1):39–45.
144. Abdulahad WH, Stegeman CA, Kallenberg CG. Review article: The role of CD4(+) T cells in ANCA-associated systemic vasculitis. Nephrology (Carlton). 2009;14(1):26–32.
145. Xiao H, Heeringa P, Hu P, Liu Z, Zhao M, Aratani Y, et al. Antineutrophil cytoplasmic autoantibodies specific for myeloperoxidase cause glomerulonephritis and vasculitis in mice. J Clin Invest. 2002;110(7):955–63.
146. Primo VC, Marusic S, Franklin CC, Goldmann WH, Achaval CG, Smith RN, et al. Anti-PR3 immune responses induce segmental and necrotizing glomerulonephritis. Clin Exp Immunol. 2010;159(3):327–37.
147. Xing GQ, Chen M, Liu G, Heeringa P, Zhang JJ, Zheng X, et al. Complement activation is involved in renal damage in human antineutrophil cytoplasmic autoantibody associated pauci-immune vasculitis. J Clin Immunol. 2009;29(3):282–91.
148. Magro CM, Crowson AN, Harrist TJ. The use of antibody to C5b-9 in the subclassification of lupus erythematosus. Br J Dermatol. 1996;134(5):855–62.

149. Vassallo G, Newton RW, Chieng SE, Haeney MR, Shabani A, Arkwright PD. Clinical variability and characteristic autoantibody profile in primary C1q complement deficiency. Rheumatology. 2007;46(10):1612–4.
150. Jönsson G, Sjöholm AG, Truedsson L, Bengtsson AA, Braconier JH, Sturfelt G. Rheumatological manifestations, organ damage and autoimmunity in hereditary C2 deficiency. Rheumatology. 2007;46(7):1133–9.
151. Schreiber A, Xiao H, Jennette JC, Schneider W, Luft FC, Kettritz R. C5a receptor mediates neutrophil activation and ANCA-induced glomerulonephritis. J Am Soc Nephrol. 2009; 20(2):289–98.
152. Huugen D, van Esch A, Xiao H, Peutz-Kootstra CJ, Buurman WA, Tervaert JW, et al. Inhibition of complement factor C5 protects against anti-myeloperoxidase antibody-mediated glomerulonephritis in mice. Kidney Int. 2007;71(7):646–54.
153. Xiao H, Schreiber A, Heeringa P, Falk RJ, Jennette JC. Alternative complement pathway in the pathogenesis of disease mediated by anti-neutrophil cytoplasmic autoantibodies. Am J Pathol. 2007;170(1):52–64.
154. Brogan PA, Dillon MJ. Endothelial microparticles and the diagnosis of the vasculitides. Intern Med. 2004;43(12):1115–9.
155. Erdbruegger U, Grossheim M, Hertel B, Wyss K, Kirsch T, Woywodt A, et al. Diagnostic role of endothelial microparticles in vasculitis. Rheumatology (Oxford). 2008;47(12):1820–5.
156. Daniel L, Fakhouri F, Joly D, Mouthon L, Nusbaum P, Grunfeld JP, et al. Increase of circulating neutrophil and platelet microparticles during acute vasculitis and hemodialysis. Kidney Int. 2006;69(8):1416–23.
157. Hong Y, Eleftheriou D, Hussain AA, Price-Kuehne FE, Savage CO, Jayne D, et al. Anti-neutrophil cytoplasmic antibodies stimulate release of neutrophil microparticles. J Am Soc Nephrol. 2012;23(1):49–62.
158. Mesri M, Altieri DC. Endothelial cell activation by leukocyte microparticles. J Immunol. 1998;161(8):4382–7.
159. Pitanga TN, de Aragao FL, Rocha VC, Meirelles T, Borges VM, Goncalves MS, et al. Neutrophil-derived microparticles induce myeloperoxidase-mediated damage of vascular endothelial cells. BMC Cell Biol. 2014;15:21.
160. Wong CS, Mooney JC, Brandt JR, Staples AO, Jelacic S, Boster DR, et al. Risk factors for the hemolytic uremic syndrome in children infected with *Escherichia coli* O157:H7: a multivariable analysis. Clin Infect Dis. 2012;55(1):33–41.
161. Menne J, Nitschke M, Stingele R, Abu-Tair M, Beneke J, Bramstedt J, et al. Validation of treatment strategies for enterohaemorrhagic *Escherichia coli* O104:H4 induced haemolytic uraemic syndrome: case-control study. BMJ. 2012;345, e4565.
162. Karpman D. Management of Shiga toxin-associated *Escherichia coli*-induced haemolytic uraemic syndrome: randomized clinical trials are needed. Nephrol Dial Transplant. 2012;27(10):3669–74.
163. Lapeyraque AL, Malina M, Fremeaux-Bacchi V, Boppel T, Kirschfink M, Oualha M, et al. Eculizumab in severe Shiga-toxin-associated HUS. N Engl J Med. 2011;364(26):2561–3.
164. Loos S, Ahlenstiel T, Kranz B, Staude H, Pape L, Hartel C, et al. An Outbreak of Shiga Toxin-Producing *Escherichia coli* O104:H4 Hemolytic Uremic Syndrome in Germany: Presentation and Short-term Outcome in Children. Clin Infect Dis. 2012;55(6):753–9.
165. Kielstein JT. Best supportive care and therapeutic plasma exchange with or without eculizumab in Shiga toxin-producing *E. coli* O104:H4-induced hemolytic uremic syndrome: An analysis of the German STEC-HUS registry. Nephrol Dial Transplant. 2012; 27(10):3807–15.
166. Legendre CM, Licht C, Muus P, Greenbaum LA, Babu S, Bedrosian C, et al. Terminal complement inhibitor eculizumab in atypical hemolytic-uremic syndrome. N Engl J Med. 2013;368(23):2169–81.
167. Zuber J, Le Quintrec M, Krid S, Bertoye C, Gueutin V, Lahoche A, et al. Eculizumab for atypical hemolytic uremic syndrome recurrence in renal transplantation. Am J Transplant. 2012;12(12):3337–54.

168. Tsai E, Chapin J, Laurence JC, Tsai HM. Use of eculizumab in the treatment of a case of refractory, ADAMTS13-deficient thrombotic thrombocytopenic purpura: additional data and clinical follow-up. Br J Haematol. 2013;162(4):558–9.
169. Coppo R, Peruzzi L, Amore A, Martino S, Vergano L, Lastauka I, et al. Dramatic effects of eculizumab in a child with diffuse proliferative lupus nephritis resistant to conventional therapy. Pediatr Nephrol. 2015;30(1):167–72.
170. El-Husseini A, Hannan S, Awad A, Jennings S, Cornea V, Sawaya BP. Thrombotic microangiopathy in systemic lupus erythematosus: efficacy of eculizumab. Am J Kidney Dis. 2015;65(1):127–30.
171. Ricklin D, Lambris JD. Progress and trends in complement therapeutics. Adv Exp Med Biol. 2013;735:1–22.

Chapter 3
Role of Complement on Broken Surfaces After Trauma

Markus Huber-Lang, Anita Ignatius, and Rolf E. Brenner

Abstract Activation of both the complement and coagulation cascade after trauma and subsequent local and systemic inflammatory response represent a major scientific and clinical problem. After severe tissue injury and bone fracture, exposure of innate immunity to damaged cells and molecular debris is considered a main trigger of the posttraumatic danger response. However, the effects of cellular fragments (e.g., histones) on complement activation remain enigmatic. Furthermore, direct effects of "broken" bone and cartilage surfaces on the fluid phase response of complement and its interaction with key cells of connective tissues are still unknown. Here, we summarize data suggesting direct and indirect complement activation by extracellular and cellular danger associated molecular patterns. In addition, key complement components and the corresponding receptors (such as C3aR, C5aR) have been detected on "exposed surfaces" of the damaged regions. On a cellular level, multiple effects of complement activation products on osteoblasts, osteoclasts, chondrocytes and mesenchymal stem cells have been found.

In conclusion, the complement system may be activated by trauma-altered surfaces and is crucially involved in connective tissue healing and posttraumatic systemic inflammatory response.

Keywords Mesenchymal stem cells • Complement • Trauma • Broken surfaces

M. Huber-Lang (✉)
Department of Orthopedic Trauma, Hand-, Plastic-, and Reconstructive Surgery, University of Ulm, Ulm, Germany
e-mail: markus.huber-lang@uniklinik-ulm.de

A. Ignatius
Institute of Orthopedic Research and Biomechanics, University of Ulm, Ulm, Germany

R.E. Brenner
Division for Biochemistry of Joint and Connective Tissue Diseases, Department of Orthopedics, University of Ulm, Ulm, Germany

J.D. Lambris et al. (eds.), *Immune Responses to Biosurfaces*,
Advances in Experimental Medicine and Biology 865,
DOI 10.1007/978-3-319-18603-0_3

3.1 Introduction

Trauma results in an immediate disruption and damage of various organs, tissues, cells and molecules. Depending on the type of trauma and injury severity the organism is challenged by cellular and molecular debris often functioning as damage associated molecular patterns (DAMPS). Furthermore, injured tissue is often exposed to a vast amount of external or internal pathogen-associated molecular patterns (PAMPS). The resulting immediate inflammatory response includes the "serine protease system" of the coagulation and complement system [1] and pattern recognition receptors of the "first line of cellular defense". Whereas the resulting "fluid phase damage control" induces clotting and walls off injured or infectious tissue, the "cellular damage control" phagocytoses and clears up damaged tissue. Micro-fragments of cartilage and bones may also be resorbed by activated phagocytes and osteoclasts.

The clinical management of damaged and broken surfaces mainly includes a careful surgical debridement of damaged tissues, electrocoagulation/ligation or revascularization of damaged vessels, and osteosynthesis of broken bones. If the tissue defects are too extended, implantation of grafts or metal implants is often necessary. Synchronic to and far beyond the surgical "macro-management" there is a constant body's own "micro-management" by the clotting cascade and immune system. However, there are only a few reports in defining the role of the complement system in detecting and clearing broken bio-surfaces.

Therefore, this presentation aims to summarize possible underlying mechanisms and finally adverts to the importance of the complement system in handling damaged surfaces. Furthermore, an outlook for future research and clinical avenues in treatment of broken surfaces are provided.

3.2 Response to Soft Tissue Disruption

Soft tissue damage mainly involving skin, adipose and muscle tissues leads to a release of multiple cellular content such as mitochondria, nucleosomes, histones, RNA/DNA, autophagosome etc. A significant release of intact or fragmented mitochondria has been shown in serum of trauma patients [2, 3]. Based on the endosymbiotic evolutionary development theory of mitochondria origin, a significant complement activation has been proposed triggered by mitochondria (released after trauma). Especially mannan-binding lectin (MBL), L-ficolin, and M-ficolin seem to sense mitochondrial components and may activate complement via the lectin pathway to clear the mitochondria—then considered as "enemies" [4]. Natural dangerous microsurfaces of non-symbiotic microorganisms function as PAMPS and may thereby massively activate the complement cascade. However, surfaces of microorganism are not covered in this review and may certainly involve the intense crosstalk of toll-like receptors and complement [5, 6].

Concerning histone release it is still a matter of debate whether—and if so, to what extent—they are capable to activate complement. A recent trauma study suggested histone-complexed DNA fragments in plasma to be associated with trauma-induced coagulopathy [5–7] which via an intense serine protease cross-talk [1] is itself associated with trauma-induced complementopathy [8]. In multiple injured patients, own studies have shown a peak appearance of nucleosomes in plasma of patients within the first hour after trauma impact remaining significantly enhanced for 12 h after injury [9]. The nucleosomes correlated with activation of factor seven-activating serine protease (FSAP), a phylogenetically old activation pathway of the coagulation system [9]. In turn, FSAP is capable to cleave both, C3 and C5, thereby generating biologically active anaphylatoxins C3a and C5a, respectively. It is noteworthy, that major activators of FSAP itself are both RNA and DNA fragments, which can be detected after trauma in the experimental and clinical setting. It is still unclear whether complement is directly activated by nucleosomes or histones or via the FSAP pathway. However, histones were capable to generate in human plasma 3 ng/ml C5a whereas at least ten times more was produced by zymosan stimulation [9]. Further differentiation between the various histone forms (e.g., citrullinated, acytylated, methylated) as a potential trigger for complement activation needs to be studied in further detail.

Soft tissue injury excessively releases adipokines including adiponectin which in turn may adhere to damaged tissue surfaces such as the endothelium. Adiponectin as adipose tissue-specific protein has some structural homology with C1q and may theoretically interact with and enforce the complement system [10]. In this regard, it has been shown that adiponectin in partnership with C1q and MBL is capable to induce a shift of macrophages from the pro-inflammatory M1—phenotype to the pro-regenerative M2 phenotype [11] to induced repair processes.

Soft tissue disruption also results in a local and remote ischemia-reperfusion damage activating complement in an early phase by exposition of neo-epitopes in ischemic membranes [12]. In turn, complement-dependent up-regulation of endothelial adhesion molecules such as p-selectin triggers recruitment of additional inflammatory cells. In addition, the stress response to broken surfaces pathophysiologically results in a prolonged vaso-constriction of soft tissue areas thereby further aggravating ischemia in a vicious circle.

Although several experimental studies suggest some promising effects of a specific complement blockade (e.g., C1q-inhibition, C5a-blockade) in improving local and systemic ischemia-reperfusion injury [13–16], its protective effects in the clinical setting remains to be proven [12].

3.3 Acute Response to Broken to Surfaces of Cartilage Tissue

Cartilage injury represents a major risk factor for the development of posttraumatic arthritis and subsequently degenerative joint disease [17]. In the USA about 12 % of all cases with osteoarthritis of the hip-, knee- and ankle-joint have been attributed to a preceeding major traumatic event [18]. Different kinds of tissue damage such as direct cartilage trauma, intra-articular fracture and injury of menisci, ligaments or joint capsule could be involved in varying combination [19]. There is experimental evidence for an additive morphological effect of cartilage degeneration by combined lesions as shown for a rabbit model of blunt cartilage trauma with and without radial transection of the medial meniscus [20]. Among the different types of joint injuries articular fracture carries the highest risk for development of posttraumatic osteoarthritis [19, 21]. It has been reported that between 23 and 44 % of patients with intraarticular fractures of the knee joint develop degenerative joint disease [22, 23]. Joint injuries—especially those with intra-articular fractures—may also lead to hemarthrosis. Its influence on the long-term prognosis of a traumatized joint is not completely clarified. In vitro, however, blood exposure of cartilage has adverse effects and leads to chondrocyte apoptosis and loss of proteoglycans [24]. In an animal study on dogs the coagulation system further increased the resulting cartilage damage after repeated intra-articular blood application [25]. However, possible detrimental effects of complement activation within hemathrosis have not been clarified so far. During the last years the concept arose that synovitis and the innate immune system play an essential role in the pathogenesis of osteoarthritis [26, 27]. This is based on the notion that mechanical or other primary damage to cartilage tissue causes the release DAMPs inducing toll-like receptor (TLR)-activation, synovial inflammation, recruitment of inflammatory cells, enhanced secretion of cytokines, release and activation of cartilage degrading proteases like MMP13, ADAMTS 4 and ADAMTS 5 [26]. As mentioned above, DAMPs are heterogeneous molecules originating either from various compartments inside the cell or the extracellular matrix [28]. Since cartilage is characterized by a low cell density, the release of intracellular DAMPs may play a minor role compared to other tissues (especially in comparison to the previously discussed soft tissue). On the other hand it is known that chondrocyte death is highest around matrix cracks and that articular fracture causes significant chondrocyte death at the fracture edge [29, 30]. Therefore, DAMPs released from necrotic or apoptotic chondrocytes may contribute to an activation of the innate immune system. Binding of these ligands to TLRs of synovial cells or chondrocytes leads to the secretion of pro-inflammatory cytokines and chemokines, cellular infiltration of the synovial tissue and generation of proteases with the capacity to degrade the extracellular matrix of the cartilage [27, 31]. This extracellular matrix determines the functional properties of articular cartilage and has a highly complex molecular organization [32]. By structural mechanical disintegration or proteolytic activity single components or fragments of matrix components may also interact with TLRs further enhancing the inflammatory process. Furthermore, there is growing evidence that besides these TLR-transmitted processes the complement system is crucially involved in the pathogenetic processes induced by broken cartilage surfaces generated by a severe joint trauma [33] (Fig. 3.1).

Complement factors and complement regulatory proteins in the synovial fluid may originate from synovial blood supply, different types of synovial cells, chondrocytes or hemarthrosis in the course of a joint trauma. Overall, relevant complement components are present in the synovial fluid in sufficient amounts [31, 33].

In chondrocytes the synthesis of C1q, C1s, C2 and C4 as well as membrane bound complement regulatory proteins CD46, CD55 and CD59 has been reported [34, 35]. Furthermore, the expression of receptors for the anaphylatoxins C3a and C5a (C3aR and C5aR, respectively) has been described for human chondrocytes [35, 36]. Although C3aR and C5aR expression was found to be regulated by TNF, IL1beta and IL10 [35, 36] their functional relevance in cartilage homeostasis and pathology has not been elucidated in detail so far. Nevertheless, their presence at the chondrocyte surface and the identification of terminal complement complex formation (TCC, sC5b-9) on the surface of chondrocytes from osteoarthritic cartilage [37] indicates that chondrocytes are potential targets for complement-mediated processes. Interestingly, several of the key matrix components of cartilage itself have been shown to directly influence complement activation either through a stimulatory or inhibitory function [38].

The extracellular matrix of cartilage mainly consists of a network of collagen heterofibrils composed of collagen types II, IX and XI, collagen type VI, negatively charged proteoglycans (mostly aggrecan and members of the small leucine-rich repeat protein family like decorin, biglycan, fibromodulin, asporin, chondroadherin and PRELP), hyaluronan and additional molecules specifically interacting with other matrix components like matrilin 1 and 3, cartilage oligomeric matrix protein (COMP), fibronectin or link protein [32]. For several of these components like decorin, biglycan, chondroadherin or fibronectin and some glycosaminoglycans (present in cartilage proteoglycans) binding to complement factors e.g., C1q has been shown [39–43]. Distinct functionalities with respect to activation or inhibition of complement have been reported for the NC4-domain of collagen type IX (e.g., inhibition of C9 polymerization and thereby TCC formation), the C-type lectin part of the aggrecan G3 domain (e.g., activation of the classical and to a lesser extent the alternative pathway through binding of C1q and C3), fibromodulin (e.g., activation of the classical pathway by binding to C1q), PRELP (e.g., inhibition of TCC formation by prevention of C9 polymerization), decorin and biglycan (e.g., inhibition of the classical or classical and lectin pathway by binding of C1q and MBL) as well as COMP (e.g., activation of the alternative pathway and inhibition of the classical and lectin pathway) [44–49]. Interestingly, some of the matrix proteins that activate complement by binding of C1q also interact with complement inhibitors C4-binding protein and CFH and thus contribute to a limitation of more terminal steps of complement activation [31]. This is also the case for DNA—one of the ligands released from dying chondrocytes—which may contribute to complement activation at broken cartilage surfaces.

In the past, the role of complement received much interest in the context of primary immune-mediated rheumatoid arthritis for which an important role of C5a has been described in mouse models [50, 51]. Subsequent attempts to translate this knowledge into pharmacological therapeutic approaches in men by blocking the receptor for C5a were not successful so far [52]. More recently, a central role of complement was postulated based on a surgically induced model of osteoarthritis [37]. The authors found that C5- and C6-deficient mice were somehow protected from osteoarthritis, while mice deficient in CD59a which is an inhibitor of TCC formation were more severely affected. In line with these results mice with a deficiency of cyclopeptidase B which inhibits complement activation by inactivating C5a and reduces TCC-formation developed more severe cartilage damage compared to wildtype mice in the same surgical model of medial meniscectomy [37, 53]. These observations strongly indicate an involvement of the TCC in the pathogenetic process. This is further supported by overall elevated levels of C3a and sC5b-9 in synovial fluids of patients with osteoarthritis [37]. In this report complement activation with TCC-formation was also shown for pulverized human osteoarthritic cartilage, pulverized human osteoarthritic synovium, and for fibromodulin as well as aggrecan but not for collagen type II or matrilin 3. Furthermore, in vitro generation of TCC on human chondrocytes induced elevated expression of several genes usually associated with osteoarthritis (e.g., MMP13, ADAMTS4, ADAMTS5, PTGS2). Nevertheless, the question of relevance in human patients with joint injury remains an important point. Recently, in the synovial fluids of patients with ankle fractures significantly higher levels of C3a, C5a and sC5b-9 (TCC) were found at a mean time point of about 2 days after injury compared to synovial fluids from patients with osteochondrosis dissecans of the same joint [54]. This indicates early complement activation after intra-articular fracture in a patient cohort at high risk for development of posttraumatic osteoarthritis.

The current knowledge therefore indicates that in synovial joints a delicate balance of inhibitory and stimulatory factors of complement activation exists. By creating cartilage damage and artificial cartilage surfaces through mechanical trauma DAMPs are presented at the broken surface or released into the synovial fluid which have the potential to activate TLRs and the complement system as major functional cross-talking parts of the innate immune system. The induced inflammatory reaction increases the proteolytic activity for cartilage matrix components creating fragments thereof which can further trigger the innate immune response. Finally, trauma-associated hemarthrosis may not only provide complement factors but can influence complement activation by interaction during clotting processes [1]. At present, the relevant conditions leading to such an in vivo scenario are not sufficiently understood. Therefore, further studies are needed with different well-defined joint injury models all of which may not be possible in mice due to limitations in size. Moreover, translational investigations of intra-articular complement activation after joint trauma in human patients with different injury types and severity are necessary. For specific trauma situations characterized by deleterious complement activation early pharmacological inhibition of this process [55] might become a realistic option for limiting the resulting damage in the future.

3.4 Early Response to Broken Surfaces After Fracture

A recent clinical study showed in haematoma of ankle fractures generation of C3a, C5a and TCC [54]. In several own experimental studies complement has been shown to be involved in fracture healing (Fig. 3.1). Since the early phase of fracture healing is dominated by a local inflammatory response complement may especially

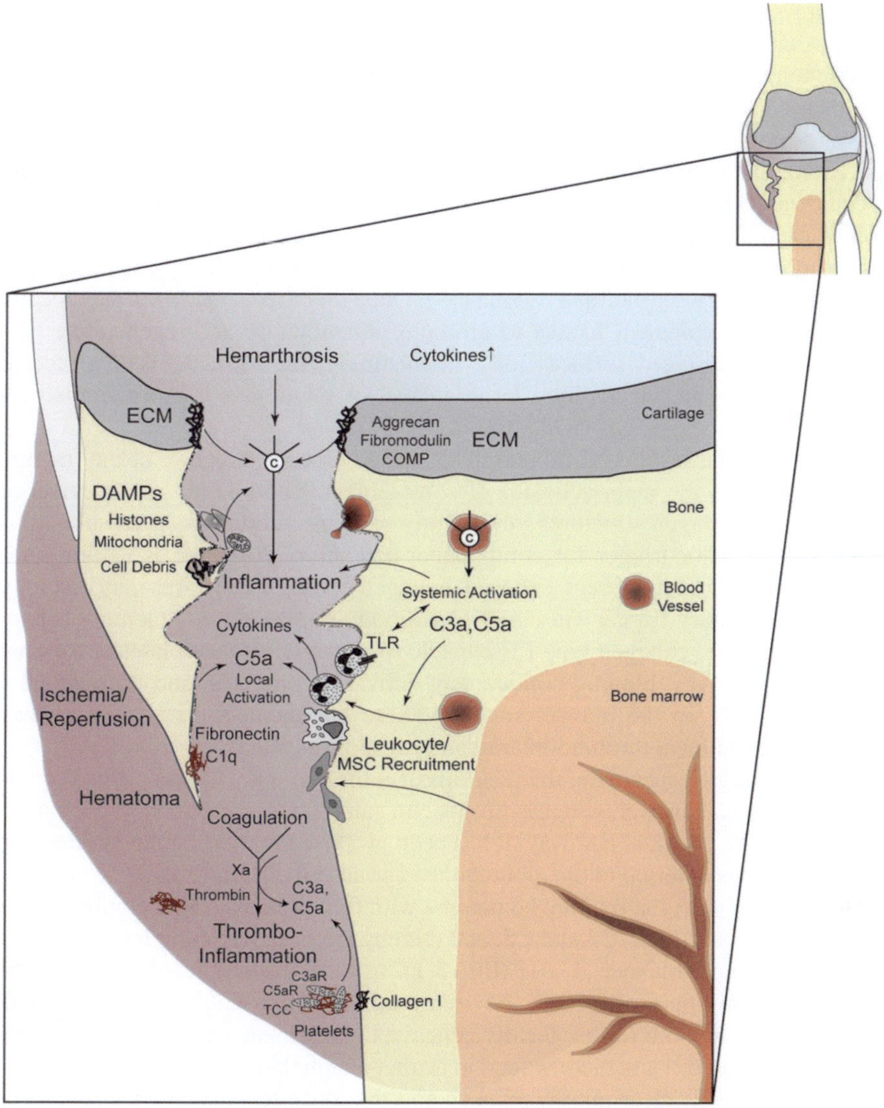

Fig. 3.1 Role of the complement and coagulation systems after fracture of bone and cartilage tissue. *ECM* extracelluar matrix, *COMP* cartilage oligomeric matrix protein, *MSC* mesenchymal stem cells, *TCC* terminal complement complex (sC5b-9), *TLR* toll-like receptor, *DAMPs* danger-associated molecular patterns, *Xa* activated clotting factor X

modulate the neutrophil and macrophage response, known as "first line of defense". Furthermore, systemic effects of complement activation may significantly contribute to regenerative processes of broken surfaces. In support, experimental blunt chest trauma-induced systemic activation of complement with generation of C3a and C5a significantly compromised fracture healing in rodents [56, 57]. It is noteworthy, that in presence of a concomitant blunt chest trauma, blockade of C5a was capable to normalize fracture healing processes therefore representing a promising therapeutic strategy [56, 57]. On a cellular level, direct effects of complement factors on osteoblast and osteoclast functions have recently been shown [58]. In addition, osteoblasts were capable of producing C3 and C5, the latter could be activated by osteoclasts. There is also evidence, that anaphylatoxin C3a plays a crucial role in mobilization of stem cells from the bone marrow to broken and defective surfaces. For example, mesenchymal stem cells (MSC) migrate along C3a gradients in concentrations found after multiple trauma [59]. Once mobilized, the MSC may locally generate key complement factors to promote inflammatory and regenerative processes on the damaged surfaces [58]. Interestingly, the MSC is itself protected against complement attacks by a large arsenal of membrane-bound complement regulatory proteins (such as CD35, CD46, CD59) [58].

On a molecular level it is still enigmatic which osseous structure of the fracture region represents the main activator of complement. There is no strong evidence that damaged collagen structures (abundantly existing in bone tissue as subtype I) may act as a major trigger for complement activation. However, if natural anti-collagen antibodies are present, exposed damaged collagen structures may certainly act as inflammatory trigger via classical antibody-dependent complement activation. Furthermore, collagen type I is capable to strongly activate platelets forming the sC5b-9 complex, binding complement activation products, and up-regulating C3aR and C5aR on the platelet surface. These complement-driven effects may result in a strong local thromboinflammatory response [60].

It is tempting to speculate, that also fibronectin may play an important role on broken surfaces. However, whereas fibronectin can bind C1q it does not result in an activation of the classical pathway. It has been proposed that the active binding site for fibronectin is external to the globular head structure of C1q [39].

On the other side, the fracture hematoma with thrombin as a main clotting factor is capable to cleave both, C3 and C5, and thereby may activate complement locally in a non-canonical manner [1, 61] (Fig. 3.1). In turn, the generated C3a and C5a represent potent chemoattractants for many inflammatory cells. Neutrophils and macrophages may both further locally generate complement activation products by a serine-protease [62] and thereby feed in further anaphylatoxins to drive the inflammatory response, to clear mini-fragments and to induce regenerative processes for successfully stabilizing the fracture region.

3.5 Role of Complement in Broken Artificial Surfaces

Surgical implantation of a biomaterial always goes along with tissue injury. Therefore, the artificial surface gets in tight contact with broken natural surfaces of soft tissue, cartilage or bone. Although surgical procedures for implantation maximally reduce the extent of tissue damage—e.g., through minimal invasive surgery—a contribution to the local inflammatory response can be assumed. On the other hand the biomaterial itself with its specific chemical, physical and (nano-) topographical characteristics determines the initial adsorption of proteins from blood or interstitial fluids which occurs within a timespan of nanoseconds [63]. The composition and amount of the bound proteins determines the local inflammatory response, blood clotting processes and complement activation mostly through the classical and alternative, partly also the lectin mediated pathway [63, 64]. The amount of inflammation and complement activation defines not only the primary biocompatibility of a biomaterial but may either support subsequent tissue integration if it is well controlled, or lead to a failure of this process if it goes out of control. In the case of biomaterial degradation or particle release complement activation triggered by newly presented artificial surfaces may become relevant and may again lead to complement activation. With respect to materials used for joint replacement it was reported that particles consisting of pure titanium, high density and ultrahigh molecular weight polyethylene (UHMWPE) as well as polymethylmethacrylate (PMMA) which is present in bone cement partly used for implantation generate C3a in a standardized in vitro assay [65]. Erythrocyte lysis-assays indicated that polyethylene particles activate the alternative pathway of complement activation and adsorb activated complement components [66]. Furthermore, in synovial tissue from patients with aseptic loosening of hip prostheses C3a, iC3b and soluble C5-9 was localized by immunostaining at the periphery of polyethylene particles [66]. Therefore, complement may also play a role in aseptic loosening of implants which includes activation of macrophages and is associated with peri-implant osteolysis.

3.6 Outlook

Complement acts as an important modulator of the acute immune response on broken surfaces and thereby represents an interesting target for specific inhibitory strategies to support clearance of molecular and cellular debris, to pursue regenerative processes, and to improve long-lasting integration of artificial surfaces of implants. Especially blockade of key complement components, such as C3 or C5 and the related anaphylatoxins may help in healing processes of broken tissue surfaces. Furthermore, coating of artificial surfaces with complement-regulatory proteins to avoid local complement activation may lead to an improved integration of implants not only in the physiological environment but also in inflamed or infected environments (e.g., in osteomyelitis). To realistically assess the possible benefit of a complement modulatory approach for these indications, more valid experimental and clinical studies are needed.

Acknowledgments The cited author's own work was funded in part by the State of Baden-Württemberg (Perspektivförderung), the DFG Clinical Research Unit KFO200 TP2 and TP4, and the SFB1149.

References

1. Huber-Lang M, Sarma JV, Zetoune FS, Rittirsch D, Neff TA, McGuire SR, Lambris JD, Warner RL, Flierl MA, Hoesel LM, Gebhard F, Younger JG, Drouin SM, Wetsel RA, Ward PA. Generation of C5a in the absence of C3: a new complement activation pathway. Nat Med. 2006;12:682–7.
2. Zhang Q, Raoof M, Chen Y, Sumi Y, Sursal T, Junger W, Brohi K, Itagaki K, Hauser CJ. Circulating mitochondrial DAMPs cause inflammatory responses to injury. Nature. 2010;464:104–7.
3. Zhao C, Itagaki K, Gupta A, Odom S, Sandler N, Hauser CJ. Mitochondrial damage-associated molecular patterns released by abdominal trauma suppress pulmonary immune responses. J Trauma Acute Care Surg. 2014;76:1222–7.
4. Brinkmann CR, Jensen L, Dagnaes-Hansen F, Holm IE, Endo Y, Fujita T, Thiel S, Jensenius JC, Degn SE. Mitochondria and the lectin pathway of complement. J Biol Chem. 2013;288:8016–27.
5. Huber-Lang M, Barratt-Due A, Pischke SE, Sandanger O, Nilsson PH, Nunn MA, Denk S, Gaus W, Espevik T, Mollnes TE. Double blockade of CD14 and complement C5 abolishes the cytokine storm and improves morbidity and survival in polymicrobial sepsis in mice. J Immunol. 2014;192:5324–31.
6. Song WC. Crosstalk between complement and toll-like receptors. Toxicol Pathol. 2012;40:174–82.
7. Johansson PI, Windelov NA, Rasmussen LS, Sorensen AM, Ostrowski SR. Blood levels of histone-complexed DNA fragments are associated with coagulopathy, inflammation and endothelial damage early after trauma. J Emerg Trauma Shock. 2013;6:171–5.
8. Burk AM, Martin M, Flierl MA, Rittirsch D, Helm M, Lampl L, Bruckner U, Stahl GL, Blom AM, Perl M, Gebhard F, Huber-Lang M. Early complementopathy after multiple injuries in humans. Shock. 2012;37:348–54.
9. Kanse SM, Gallenmueller A, Zeerleder S, Stephan F, Rannou O, Denk S, Etscheid M, Lochnit G, Krueger M, Huber-Lang M. Factor VII-activating protease is activated in multiple trauma patients and generates anaphylatoxin C5a. J Immunol. 2012;188:2858–65.
10. Diez JJ, Iglesias P. The role of the novel adipocyte-derived hormone adiponectin in human disease. Eur J Endocrinol. 2003;148:293–300.
11. Bohlson SS, O'Conner SD, Hulsebus HJ, Ho MM, Fraser DA. Complement, c1q, and c1q-related molecules regulate macrophage polarization. Front Immunol. 2014;5:402.
12. Diepenhorst GM, van Gulik TM, Hack CE. Complement-mediated ischemia-reperfusion injury: lessons learned from animal and clinical studies. Ann Surg. 2009;249:889–99.
13. Bless NM, Warner RL, Padgaonkar VA, Lentsch AB, Czermak BJ, Schmal H, Friedl HP, Ward PA. Roles for C-X-C chemokines and C5a in lung injury after hindlimb ischemia-reperfusion. Am J Physiol. 1999;276:L57–63.
14. Duehrkop C, Banz Y, Spirig R, Miescher S, Nolte MW, Spycher M, Smith RA, Sacks SH, Rieben R. C1 esterase inhibitor reduces lower extremity ischemia/reperfusion injury and associated lung damage. PLoS One. 2013;8, e72059.
15. Lindsay TF, Hill J, Ortiz F, Rudolph A, Valeri CR, Hechtman HB, Moore Jr FD. Blockade of complement activation prevents local and pulmonary albumin leak after lower torso ischemia-reperfusion. Ann Surg. 1992;216:677–83.
16. Woodruff TM, Arumugam TV, Shiels IA, Reid RC, Fairlie DP, Taylor SM. Protective effects of a potent C5a receptor antagonist on experimental acute limb ischemia-reperfusion in rats. J Surg Res. 2004;116:81–90.
17. Lotz MK, Kraus VB. New developments in osteoarthritis. Posttraumatic osteoarthritis: pathogenesis and pharmacological treatment options. Arthritis Res Ther. 2010;12:211.
18. Brown TD, Johnston RC, Saltzman CL, Marsh JL, Buckwalter JA. Posttraumatic osteoarthritis: a first estimate of incidence, prevalence, and burden of disease. J Orthop Trauma. 2006;20:739–44.

19. Furman BD, Olson SA, Guilak F. The development of posttraumatic arthritis after articular fracture. J Orthop Trauma. 2006;20:719–25.
20. Brophy RH, Martinez M, Borrelli Jr J, Silva MJ. Effect of combined traumatic impact and radial transection of medial meniscus on knee articular cartilage in a rabbit in vivo model. Arthroscopy. 2012;28:1490–6.
21. McKinley TO, Borrelli Jr J, D'Lima DD, Furman BD, Giannoudis PV. Basic science of intra-articular fractures and posttraumatic osteoarthritis. J Orthop Trauma. 2010;24:567–70.
22. Honkonen SE. Degenerative arthritis after tibial plateau fractures. J Orthop Trauma. 1995;9:273–7.
23. Volpin G, Dowd GS, Stein H, Bentley G. Degenerative arthritis after intra-articular fractures of the knee. Long-term results. J Bone Joint Surg Br. 1990;72:634–8.
24. Jansen NW, Roosendaal G, Bijlsma JW, Degroot J, Lafeber FP. Exposure of human cartilage tissue to low concentrations of blood for a short period of time leads to prolonged cartilage damage: an in vitro study. Arthritis Rheum. 2007;56:199–207.
25. van Meegeren ME, Roosendaal G, Barten-van Rijbroek AD, Schutgens RE, Lafeber FP, Mastbergen SC. Coagulation aggravates blood-induced joint damage in dogs. Arthritis Rheum. 2012;64:3231–9.
26. Scanzello CR, Goldring SR. The role of synovitis in osteoarthritis pathogenesis. Bone. 2012;51:249–57.
27. Orlowsky EW, Kraus VB. The role of innate immunity in osteoarthritis: when our first line of defense goes on the offensive. J Rheumatol. 2015;2(3):363–71.
28. Schaefer L. Complexity of danger: the diverse nature of damage-associated molecular patterns. J Biol Chem. 2014;289:35237–45.
29. Backus JD, Furman BD, Swimmer T, Kent CL, McNulty AL, Defrate LE, Guilak F, Olson SA. Cartilage viability and catabolism in the intact porcine knee following transarticular impact loading with and without articular fracture. J Orthop Res. 2011;29:501–10.
30. Lewis JL, Deloria LB, Oyen-Tiesma M, Thompson Jr RC, Ericson M, Oegema Jr TR. Cell death after cartilage impact occurs around matrix cracks. J Orthop Res. 2003;21:881–7.
31. Sjoberg AP, Trouw LA, Blom AM. Complement activation and inhibition: a delicate balance. Trends Immunol. 2009;30:83–90.
32. Heinegard D, Saxne T. The role of the cartilage matrix in osteoarthritis. Nat Rev Rheumatol. 2011;7:50–6.
33. John T, Stahel PF, Morgan SJ, Schulze-Tanzil G. Impact of the complement cascade on post-traumatic cartilage inflammation and degradation. Histol Histopathol. 2007;22:781–90.
34. Bradley K, North J, Saunders D, Schwaeble W, Jeziorska M, Woolley DE, Whaley K. Synthesis of classical pathway complement components by chondrocytes. Immunology. 1996;88:648–56.
35. Schulze-Tanzil G, Kohl B, El SK, Arens S, Ertel W, Stolzel K, John T. Anaphylatoxin receptors and complement regulatory proteins in human articular and non-articular chondrocytes: interrelation with cytokines. Cell Tissue Res. 2012;350:465–75.
36. Onuma H, Masuko-Hongo K, Yuan G, Sakata M, Nakamura H, Kato T, Aoki H, Nishioka K. Expression of the anaphylatoxin receptor C5aR (CD88) by human articular chondrocytes. Rheumatol Int. 2002;22:52–5.
37. Wang Q, Rozelle AL, Lepus CM, Scanzello CR, Song JJ, Larsen DM, Crish JF, Bebek G, Ritter SY, Lindstrom TM, Hwang I, Wong HH, Punzi L, Encarnacion A, Shamloo M, Goodman SB, Wyss-Coray T, Goldring SR, Banda NK, Thurman JM, Gobezie R, Crow MK, Holers VM, Lee DM, Robinson WH. Identification of a central role for complement in osteoarthritis. Nat Med. 2011;17:1674–9.
38. Happonen KE, Heinegard D, Saxne T, Blom AM. Interactions of the complement system with molecules of extracellular matrix: relevance for joint diseases. Immunobiology. 2012;217:1088–96.
39. Sjoberg AP, Manderson GA, Morgelin M, Day AJ, Heinegard D, Blom AM. Short leucine-rich glycoproteins of the extracellular matrix display diverse patterns of complement interaction and activation. Mol Immunol. 2009;46:830–9.

40. Krumdieck R, Hook M, Rosenberg LC, Volanakis JE. The proteoglycan decorin binds C1q and inhibits the activity of the C1 complex. J Immunol. 1992;149:3695–701.
41. Bing DH, Almeda S, Isliker H, Lahav J, Hynes RO. Fibronectin binds to the C1q component of complement. Proc Natl Acad Sci U S A. 1982;79:4198–201.
42. Almeda S, Rosenberg RD, Bing DH. The binding properties of human complement component C1q. Interaction with mucopolysaccharides. J Biol Chem. 1983;258:785–91.
43. Clark SJ, Bishop PN, Day AJ. The proteoglycan glycomatrix: a sugar microenvironment essential for complement regulation. Front Immunol. 2013;4:412.
44. Kalchishkova N, Furst CM, Heinegard D, Blom AM. NC4 Domain of cartilage-specific collagen IX inhibits complement directly due to attenuation of membrane attack formation and indirectly through binding and enhancing activity of complement inhibitors C4B-binding protein and factor H. J Biol Chem. 2011;286:27915–26.
45. Melin FC, Morgelin M, Vadstrup K, Heinegard D, Aspberg A, Blom AM. The C-type lectin of the aggrecan G3 domain activates complement. PLoS One. 2013;8, e61407.
46. Sjoberg A, Onnerfjord P, Morgelin M, Heinegard D, Blom AM. The extracellular matrix and inflammation: fibromodulin activates the classical pathway of complement by directly binding C1q. J Biol Chem. 2005;280:32301–8.
47. Happonen KE, Furst CM, Saxne T, Heinegard D, Blom AM. PRELP protein inhibits the formation of the complement membrane attack complex. J Biol Chem. 2012;287:8092–100.
48. Happonen KE, Saxne T, Aspberg A, Morgelin M, Heinegard D, Blom AM. Regulation of complement by cartilage oligomeric matrix protein allows for a novel molecular diagnostic principle in rheumatoid arthritis. Arthritis Rheum. 2010;62:3574–83.
49. Acharya C, Yik JH, Kishore A, Van DV, Di Cesare PE, Haudenschild DR. Cartilage oligomeric matrix protein and its binding partners in the cartilage extracellular matrix: interaction, regulation and role in chondrogenesis. Matrix Biol. 2014;37:102–11.
50. Wang Y, Rollins SA, Madri JA, Matis LA. Anti-C5 monoclonal antibody therapy prevents collagen-induced arthritis and ameliorates established disease. Proc Natl Acad Sci U S A. 1995;92:8955–9.
51. Okroj M, Heinegard D, Holmdahl R, Blom AM. Rheumatoid arthritis and the complement system. Ann Med. 2007;39:517–30.
52. Vergunst CE, Gerlag DM, Dinant H, Schulz L, Vinkenoog M, Smeets TJ, Sanders ME, Reedquist KA, Tak PP. Blocking the receptor for C5a in patients with rheumatoid arthritis does not reduce synovial inflammation. Rheumatology (Oxford). 2007;46:1773–8.
53. Lepus CM, Song JJ, Wang Q, Wagner CA, Lindstrom TM, Chu CR, Sokolove J, Leung LL, Robinson WH. Brief report: carboxypeptidase B serves as a protective mediator in osteoarthritis. Arthritis Rheumatol. 2014;66:101–6.
54. Schmal H, Salzmann GM, Niemeyer P, Langenmair E, Guo R, Schneider C, Habel M, Riedemann N. Early intra-articular complement activation in ankle fractures. Biomed Res Int. 2014;2014:426893.
55. Ricklin D, Lambris JD. Complement in immune and inflammatory disorders: therapeutic interventions. J Immunol. 2013;190:3839–47.
56. Huber-Lang M, Kovtun A, Ignatius A. The role of complement in trauma and fracture healing. Semin Immunol. 2013;25:73–8.
57. Recknagel S, Bindl R, Brochhausen C, Gockelmann M, Wehner T, Schoengraf P, Huber-Lang M, Claes L, Ignatius A. Systemic inflammation induced by a thoracic trauma alters the cellular composition of the early fracture callus. J Trauma Acute Care Surg. 2013;74:531–7.
58. Ignatius A, Schoengraf P, Kreja L, Liedert A, Recknagel S, Kandert S, Brenner RE, Schneider M, Lambris JD, Huber-Lang M. Complement C3a and C5a modulate osteoclast formation and inflammatory response of osteoblasts in synergism with IL-1beta. J Cell Biochem. 2011;112:2594–605.
59. Hengartner NE, Fiedler J, Schrezenmeier H, Huber-Lang M, Brenner RE. Crucial role of IL1beta and C3a in the in vitro-response of multipotent mesenchymal stromal cells to inflammatory mediators of polytrauma. PLoS One. 2015;10, e0116772.

60. Martel C, Cointe S, Maurice P, Matar S, Ghitescu M, Theroux P, Bonnefoy A. Requirements for membrane attack complex formation and anaphylatoxins binding to collagen-activated platelets. PLoS One. 2011;6, e18812.
61. Amara U, Flierl MA, Rittirsch D, Klos A, Chen H, Acker B, Bruckner UB, Nilsson B, Gebhard F, Lambris JD, Huber-Lang M. Molecular intercommunication between the complement and coagulation systems. J Immunol. 2010;185:5628–36.
62. Huber-Lang M, Younkin EM, Sarma JV, Riedemann N, McGuire SR, Lu KT, Kunkel R, Younger JG, Zetoune FS, Ward PA. Generation of C5a by phagocytic cells. Am J Pathol. 2002;161:1849–59.
63. Franz S, Rammelt S, Scharnweber D, Simon JC. Immune responses to implants – a review of the implications for the design of immunomodulatory biomaterials. Biomaterials. 2011;32:6692–709.
64. Nilsson B, Ekdahl KN, Mollnes TE, Lambris JD. The role of complement in biomaterial-induced inflammation. Mol Immunol. 2007;44:82–94.
65. Noordin S, Shortkroff S, Sledge CB, Spector M. Investigation of the activation of a human serum complement protein, C3, by orthopedic prosthetic particulates. Biomaterials. 2004;25:5347–52.
66. DeHeer DH, Engels JA, DeVries AS, Knapp RH, Beebe JD. In situ complement activation by polyethylene wear debris. J Biomed Mater Res. 2001;54:12–9.

Chapter 4
Complement Involvement in Periodontitis: Molecular Mechanisms and Rational Therapeutic Approaches

George Hajishengallis, Tomoki Maekawa, Toshiharu Abe, Evlambia Hajishengallis, and John D. Lambris

Abstract The complement system is a network of interacting fluid-phase and cell surface-associated molecules that trigger, amplify, and regulate immune and inflammatory signaling pathways. Dysregulation of this finely balanced network can destabilize host-microbe homeostasis and cause inflammatory tissue damage. Evidence from clinical and animal model-based studies suggests that complement is implicated in the pathogenesis of periodontitis, a polymicrobial community-induced chronic inflammatory disease that destroys the tooth-supporting tissues. This review discusses molecular mechanisms of complement involvement in the dysbiotic transformation of the periodontal microbiome and the resulting destructive inflammation, culminating in loss of periodontal bone support. These mechanistic studies have additionally identified potential therapeutic targets. In this regard, interventional studies in preclinical models have provided proof-of-concept for using complement inhibitors for the treatment of human periodontitis.

Keywords Complement • C3 • C5a receptor • Periodontitis • Dysbiosis • Inflammation • *P. gingivalis* • Therapeutics

---

G. Hajishengallis (✉) • T. Maekawa • T. Abe
Department of Microbiology, Penn Dental Medicine, University of Pennsylvania, Philadelphia, PA 19104, USA
e-mail: geoh@upenn.edu

E. Hajishengallis
Department of Restorative and Preventive Sciences, Penn Dental Medicine, University of Pennsylvania, Philadelphia, PA 19104, USA

J.D. Lambris
Perelman School of Medicine, University of Pennsylvania, Philadelphia, PA, USA

J.D. Lambris et al. (eds.), *Immune Responses to Biosurfaces*, Advances in Experimental Medicine and Biology 865,
DOI 10.1007/978-3-319-18603-0_4

4.1 Introduction

Periodontitis is a chronic inflammatory disease that compromises the integrity of the periodontium, i.e., the tooth-supporting structures such as the gingiva, periodontal ligament, and the alveolar bone [1]. The disease is initiated by inflammation caused by dysbiotic bacterial communities forming on subgingival tooth sites [2]. Similarly to other chronic diseases, periodontitis requires a susceptible host. Susceptibility to periodontitis is determined by genetic factors that may predispose to hyperinflammatory responses or by environmental factors (e.g., diet and stress) and risk-related behavior (e.g., smoking) that can modify the host immune response in a destructive direction [3–7]. Regardless of the complexity underlying periodontal disease susceptibility, the control of the host periodontal inflammatory response is considered to be central to the treatment of the disease [1]. Therefore, identifying key inflammatory pathways that mediate periodontal tissue destruction has important translational implications.

Chronic periodontitis affects >47 % of U.S. adults [8] and has been prevalent since antiquity [9]. Severe periodontitis, which affects 8.5 % of adults [8], is not only a common cause of tooth loss, but is also associated with increased risk for atherosclerosis, diabetes, rheumatoid arthritis, and adverse pregnancy outcomes [10–13]. The high prevalence of periodontitis [8], its significant economic burden [14, 15], and the fact that many clinical cases are refractory to standard modes of treatment (combined mechanical and antimicrobial therapy, including scaling and root planning, surgery, and systemically administered antibiotics) [16, 17] underscore the importance of implementing innovative and cost-effective therapeutic interventions. In this review, we summarize published evidence that the destructive host inflammatory response in periodontitis is heavily dependent on the activation of the complement system. Moreover, we discuss recent studies that provided proof-of-concept that complement inhibition is a promising therapeutic strategy for the treatment of this oral disease.

4.2 Complement

Traditionally known as a cascade of antimicrobial proteins in the blood, complement is now recognized as a key system for immune surveillance and homeostasis and a major link between the innate and the adaptive arms of the host immune response [18]. In addition to the classic serum proteins (C1-9), the integrated complement system comprises pattern-recognition molecules, convertases and other proteases, regulators, and receptors for interactions with immune mediators [18]. The complement cascade can be triggered by distinct mechanisms (classical, lectin, or alternative), all of which converge at the third complement component (C3) and lead to the generation of effectors that mediate diverse functions. These include the recruitment and activation of inflammatory cells (via the C3a and C5a anaphylatoxins that activate specific G-protein-coupled receptors, C3aR and C5aR [CD88], respectively), microbial opsonization and phagocytosis (e.g., through the C3b or C4b opsonins), and direct lysis of susceptible targeted microbes (by means of the C5b-9 membrane attack complex) [18]. It should be noted that the activities of complement are not restricted to a linear cascade of events but rather involve a network of interactions with other systems, which together coordinate the host response to infection or tissue injury. These complement interactions can amplify innate immune and inflammatory responses through synergy with Toll-like receptors (TLRs) [19], provide a barrier against the spread of invading bacteria by potentiating local clotting [20], replenish the immune system through mobilization of hematopoietic stem/progenitor cells from the bone marrow [21] and regulate the activation and differentiation of T-cell subsets [22, 23].

Owing to the operation of a sophisticated system of negative regulators (e.g., the fluid-phase regulators factor H and C4-binding protein and the cell-associated regulators CD46 and CD59), complement is not normally activated on the surface of host cells and tissues [24]. However, disruption of these regulatory mechanisms by specific complement gene mutations or by subversive pathogens can lead to complement over-activation and hence unwarranted inflammation and possibly damage to host tissues. Indeed, genetic defects in complement regulators have been implicated in the development of local or systemic diseases, such as age-related macular degeneration and systemic lupus erythematosus [18, 24, 25]. From a microbial perspective, several pathogens not only hijack soluble negative regulators to protect themselves against complement attack but can also degrade cell-associated regulatory molecules that would otherwise protect host tissues or cells [26–29]. Moreover, it is plausible that complement over-activation can occur for reasons unrelated to compromised regulatory mechanisms, such as when the host fails to clear infections [30, 31]. In such cases, the infection could become chronic providing a persistent stimulus for complement activation.

4.3 Role of Complement in Periodontal Dysbiosis and Inflammation

In order to better understand the role of complement in periodontitis, it is instructive to first discuss the role of bacteria in periodontal disease pathogenesis. Until fairly recently, the prevailing paradigm was that specific organisms were involved in the etiology of periodontitis, the most prominent being a troika of bacteria known as the "red complex," namely, *Porphyromonas gingivalis*, *Treponema denticola*, and *Tannerella forsythia* [32, 33]. This notion was in part fueled by the bias of culture-based methods to overestimate the abundance of the easily grown species (such as *P. gingivalis*) in the periodontitis-associated biofilms, while neglecting the presence of uncultivable bacteria. However, culture-independent molecular methods used in recent metagenomic studies have revealed a more heterogeneous and diverse periodontitis-associated microbiota than previously known from cultural studies [34–39]. Many of the newly recognized organisms (e.g., certain gram-positive bacteria and other species from the gram-negative genera *Prevotella*, *Megasphaera*,

Selenomonas, *Desulfobulbus*, *Dialister*, *and Synergistes*) show as good or better a correlation with disease than the red complex bacteria [34–39]. Moreover, a recent metatranscriptomic study revealed that the majority of virulence factors that are upregulated in the microbiome of periodontitis patients is primarily derived from the previously underappreciated species that were not traditionally associated with periodontitis [40]. These recent human microbiome analyses and animal model-based mechanistic studies collectively suggest that the pathogenesis of periodontitis involves polymicrobial synergy and dysbiosis [2, 34, 36, 37, 40–49].

The dysbiosis of the periodontal microbiota represents an alteration in the relative abundance or influence of individual components of the bacterial community (relative to their abundance or influence in health) leading to dysregulated host-microbial crosstalk sufficient to induce destructive inflammation and bone loss [1]. Dysbiotic communities exhibit synergistic interactions that can enhance colonization, persistence, or virulence; bacteria known as keystone pathogens are involved in the breakdown of periodontal tissue homeostasis, whereas other, known as pathobionts, can trigger destructive inflammation once homeostasis is disrupted [2]. Certain commensals, though non-pathogenic by themselves in the oral environment, can promote keystone pathogen colonization and, as such, are implicated as accessory pathogens [2]. Briefly stated, according to the polymicrobial synergy and dysbiosis (PSD) model, the host immune response is initially subverted by keystone pathogens with the help of accessory pathogens and is subsequently over-activated by pathobionts, leading to destructive inflammation in susceptible hosts (Fig. 4.1). Therefore, according to the PSD model, periodontitis is not a bacterial infection in the classical sense (i.e., not caused by a single or a select few pathogens) but, rather, represents a polymicrobial community-induced perturbation of host homeostasis that leads to destructive inflammation in susceptible individuals [2].

These recent advances should not be interpreted to suggest that *P. gingivalis* or other red complex bacteria are not important in periodontal disease pathogenesis; simply, their roles need to be re-interpreted in a manner consistent with emerging new evidence. In this regard, it was recently shown that *P. gingivalis* acts as a keystone pathogen at low colonization levels. Specifically, *P. gingivalis* induces the conversion from a symbiotic community structure to a dysbiotic one capable of causing destructive inflammation and periodontal bone loss [44, 50, 51]. In line with this concept, *P. gingivalis* cannot cause disease in germ-free mice despite colonizing this host, that is, it cannot cause inflammatory bone loss in the absence of other bacteria [44]. Contrary to the findings of some of the early culture-based microbiological studies, the recent metagenomic studies using culture-independent molecular methods show that *P. gingivalis* constitutes a quantitatively minor constituent of human periodontitis-associated biofilms [36, 38, 52]. Moreover, in non-human primates where *P. gingivalis* is a natural inhabitant of the subgingival biofilm, a specific vaccine (against a key virulence factor, the gingipain proteases) causes a reduction both in *P. gingivalis* counts and in the total subgingival bacterial load, in addition to inhibiting bone loss [53]. These findings suggest that the presence of *P. gingivalis* benefits the entire biofilm, as predicted by the keystone-pathogen concept [50]. It should be clarified that the mere presence of *P. gingivalis* does not

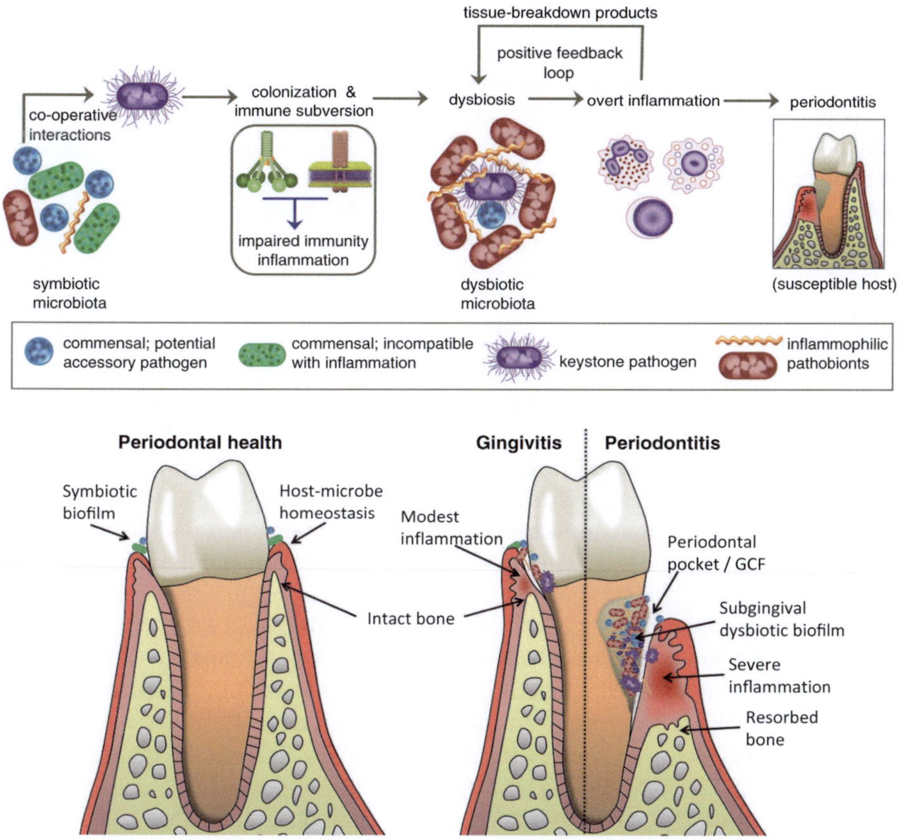

Fig. 4.1 The polymicrobial synergy and dysbiosis (PSD) model of periodontal disease pathogenesis. Periodontitis is induced by a polymicrobial bacterial community, wherein different members have distinct roles that synergize to cause destructive inflammation. Keystone pathogens, the colonization of which is facilitated by accessory pathogens, manipulate the host response leading from a symbiotic to a dysbiotic microbiota, in which pathobionts over-activate the inflammatory response and cause destructive resorption of the supporting bone. Inflammation and dysbiosis reinforce each other by engaging in a positive feedback loop (inflammatory tissue breakdown products are used as nutrients by the dysbiotic microbiota, which further exacerbates inflammation). The *lower panel* shows the progression from periodontal health to gingivitis (gingival inflammation without bone loss) to periodontitis (loss of epithelial attachment, formation of deep periodontal pockets, and inflammatory bone loss). Periodontal pockets serve as a niche that can harbor dysbiotic bacterial communities feeding on the inflammatory spoils (e.g., degraded collagen peptides, haem-containing compounds) transferred with the gingival crevicular fluid (GCF) that bathes the pockets. Redrawn from Ref. [13]. Used by permission

necessarily trigger a transition toward periodontitis. Indeed, *P. gingivalis* can be detected, albeit with reduced frequency, also in periodontally healthy individuals [36, 54]. In this regard, there is considerable strain and virulence diversity within the population structure of *P. gingivalis*. Moreover, key virulence factors (e.g., gingipains and lipid A phosphatases) of this bacterium are regulated by local environmental conditions that likely differ among different individuals [51]. Another potential explanation is that there might be individuals who can resist the capacity of *P. gingivalis* to convert a symbiotic microbiota into a dysbiotic one by virtue of their intrinsic immune status (e.g., alterations in signaling pathways required for immune subversion by *P. gingivalis*). In other words, *P. gingivalis* does not necessarily initiate disease but rather signifies a risk factor for periodontitis [13, 55]

Recent studies in mice and non-human primates indicate that complement is involved in both the dysbiotic transformation of the periodontal microbiota and the inflammatory response that leads to destruction of periodontal bone [44, 47, 56–59]. In this model of periodontal disease pathogenesis, C5aR (CD88) is a target of immune subversion by *P. gingivalis* leading to the dysbiotic transformation of the microbiota, which in turn causes destructive inflammation that is largely dependent on C3 activation (Fig. 4.2). This involvement of C3 may entail synergism with TLRs, as suggested by previous findings on the interactions of complement and the TLR signaling system in the periodontium and other tissues [19, 57, 60].

Intriguingly, whereas *P. gingivalis* can impair the killing capacity of leukocytes such as neutrophils and macrophages, it does not block their ability to induce inflammatory responses [47, 59, 61]. For instance, in human and mouse neutrophils, *P. gingivalis* instigates a C5aR-TLR2 crosstalk which disarms and disassociates a host-protective TLR2–MyD88 pathway from a proinflammatory and immune-evasive TLR2–MyD88 adaptor-like (Mal)–phosphoinositide 3-kinase (PI3K) pathway that prevents phagocytosis of *P. gingivalis* and bystander bacteria [47]. The ability of *P. gingivalis* to exploit C5aR in leukocytes to impair their antimicrobial but not their proinflammatory responses allows uncontrolled growth and altered composition of the microbiota in an inflammatory environment [44, 47, 59]. This documented concept has resolved a long-standing conundrum: on the one hand, periodontal bacteria need to evade immune-mediated killing; on the other hand, they require inflammation as this generates nutrients (e.g., degraded collagen peptides and haem-containing compounds) that periodontitis-associated bacteria need to thrive [62]. In other words, periodontal bacteria cannot afford to evade killing via immunosuppression, even though this represents a common evasion strategy of many other pathogens [63].

It should be noted that *P. gingivalis* can activate C5aR independently of the immunologically activated complement cascade, as this bacterium can release biologically active C5a from C5 through the action of its Arg-specific gingipains [59, 61, 64]. Consistent with this, *P. gingivalis* was shown to retain its capacity to colonize the periodontium of C3-deficient ($C3^{-/-}$) mice, since these mice express normal levels of C5 and C5aR that are required for *P. gingivalis* colonization [56]. Intriguingly, although *P. gingivalis* can colonize $C3^{-/-}$ mice, its dysbiotic effect is transient in this host and the periodontal microbiota cannot be sustained at high levels throughout the experimental period as seen in wild-type [56]. Moreover,

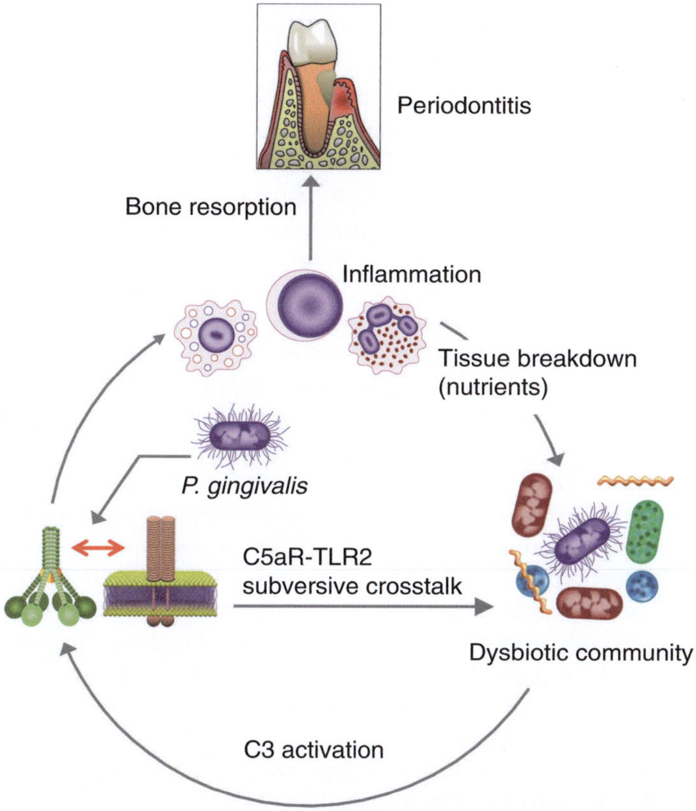

Fig. 4.2 Complement involvement in periodontal dysbiosis and inflammation. Colonization of the periodontium by *P. gingivalis* impairs innate host defense by instigating a subversive C5aR-TLR2 crosstalk, which leads to the dysbiotic transformation of the periodontal microbiota. The dysbiotic microbial community in turn causes C3-dependent inflammatory bone loss, the hallmark of periodontitis. The resulting inflammatory environment selects for inflammophilic bacteria that feed on inflammatory breakdown products, thereby promoting further bacterial growth and dysbiosis. These pathologic interactions generate and perpetuate a vicious cycle of periodontal tissue destruction. Modified from Ref. [50] on the basis of recent studies [47, 56]. Used by permission

P. gingivalis-colonized $C3^{-/-}$ mice exhibit significantly less periodontal inflammation and bone loss than *P. gingivalis*-colonized wild-type mice [56]. Therefore, C3 is crucial for the long-term sustenance of the dysbiotic microbiota and for maximal inflammatory bone loss. The reason why *P. gingivalis*-induced dysbiosis cannot be sustained in $C3^{-/-}$ mice is likely related to the diminished periodontal inflammation, which—as alluded to above—is required for nutrient acquisition. Consistent with the notion that periodontitis-associated bacteria are "inflammo-philic" (from the Greek suffix *philic* indicating fondness), the bacterial biomass of human periodontitis-associated biofilms was shown to increase with increasing periodontal inflammation [36], and anti-inflammatory treatments in animal models suppress the periodontal bacterial load [65–67].

4.4 Clinical Evidence Linking Complement to Periodontitis

The space between the free gingiva and the tooth surfaces is known as the gingival crevice. This anatomical location is bathed with an inflammatory exudate termed gingival crevicular fluid (GCF) [68]. When deepened due to periodontal disease activity, the gingival crevice is referred to as periodontal pocket, which is a niche heavily populated with periodontitis-associated microbial communities [13] (Fig. 4.1, lower panel). The enhanced host inflammatory response in periodontitis is correlated with elevated flow of GCF, in part owing to the increased vascular permeability of the subepithelial blood vessels [69]. Under inflammatory conditions, the GCF contains complement at up to 70–80 % of its concentration in serum, although the serum is not the only source of periodontal complement since it is also produced locally [70–73]. For instance, recruited leukocytes and especially macrophages constitute a source of local production of complement components [74, 75].

The periodontal pockets contain a functional complement system as attested by analyses of collected GCF samples showing robust complement-dependent hemolytic activity [76, 77]. Moreover, GCF collected from periodontitis patients contains activated complement fragments at higher concentrations than in GCF from healthy individuals [72, 78–81]. Consistent with this, complement components and cleavage products are readily detected in chronically inflamed gingiva, whereas complement is undetectable or present at lower levels in healthy gingival biopsy samples [71, 73, 76, 77, 82–84]. An immunohistochemical study revealed weaker expression of CD59 in periodontitis-involved gingiva than in healthy gingival tissue, implying impaired protection of diseased tissues against potential tissue damage by autologous membrane attack complex formation [84].

Using an integrative gene prioritization method and databases from genome-wide association studies and microarray experiments, a recent study has identified C3 among the top 21 most promising candidate genes involved in periodontitis [85]. A genetic basis for periodontal disease is suggested by twin studies and familial aggregation of severe forms of the disease [3, 4, 6]. Although a number of candidate susceptibility genes have been proposed, it remains uncertain whether individual genes play important roles in periodontal disease pathogenesis [3, 4, 6]. In this regard, chronic (or adult-type) periodontitis is a polygenic disease, where multiple genes contribute cumulatively to the overall disease risk (or protection) by influencing the host immune response and the microbiota. Nevertheless, a role for C3 is supported by additional evidence: Induction of experimental gingivitis in human volunteers causes progressive elevation of complement activation (as determined by C3 conversion) correlating with increased clinical inflammatory parameters [81]. Conversely, the resolution of inflammation in periodontitis patients undergoing therapy leads to decreased complement activity, as revealed by reduced C3-to-C3c conversion in the GCF [86]. In a similar context, C3 is among the top 5 % of genes that are most strongly downregulated following periodontal therapy [87]. Importantly, local inhibition of C3 blocks experimental periodontitis in non-human primates [56].

It is of interest to note that despite excessive complement activation in periodontitis, periodontal bacteria have a number of protective mechanisms against complement-mediated killing. For instance, *P. gingivalis* and *Prevotella intermedia* can capture and co-opt physiological soluble inhibitors of the complement cascade, such as the C4b-binding protein [88, 89] (Fig. 4.3). In a similar context, *T. denticola* expresses an 11.4-kDa cell surface lipoprotein which binds complement factor H (hence known as factor H-binding protein) [90]. Moreover, whereas certain bacterial proteases (*P. gingivalis* Arg-specific gingipains and *T. forsythia* karilysin) cleave C5 to release biologically active C5a, the same proteases readily destroy the C5b component, thereby preventing the generation of the membrane attack complex [70, 91] (Fig. 4.3).

In summary, clinical and histological studies in human patients are consistent with the involvement of complement in local tissue destruction in periodontitis.

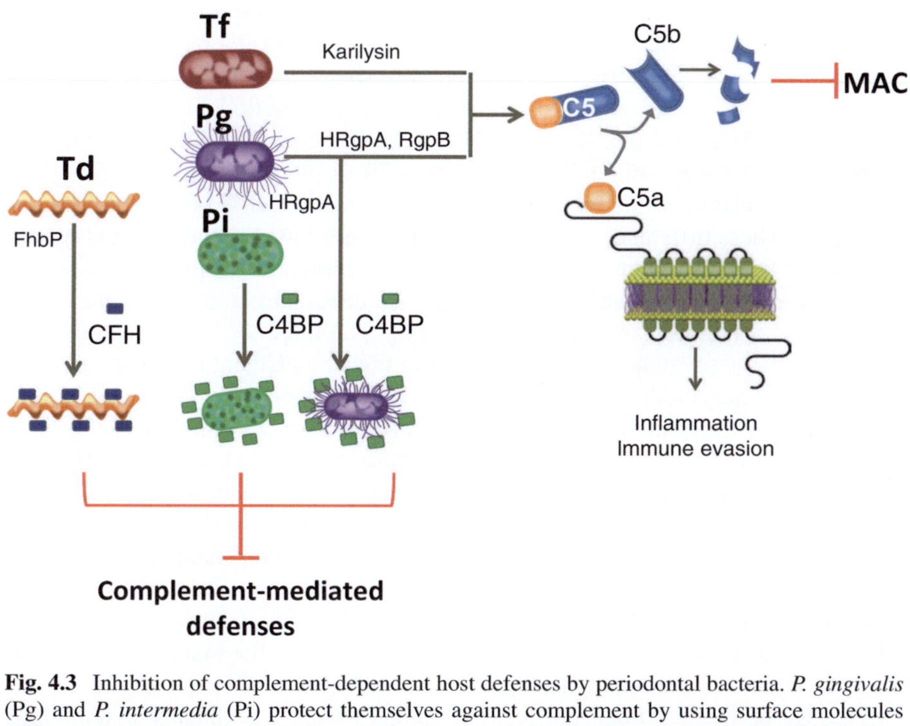

Fig. 4.3 Inhibition of complement-dependent host defenses by periodontal bacteria. *P. gingivalis* (Pg) and *P. intermedia* (Pi) protect themselves against complement by using surface molecules (HRgpA gingipain for *P. gingivalis*, undefined molecule for *P. intermedia*) to capture the circulating C4b-binding protein (C4BP), a physiological negative regulator of the classical and lectin pathways. *Treponema denticola* (Td) hijacks another regulator, the complement factor H (CFH), using a lipoprotein known as factor H-binding protein (FhbP). In this way, the bacteria can prevent complement-dependent opsonophagocytosis and the formation of the membrane attack complex (MAC). Moreover, although *P. gingivalis* and *T. forsythia* proteases can release biologically active C5a from C5 (which leads to immune evasion and inflammation), the generated C5b component is degraded by the same proteases (Arg-specific gingipains HRgpA and RgpB and karilysin), thereby preventing the generation of MAC

This notion is supported by interventional studies in preclinical models, which additionally offer promising targets for treating human periodontitis (below).

4.5 Complement as a Therapeutic Target in Periodontitis

The above-discussed mechanistic studies in mice have implicated both C3 and C5aR in periodontal disease pathogenesis, thereby offering two novel targets for therapeutic intervention in this oral inflammatory disease. In a proof-of-concept study, local intragingival injection of PMX-53, a C5aR antagonist, blocked periodontal inflammation and bone loss in a model of *P. gingivalis*-induced periodontitis, regardless of whether it was administered before or after disease initiation [57]. The same inhibitor inhibited inflammatory periodontal bone loss also in a mouse model of ligature-induced periodontitis where the disease is induced independently of *P. gingivalis* [57]. In this model, a silk ligature is placed around molar teeth, resulting in massive local accumulation of bacteria and development of inflammation and bone loss in conventional (but not germ-free) mice or rats [45, 92]. Work by an independent group using a similar ligature-induced periodontitis model in rats showed that PMX205 (an analog of PMX53) inhibits bone loss when administered in the drinking water, although the efficacy (<20 % protection vs. controls) [93], was reduced relative to the local administration method (50 % protection vs. controls) [57]. These differences in efficacy might be attributed to the different modes of inhibitor administration and/or to the use of different animal species.

More recently, the suitability of C3 as a therapeutic target in periodontitis was evaluated in a non-human primate model [56]. The inhibitor used was Cp40, an improved analog of compstatin, which is a peptidic complement inhibitor acting on C3 [94, 95]. Thus, unlike physiological negative regulators of complement, compstatin and its analogs do not only bind the C3 convertase but also bind and protect C3 from both classical and alternative convertase-mediated cleavage [94, 95]. C3 inhibition is advantageous in that it blocks the generation of downstream effector molecules regardless of the initiation mechanism of complement activation. Moreover, by inhibiting complement at the level of C3, compstatin and its analogs do not interfere with C4b opsonization induced via the classical and lectin pathways. The improved analog Cp40 exhibits plasma half-life values more than 50 h, which exceeds expectations for most peptidic drugs, and is the first compstatin analog with subnanomolar target affinity (K_D=0.5 nM) [96, 114].

The use of a non-human primate preclinical model (specifically cynomolgus monkeys; *Macaca fascicularis*) was necessary for an initial evaluation of the efficacy of Cp40 in periodontitis due to its exclusive specificity for C3 of humans and non-human primates. Importantly, the immune system and periodontal anatomy of the cynomolgus monkey is very similar to that of humans, and periodontitis in this model displays bacteriological, immuno-histological, and clinical features that are highly similar to those observed in human periodontitis [97–101]. The cynomolgus model is thus considerably more predictive of drug efficacy in human periodontitis compared to widely used models such as those in rodents, rabbits, or dogs. In the Cp40 intervention study, which had a 6-week duration, silk ligatures were placed around posterior teeth on both halves of the lower jaw (mandible) for a split-mouth experimental design. Specifically, one side was treated with active drug (Cp40) and the other with inactive analog (control peptide), therefore, each animal served as its own control. Treatment with Cp40 resulted in decreased clinical indices that measure periodontal inflammation and tissue destruction. The decreased clinical inflammation correlated with lower GCF levels of proinflammatory cytokines (e.g., TNF, IL-1β, IL-17, and RANKL, a key osteoclastogenic factor) and decreased numbers of osteoclasts in bone biopsy specimens [56]. Consistent with the latter, radiographic analysis showed that Cp40 caused a significant inhibition of periodontal bone loss. Interestingly, the GCF levels of osteoprotegerin (OPG), a natural inhibitor of RANKL, were maintained at higher levels in Cp40-treated sites than control sites during the course of the study. Therefore, Cp40 caused a favorable reversal of the RANKL/OPG ratio, which is thought to be a useful biomarker of human periodontitis [102]. This study therefore supports the therapeutic potential of Cp40 in human periodontitis and marks the first time, for any disease, that complement inhibition was shown to inhibit inflammatory processes that lead to bone loss in non-human primates. More recently, locally administered Cp40 was shown to inhibit preexisting, naturally occurring chronic periodontitis in non-human primates (Maekawa et al., submitted).

The mechanism(s) by which C3 inhibition blocks periodontal inflammation may not be restricted to mere suppression of the complement cascade. This is because complement pathways (e.g., C3a or C5a receptor signaling) cross-talk with and amplify TLR-mediated inflammatory responses in both systemic and mucosal settings [19, 60] including the periodontium [57]. Complement inhibition may thus also suppress inflammation that is initiated by TLR activation in response to microbial ligands such as lipopolysaccharide, lipoproteins, and bacterial DNA [1, 103]. Moreover, TLR activation can be triggered by endogenous molecules (e.g., biglycan, hyaluronan fragments, and heparan sulfate fragments) that are released upon inflammatory tissue damage and act as danger signals [104, 105]. This suggests that complement may also be involved in the progression of periodontal inflammation; hence complement inhibitors may additionally interfere with this stage of the disease.

4.6 Conclusions and Perspective

There is currently an unmet need for efficacious and safe therapeutics in periodontitis, which is often unresponsive to conventional periodontal treatment [17, 106–108]. At present, there is no satisfactory adjunctive therapy to scaling and root planing for the treatment of chronic periodontitis. The use of antimicrobials and generic antibiotics as adjunctive therapies has met with limited success at best [109]. Therefore, the treatment of periodontal disease should benefit from safe and effective products appropriate for chronic administration. On the basis of evidence from preclinical models, locally applied complement inhibitors can potentially block periodontal inflammation and thereby provide protection as an adjunctive therapy to standard periodontal treatment. Being a host modulation-based approach, complement inhibition is advantageous to antimicrobial approaches since it is the host response that primarily inflicts damage upon the periodontal tissues. Moreover, as discussed above, the inhibition of periodontal inflammation also exerts indirect antimicrobial effects, since the periodontitis-associated microbiota requires an inflammatory environment to obtain nutrients for its growth and sustenance [56, 65–67].

Compstatin-derived compounds with improved inhibitory potency and pharmacokinetic properties have shown safety and efficacy in several other clinically relevant non-human primate disease models. These involve treatment of age-related macular degeneration, sepsis, hemodialysis-induced inflammation, and paroxysmal nocturnal hemoglobinuria [110–113]. A Cp40-based drug (AMY-101; Amyndas Pharmaceuticals) is currently being evaluated as a novel therapeutic approach to treat complications of ABO-incompatible kidney transplantation and paroxysmal nocturnal hemoglobinuria [114]. The recent interventional periodontal studies in non-human primates suggest that periodontitis may be a promising clinical application for Cp40 and the clinically developed drug candidate AMY-101, a possibility that can be pursued in future clinical trials.

Acknowledgments The authors are supported by grants from the U.S. National Institutes of Health: DE015254, DE017138, DE021685, and DE024716 (G.H.); AI003040, AI068730, EY020633, and GM097747 (J.D.L.) and from the European Community's Seventh Framework Programme under grant agreement number 602699 (DIREKT) (J.D.L.).

References

1. Hajishengallis G. Immunomicrobial pathogenesis of periodontitis: keystones, pathobionts, and host response. Trends Immunol. 2014;35(1):3–11. http://dx.doi.org/10.1016/j.it.2013.09.001.
2. Lamont RJ, Hajishengallis G. Polymicrobial synergy and dysbiosis in inflammatory disease. Trends Mol Med. 2014; Epub ahead of print: doi: 10.1016/j.molmed.2014.11.004.
3. Stabholz A, Soskolne WA, Shapira L. Genetic and environmental risk factors for chronic periodontitis and aggressive periodontitis. Periodontol 2000. 2010;53:138–53. doi:10.1111/j.1600-0757.2010.00340.x..
4. Laine ML, Crielaard W, Loos BG. Genetic susceptibility to periodontitis. Periodontol 2000. 2012;58(1):37–68. doi:10.1111/j.1600-0757.2011.00415.x.
5. Zhou Q, Leeman SE, Amar S. Signaling mechanisms in the restoration of impaired immune function due to diet-induced obesity. Proc Natl Acad Sci U S A. 2011;108(7):2867–72. doi:10.1073/pnas.1019270108.
6. Divaris K, Monda KL, North KE, Olshan AF, Reynolds LM, Hsueh WC, et al. Exploring the genetic basis of chronic periodontitis: a genome-wide association study. Hum Mol Genet. 2013;22(11):2312–24. doi:10.1093/hmg/ddt065.
7. Palmer RM, Wilson RF, Hasan AS, Scott DA. Mechanisms of action of environmental factors–tobaccosmoking.JClinPeriodontol.2005;32Suppl6:180–95.doi:10.1111/j.1600-051X.2005.00786.x.

8. Eke PI, Dye BA, Wei L, Thornton-Evans GO, Genco RJ. Prevalence of periodontitis in adults in the United States: 2009 and 2010. J Dent Res. 2012;91(10):914–20. doi:10.1177/0022034512457373.
9. Tsilivakos MG, Manolis SK, Vikatou O, Papagrigorakis MJ. Periodontal disease in the Mycenean (1450–1150 BC) population of Aghia Triada, W. Peloponnese, Greece. Int J Anthropol. 2002;17(2):91–9. doi:10.1007/bf02447400.
10. Lalla E, Papapanou PN. Diabetes mellitus and periodontitis: a tale of two common interrelated diseases. Nat Rev Endocrinol. 2011;7(12):738–48. doi:10.1038/nrendo.2011.106. doi:nrendo.2011.106 [pii].
11. Kebschull M, Demmer RT, Papapanou PN. "Gum bug leave my heart alone": epidemiologic and mechanistic evidence linking periodontal infections and atherosclerosis. J Dent Res. 2010;89:879–902. doi:10.1177/0022034510375281. 0022034510375281 [pii].
12. Han YW, Houcken W, Loos BG, Schenkein HA, Tezal M. Periodontal disease, atherosclerosis, adverse pregnancy outcomes, and head-and-neck cancer. Adv Dent Res. 2014;26(1):47–55. doi:10.1177/0022034514528334.
13. Hajishengallis G. Periodontitis: from microbial immune subversion to systemic inflammation. Nat Rev Immunol. 2015;15(1):30–44. doi:10.1038/nri3785.
14. Brown LJ, Johns BA, Wall TP. The economics of periodontal diseases. Periodontol 2000. 2002;29:223–34. prd290111 [pii].
15. Beikler T, Flemmig TF. Oral biofilm-associated diseases: trends and implications for quality of life, systemic health and expenditures. Periodontol 2000. 2011;55(1):87–103. doi:10.1111/j.1600-0757.2010.00360.x.
16. Colombo AP, Bennet S, Cotton SL, Goodson JM, Kent R, Haffajee AD, et al. Impact of periodontal therapy on the subgingival microbiota of severe periodontitis: comparison between good responders and individuals with refractory periodontitis using the human oral microbe identification microarray. J Periodontol. 2012;83(10):1279–87. doi:10.1902/jop.2012.110566.
17. Armitage GC. Classifying periodontal diseases: a long-standing dilemma. Periodontol 2000. 2002;30:9–23.
18. Ricklin D, Hajishengallis G, Yang K, Lambris JD. Complement: a key system for immune surveillance and homeostasis. Nat Immunol. 2010;11(9):785–97. doi:10.1038/ni.1923.
19. Hajishengallis G, Lambris JD. Crosstalk pathways between Toll-like receptors and the complement system. Trends Immunol. 2010;31(4):154–63. doi:10.1016/j.it.2010.01.002.
20. Markiewski MM, Nilsson B, Ekdahl KN, Mollnes TE, Lambris JD. Complement and coagulation: strangers or partners in crime? Trends Immunol. 2007;28(4):184–92. doi:10.1016/j.it.2007.02.006. S1471-4906(07)00048-8 [pii].
21. Jalili A, Shirvaikar N, Marquez-Curtis L, Qiu Y, Korol C, Lee H, et al. Fifth complement cascade protein (C5) cleavage fragments disrupt the SDF-1/CXCR4 axis: further evidence that innate immunity orchestrates the mobilization of hematopoietic stem/progenitor cells. Exp Hematol. 2010;38:321–32. doi:10.1016/j.exphem.2010.02.002. S0301-472X(10)00046-9 [pii].
22. Dunkelberger JR, Song WC. Complement and its role in innate and adaptive immune responses. Cell Res. 2010;20(1):34–50. doi:10.1038/cr.2009.139. cr2009139 [pii].
23. Friec GL, Kemper C. Complement: coming full circle. Arch Immunol Ther Exp (Warsz). 2009;57(6):393–407. doi:10.1007/s00005-009-0047-4.
24. Zipfel PF, Skerka C. Complement regulators and inhibitory proteins. Nat Rev Immunol. 2009;9(10):729–40. doi:10.1038/nri2620. nri2620 [pii].
25. Ricklin D, Lambris J, D. Complement in Immune and Inflammatory Disorders: Pathophysiological Mechanisms. J Immunol. 2013;190: in press.
26. Lambris JD, Ricklin D, Geisbrecht BV. Complement evasion by human pathogens. Nat Rev Microbiol. 2008;6(2):132–42.
27. Potempa J, Pike RN. Corruption of innate immunity by bacterial proteases. J Innate Immun. 2009;1:70–87.

28. Krauss JL, Potempa J, Lambris JD, Hajishengallis G. Complementary Tolls in the periodontium: how periodontal bacteria modify complement and Toll-like receptor responses to prevail in the host. Periodontol 2000. 2010;52(1):141–62. doi:10.1111/j.1600-0757.2009.00324.x.
29. Mahtout H, Chandad F, Rojo JM, Grenier D. Porphyromonas gingivalis mediates the shedding and proteolysis of complement regulatory protein CD46 expressed by oral epithelial cells. Oral Microbiol Immunol. 2009;24(5):396–400. doi:10.1111/j.1399-302X.2009.00532.x.
30. Hajishengallis G, Abe T, Maekawa T, Hajishengallis E, Lambris JD. Role of complement in host-microbe homeostasis of the periodontium. Semin Immunol. 2013;25:65–72. doi:10.1016/j.smim.2013.04.004.
31. Hajishengallis G, Lambris JD. Complement and dysbiosis in periodontal disease. Immunobiology. 2012;217(11):1111–6. doi:10.1016/j.imbio.2012.07.007.
32. Socransky SS, Haffajee AD, Cugini MA, Smith C, Kent Jr RL. Microbial complexes in subgingival plaque. J Clin Periodontol. 1998;25(2):134–44.
33. Holt SC, Ebersole JL. *Porphyromonas gingivalis*, *Treponema denticola*, and *Tannerella forsythia*: the "red complex", a prototype polybacterial pathogenic consortium in periodontitis. Periodontol 2000. 2005;38:72–122.
34. Dewhirst FE, Chen T, Izard J, Paster BJ, Tanner AC, Yu WH, et al. The human oral microbiome. J Bacteriol. 2010;192(19):5002–17. doi:10.1128/JB.00542-10.
35. Griffen AL, Beall CJ, Firestone ND, Gross EL, Difranco JM, Hardman JH, et al. CORE: a phylogenetically-curated 16S rDNA database of the core oral microbiome. PLoS One. 2011;6(4), e19051. doi:10.1371/journal.pone.0019051.
36. Abusleme L, Dupuy AK, Dutzan N, Silva N, Burleson JA, Strausbaugh LD, et al. The subgingival microbiome in health and periodontitis and its relationship with community biomass and inflammation. ISME J. 2013;7(5):1016–25. doi:10.1038/ismej.2012.174.
37. Griffen AL, Beall CJ, Campbell JH, Firestone ND, Kumar PS, Yang ZK, et al. Distinct and complex bacterial profiles in human periodontitis and health revealed by 16S pyrosequencing. ISME J. 2012;6(6):1176–85. doi:10.1038/ismej.2011.191.
38. Kumar PS, Leys EJ, Bryk JM, Martinez FJ, Moeschberger ML, Griffen AL. Changes in periodontal health status are associated with bacterial community shifts as assessed by quantitative 16S cloning and sequencing. J Clin Microbiol. 2006;44(10):3665–73. doi:10.1128/JCM.00317-06. 44/10/3665 [pii].
39. Perez-Chaparro PJ, Goncalves C, Figueiredo LC, Faveri M, Lobao E, Tamashiro N, et al. Newly identified pathogens associated with periodontitis: a systematic review. J Dent Res. 2014;93:846–58. doi:10.1177/0022034514542468.
40. Duran-Pinedo AE, Chen T, Teles R, Starr JR, Wang X, Krishnan K, et al. Community-wide transcriptome of the oral microbiome in subjects with and without periodontitis. ISME J. 2014;8(8):1659–72. doi:10.1038/ismej.2014.23.
41. Orth RK, O'Brien-Simpson NM, Dashper SG, Reynolds EC. Synergistic virulence of *Porphyromonas gingivalis* and *Treponema denticola* in a murine periodontitis model. Mol Oral Microbiol. 2011;26(4):229–40. doi:10.1111/j.2041-1014.2011.00612.x.
42. Ramsey MM, Rumbaugh KP, Whiteley M. Metabolite cross-feeding enhances virulence in a model polymicrobial infection. PLoS Pathog. 2011;7(3), e1002012. doi:10.1371/journal.ppat.1002012.
43. Settem RP, El-Hassan AT, Honma K, Stafford GP, Sharma A. *Fusobacterium nucleatum* and *Tannerella forsythia* induce synergistic alveolar bone loss in a mouse periodontitis model. Infect Immun. 2012;80(7):2436–43. doi:10.1128/IAI.06276-11.
44. Hajishengallis G, Liang S, Payne MA, Hashim A, Jotwani R, Eskan MA, et al. Low-abundance biofilm species orchestrates inflammatory periodontal disease through the commensal microbiota and complement. Cell Host Microbe. 2011;10(5):497–506. doi:10.1016/j.chom.2011.10.006.
45. Jiao Y, Darzi Y, Tawaratsumida K, Marchesan JT, Hasegawa M, Moon H, et al. Induction of bone loss by pathobiont-mediated nod1 signaling in the oral cavity. Cell Host Microbe. 2013;13(5):595–601. doi:10.1016/j.chom.2013.04.005.

46. Jorth P, Turner KH, Gumus P, Nizam N, Buduneli N, Whiteley M. Metatranscriptomics of the human oral microbiome during health and disease. MBio. 2014;5(2):e01012–4. doi:10.1128/mBio.01012-14.
47. Maekawa T, Krauss JL, Abe T, Jotwani R, Triantafilou M, Triantafilou K, et al. *Porphyromonas gingivalis* manipulates complement and TLR signaling to uncouple bacterial clearance from inflammation and promote dysbiosis. Cell Host Microbe. 2014;15(6):768–78. doi:10.1016/j.chom.2014.05.012.
48. Bao K, Belibasakis GN, Thurnheer T, Aduse-Opoku J, Curtis MA, Bostanci N. Role of *Porphyromonas gingivalis* gingipains in multi-species biofilm formation. BMC Microbiol. 2014;14(1):258. doi:10.1186/s12866-014-0258-7.
49. Hajishengallis G, Lamont RJ. Beyond the red complex and into more complexity: The Polymicrobial Synergy and Dysbiosis (PSD) model of periodontal disease etiology. Mol Oral Microbiol. 2012;27:409–19.
50. Hajishengallis G, Darveau RP, Curtis MA. The keystone-pathogen hypothesis. Nat Rev Microbiol. 2012;10(10):717–25. doi:10.1038/nrmicro2873.
51. Darveau RP, Hajishengallis G, Curtis MA. *Porphyromonas gingivalis* as a potential community activist for disease. J Dent Res. 2012;91(9):816–20. doi:10.1177/0022034512453589.
52. Doungudomdacha S, Rawlinson A, Douglas CW. Enumeration of *Porphyromonas gingivalis*, *Prevotella intermedia* and *Actinobacillus actinomycetemcomitans* in subgingival plaque samples by a quantitative-competitive PCR method. J Med Microbiol. 2000;49(10):861–74.
53. Page RC, Lantz MS, Darveau R, Jeffcoat M, Mancl L, Houston L, et al. Immunization of *Macaca fascicularis* against experimental periodontitis using a vaccine containing cysteine proteases purified from *Porphyromonas gingivalis*. Oral Microbiol Immunol. 2007;22(3):162–8.
54. Haffajee AD, Cugini MA, Tanner A, Pollack RP, Smith C, Kent Jr RL, et al. Subgingival microbiota in healthy, well-maintained elder and periodontitis subjects. J Clin Periodontol. 1998;25(5):346–53.
55. Darveau RP. Porphyromonas gingivalis neutrophil manipulation: risk factor for periodontitis? Trends Microbiol. 2014. doi:10.1016/j.tim.2014.06.006.
56. Maekawa T, Abe T, Hajishengallis E, Hosur KB, DeAngelis RA, Ricklin D, et al. Genetic and intervention studies implicating complement C3 as a major target for the treatment of periodontitis. J Immunol. 2014;192:6020–7. doi:10.4049/jimmunol.1400569.
57. Abe T, Hosur KB, Hajishengallis E, Reis ES, Ricklin D, Lambris JD, et al. Local complement-targeted intervention in periodontitis: proof-of-concept using a C5a receptor (CD88) antagonist. J Immunol. 2012;189(11):5442–8. doi:10.4049/jimmunol.1202339.
58. Hajishengallis G, Lambris JD. Microbial manipulation of receptor crosstalk in innate immunity. Nat Rev Immunol. 2011;11(3):187–200. doi:10.1038/nri2918.
59. Wang M, Krauss JL, Domon H, Hosur KB, Liang S, Magotti P, et al. Microbial hijacking of complement-toll-like receptor crosstalk. Sci Signal. 2010;3(109):ra11. doi:10.1126/scisignal.2000697.
60. Zhang X, Kimura Y, Fang C, Zhou L, Sfyroera G, Lambris JD, et al. Regulation of Toll-like receptor-mediated inflammatory response by complement in vivo. Blood. 2007;110(1):228–36.
61. Liang S, Krauss JL, Domon H, McIntosh ML, Hosur KB, Qu H, et al. The C5a receptor impairs IL-12-dependent clearance of *Porphyromonas gingivalis* and is required for induction of periodontal bone loss. J Immunol. 2011;186(2):869–77. doi:10.4049/jimmunol.1003252.
62. Hajishengallis G. The inflammophilic character of the periodontitis-associated microbiota. Mol Oral Microbiol. 2014;doi: 10.1111/omi.12065.
63. Cyktor JC, Turner J. Interleukin-10 and immunity against prokaryotic and eukaryotic intracellular pathogens. Infect Immun. 2011;79(8):2964–73. doi:10.1128/IAI.00047-11.
64. Wingrove JA, DiScipio RG, Chen Z, Potempa J, Travis J, Hugli TE. Activation of complement components C3 and C5 by a cysteine proteinase (gingipain-1) from *Porphyromonas (Bacteroides) gingivalis*. J Biol Chem. 1992;267(26):18902–7.

65. Hasturk H, Kantarci A, Goguet-Surmenian E, Blackwood A, Andry C, Serhan CN, et al. Resolvin E1 regulates inflammation at the cellular and tissue level and restores tissue homeostasis in vivo. J Immunol. 2007;179(10):7021–9.
66. Eskan MA, Jotwani R, Abe T, Chmelar J, Lim JH, Liang S, et al. The leukocyte integrin antagonist Del-1 inhibits IL-17-mediated inflammatory bone loss. Nat Immunol. 2012;13(5):465–73. doi:10.1038/ni.2260.
67. Moutsopoulos NM, Konkel J, Sarmadi M, Eskan MA, Wild T, Dutzan N, et al. Defective neutrophil recruitment in leukocyte adhesion deficiency type I disease causes local IL-17–driven inflammatory bone loss. Sci Transl Med. 2014;6(229):229ra40. doi:10.1126/scitranslmed.3007696.
68. Lamster IB, Ahlo JK. Analysis of gingival crevicular fluid as applied to the diagnosis of oral and systemic diseases. Ann N Y Acad Sci. 2007;1098:216–29. doi:10.1196/annals.1384.027. 1098/1/216 [pii].
69. Delima AJ, Van Dyke TE. Origin and function of the cellular components in gingival crevice fluid. Periodontol 2000. 2003;31:55–76.
70. Popadiak K, Potempa J, Riesbeck K, Blom AM. Biphasic effect of gingipains from *Porphyromonas gingivalis* on the human complement system. J Immunol. 2007;178(11):7242–50.
71. Schenkein HA, Genco RJ. Complement cleavage products in inflammatory exudates from patients with periodontal diseases. J Immunol. 1978;120(5):1796.
72. Schenkein HA, Genco RJ. Gingival fluid and serum in periodontal diseases. II. Evidence for cleavage of complement components C3, C3 proactivator (factor B) and C4 in gingival fluid. J Periodontol. 1977;48(12):778–84.
73. Lally ET, McArthur WP, Baehni PC. Biosynthesis of complement components in chronically inflamed gingiva. J Periodontal Res. 1982;17(3):257–62.
74. Hetland G, Johnson E, Royset P, Eskeland T. Human alveolar macrophages and monocytes generate the functional classical pathway of complement in vitro. Acta Pathol Microbiol Immunol Scand C. 1987;95(3):117–22.
75. Li K, Sacks SH, Zhou W. The relative importance of local and systemic complement production in ischaemia, transplantation and other pathologies. Mol Immunol. 2007;44(16):3866–74. doi:10.1016/j.molimm.2007.06.006.
76. Courts FJ, Boackle RJ, Fudenberg HH, Silverman MS. Detection of functional complement components in gingival crevicular fluid from humans with periodontal diseases. J Dent Res. 1977;56(3):327–31.
77. Boackle RJ. The interaction of salivary secretions with the human complement system–a model for the study of host defense systems on inflamed mucosal surfaces. Crit Rev Oral Biol Med. 1991;2(3):355–67.
78. Attstrom R, Laurel AB, Lahsson U, Sjoholm A. Complement factors in gingival crevice material from healthy and inflamed gingiva in humans. J Periodontal Res. 1975;10(1):19–27.
79. Schenkein HA, Genco RJ. Gingival fluid and serum in periodontal diseases. I. Quantitative study of immunoglobulins, complement components, and other plasma proteins. J Periodontol. 1977;48(12):772–7.
80. Challacombe SJ, Shirlaw PJ. Immunology of diseases of the oral cavity. In: Mestecky J, Lamm ME, Strober W, Bienenstock J, McGhee JR, Mayer L, editors. Mucosal immunology. London: Academic; 2005. p. 1517–46.
81. Patters MR, Niekrash CE, Lang NP. Assessment of complement cleavage in gingival fluid during experimental gingivitis in man. J Clin Periodontol. 1989;16(1):33–7.
82. Toto PD, Lin L, Gargiulo A. Identification of C3a, IgG, IgM in inflamed human gingiva. J Dent Res. 1978;57(5-6):696.
83. Nikolopoulou-Papaconstantinou AA, Johannessen AC, Kristoffersen T. Deposits of immunoglobulins, complement, and immune complexes in inflamed human gingiva. Acta Odontol Scand. 1987;45(3):187–93.

84. Rautemaa R, Meri S. Protection of gingival epithelium against complement-mediated damage by strong expression of the membrane attack complex inhibitor protectin (CD59). J Dent Res. 1996;75(1):568–74.
85. Zhan Y, Zhang R, Lv H, Song X, Xu X, Chai L, et al. Prioritization of candidate genes for periodontitis using multiple computational tools. J Periodontol. 2014;85(8):1059–69. doi:10.1902/jop.2014.130523.
86. Niekrash CE, Patters MR. Simultaneous assessment of complement components C3, C4, and B and their cleavage products in human gingival fluid. II Longitudinal changes during periodontal therapy. J Periodontal Res. 1985;20(3):268–75.
87. Beikler T, Peters U, Prior K, Eisenacher M, Flemmig TF. Gene expression in periodontal tissues following treatment. BMC Med Genomics. 2008;1:30. doi:10.1186/1755-8794-1-30. 1755-8794-1-30 [pii].
88. Potempa M, Potempa J, Okroj M, Popadiak K, Eick S, Nguyen KA, et al. Binding of complement inhibitor C4b-binding protein contributes to serum resistance of *Porphyromonas gingivalis*. J Immunol. 2008;181(8):5537–44.
89. Malm S, Jusko M, Eick S, Potempa J, Riesbeck K, Blom AM. Acquisition of complement inhibitor serine protease factor I and its cofactors C4b-binding protein and factor H by *Prevotella intermedia*. PLoS One. 2012;7(4), e34852. doi:10.1371/journal.pone.0034852.
90. McDowell JV, Huang B, Fenno JC, Marconi RT. Analysis of a unique interaction between the complement regulatory protein factor H and the periodontal pathogen *Treponema denticola*. Infect Immun. 2009;77(4):1417–25.
91. Jusko M, Potempa J, Karim AY, Ksiazek M, Riesbeck K, Garred P, et al. A metalloproteinase karilysin present in the majority of *Tannerella forsythia* isolates inhibits all pathways of the complement system. J Immunol. 2012;188(5):2338–49. doi:10.4049/jimmunol.1101240.
92. Graves DT, Fine D, Teng YT, Van Dyke TE, Hajishengallis G. The use of rodent models to investigate host-bacteria interactions related to periodontal diseases. J Clin Periodontol. 2008;35(2):89–105. doi:10.1111/j.1600-051X.2007.01172.x.
93. Breivik T, Gundersen Y, Gjermo P, Taylor SM, Woodruff TM, Opstad PK. Oral treatment with complement factor C5a receptor (CD88) antagonists inhibits experimental periodontitis in rats. J Periodontal Res. 2011;46(6):643–7. doi:10.1111/j.1600-0765.2011.01383.x.
94. Sahu A, Morikis D, Lambris JD. Compstatin, a peptide inhibitor of complement, exhibits species-specific binding to complement component C3. Mol Immunol. 2003;39(10):557–66.
95. Ricklin D, Lambris JD. Complement in immune and inflammatory disorders: therapeutic interventions. J Immunol. 2013;190(8):3839–47. doi:10.4049/jimmunol.1203200.
96. Qu H, Ricklin D, Bai H, Chen H, Reis ES, Maciejewski M, et al. New analogs of the clinical complement inhibitor compstatin with subnanomolar affinity and enhanced pharmacokinetic properties. Immunobiology. 2013;218(4):496–505. doi:10.1016/j.imbio.2012.06.003.
97. Brecx MC, Nalbandian J, Ooya K, Kornman KS, Robertson PB. Morphological studies on periodontal disease in the cynomolgus monkey. II. Light microscopic observations on ligature-induced periodontitis. J Periodontal Res. 1985;20(2):165–75.
98. Page RC, Schroeder HE. Periodontitis in man and other animals- a comparative review. Basel, Switzerland: Karger; 1982.
99. Kornman KS, Holt SC, Robertson PB. The microbiology of ligature-induced periodontitis in the cynomolgus monkey. J Periodontal Res. 1981;16(4):363–71.
100. Assuma R, Oates T, Cochran D, Amar S, Graves DT. IL-1 and TNF antagonists inhibit the inflammatory response and bone loss in experimental periodontitis. J Immunol. 1998;160(1):403–9.
101. Holt SC, Ebersole J, Felton J, Brunsvold M, Kornman KS. Implantation of *Bacteroides gingivalis* in nonhuman primates initiates progression of periodontitis. Science. 1988;239(4835):55–7.
102. Belibasakis GN, Bostanci N. The RANKL-OPG system in clinical periodontology. J Clin Periodontol. 2012;39(3):239–48. doi:10.1111/j.1600-051X.2011.01810.x.

103. Hajishengallis G, Sahingur SE. Novel inflammatory pathways in periodontitis. Adv Dent Res. 2014;26(1):23–9. doi:10.1177/0022034514526240.
104. Miyake K. Innate immune sensing of pathogens and danger signals by cell surface Toll-like receptors. Semin Immunol. 2007;19(1):3–10. doi:10.1016/j.smim.2006.12.002.
105. Schaefer L. Extracellular matrix molecules: endogenous danger signals as new drug targets in kidney diseases. Curr Opin Pharmacol. 2010;10(2):185–90. doi:10.1016/j.coph.2009.11.007.
106. Colombo AP, Boches SK, Cotton SL, Goodson JM, Kent R, Haffajee AD, et al. Comparisons of subgingival microbial profiles of refractory periodontitis, severe periodontitis, and periodontal health using the human oral microbe identification microarray. J Periodontol. 2009;80(9):1421–32. doi:10.1902/jop.2009.090185.
107. Hasturk H, Kantarci A, Van Dyke TE. Paradigm shift in the pharmacological management of periodontal diseases. Front Oral Biol. 2012;15:160–76. doi:10.1159/000329678.
108. Hajishengallis G, Lambris JD. Complement-targeted therapeutics in periodontitis. Adv Exp Med Biol. 2013;734:197–206. doi:10.1007/978-1-4614-4118-2_13.
109. Krayer JW, Leite RS, Kirkwood KL. Non-surgical chemotherapeutic treatment strategies for the management of periodontal diseases. Dent Clin North Am. 2010;54(1):13–33. doi:10.1016/j.cden.2009.08.010.
110. Reis ES, DeAngelis RA, Chen H, Resuello RR, Ricklin D, Lambris JD. Therapeutic C3 inhibitor Cp40 abrogates complement activation induced by modern hemodialysis filters. Immunobiology. 2014. doi:10.1016/j.imbio.2014.10.026.
111. Chi ZL, Yoshida T, Lambris JD, Iwata T. Suppression of drusen formation by compstatin, a peptide inhibitor of complement C3 activation, on cynomolgus monkey with early-onset macular degeneration. Adv Exp Med Biol. 2010;703:127–35. doi:10.1007/978-1-4419-5635-4_9.
112. Silasi-Mansat R, Zhu H, Popescu NI, Peer G, Sfyroera G, Magotti P, et al. Complement inhibition decreases the procoagulant response and confers organ protection in a baboon model of E. coli sepsis. Blood. 2010;doi: 10.1182/blood-2010-02-269746; blood-2010-02-269746 [pii].
113. Risitano AM, Ricklin D, Huang Y, Reis ES, Chen H, Ricci P, et al. Peptide inhibitors of C3 activation as a novel strategy of complement inhibition for the treatment of paroxysmal nocturnal hemoglobinuria. Blood. 2014;123(13):2094–101. doi:10.1182/blood-2013-11-536573.
114. Mastellos DC, Yancopoulou D, Kokkinos P, Huber-Lang M, Hajishengallis G, Biglarnia A, et al. Compstatin: a complement C3-based inhibitor reaching its prime for bedside intervention. Eur J Clin Invest. 2015;45(4):423–40.

Part II
Biomaterials

Chapter 5
The Lectin Pathway of Complement and Biocompatibility

Estrid Hein and Peter Garred

Abstract In modern health technologies the use of biomaterials in the form of stents, haemodialysis tubes, artificial implants, bypass circuits etc. is rapidly expanding. The exposure of synthetic, foreign surfaces to the blood and tissue of the host, calls for strict biocompatibility in respect to contact activation, the coagulation system and the complement system. The complement system is an important part of the initial immune response and consists of fluid phase molecules in the blood stream. Three different activation pathways can initiate the complement system, the lectin, the classical and the alternative pathway, all converging in an amplification loop of the cascade system and downstream reactions. Thus, when exposed to foreign substances complement components will be activated and lead to a powerful inflammatory response. Biosurface induced complement activation is a recognised issue that has been broadly documented. However, the specific role of lectin pathway and the pattern recognition molecules initiating the pathway has only been transiently investigated. Here we review the current data on the field.

Keywords Lectin pathway • Complement • Biocompatibility • Pattern recognition molecules • Mannose-binding lectin • Collectins • Ficolins • MASPs

5.1 Biocompatibility

The challenge of biocompatibility is the same for all medical devices in contact with blood or tissue, but the issue can also to some degree be extended to organ transplants. The concern is to avoid surface induced thrombosis and inflammation, which in turn will lead to even more adverse clinical effects. One of the common problem

E. Hein • P. Garred (✉)
Laboratory of Molecular Medicine, Department of Clinical Immunology, Section 7631, Rigshospitalet, Faculty of Health and Medical Sciences, University of Copenhagen, Blegdamsvej 9, 2100 Copenhagen O, Denmark
e-mail: peter.garred@regionh.dk

J.D. Lambris et al. (eds.), *Immune Responses to Biosurfaces*, Advances in Experimental Medicine and Biology 865,
DOI 10.1007/978-3-319-18603-0_5

causing biomaterials are tubing used for extracorporeal circulation e.g., during haemodialysis and coronary bypass surgery, where complement is known to be activated. The complement system is of particular importance when it comes to immune responses to biosurfaces because it is abundantly present in the blood and serves as a first line of defence in the human immune system. Activation and deposition of complement components on a wide variety of biosurfaces is a well-known issue, but the focus has mainly been on the downstream activated components. The lectin pathway of complement activation was only rather recently discovered and initiating pattern recognition molecules are still being added.

5.2 The Complement System

As an integral part of innate immunity, the complement system consists of more than 30 soluble proteins circulating the blood stream. The main function of these molecules is to activate and orchestrate the inflammatory response. The complement system is evolutionary ancient and as opposed to the somatic hypermutations characteristic for the antibodies in the adaptive part of immunity, the complement proteins are germ line encoded. The scavenging pattern recognition molecules (PRMs) bind to conserved universal structures on the surface of microorganisms, so called pathogen or danger associated molecular patterns (PAMPs/DAMPs). This generalized mode of recognition, allows the initial response to be rapid, gaining time to mobilize the slower acting but more specific and powerful adaptive response. However, the strategy is crude and increases the risk of mistakes. If not tightly controlled the cascade activation of complement may run wild leaving the host in a dangerous and life threatening situation. This is why the complement system is often referred to as a double-edged sword. The central event of complement activation is the tick-over of C3 activation and accompanying amplification loop of the cascade reaction. Upstream C3, there are three distinct initiation pathways leading to this situation: the lectin pathway, the classical pathway and the alternative pathway [1]. Figure 5.1 shows the activation pathways illustrated on a foreign biosurface.

5.2.1 Activation via the Lectin Pathway

The lectin pathway of complement is initiated by the binding of pattern recognition molecules (PRM) from the subfamilies of collectins and ficolins to the PAMPs or DAMPs on the surface of microorganisms or altered self-cells. Six recognition molecules have thus far been reported to be able to activate the lectin pathway: Mannose-Binding Lectin (MBL), Collectin-10 (Collectin liver 1, CL-L1, CL-10), Collectin-11 (Collectin kidney 1, CL-K1, CL-11), Ficolin-1 (M-ficolin), Ficolin-2 (L-ficolin) and Ficolin-3 (H-ficolin). The PRMS are found in complex with the MBL/ficolin/collectin associated serine proteases (MASPs) of which three are described: MASP-1, -2 and -3. Upon binding of a PRM to a specific ligand, MASP-2 is activated and

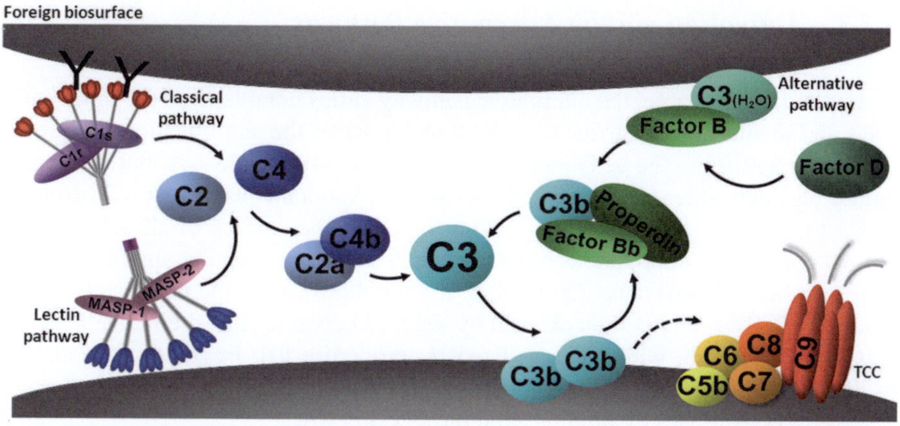

Fig. 5.1 The three activation pathways of the complement system sketched on a foreign biosurface. The classical pathway is initiated by the C1q complex and the lectin pathway is initiated by collectins or ficolins in association with the MASPs; both mediating the cleavage of C2 and C4, which then forms the C3 convertase. The alternative pathway C3 convertase is formed by activated Factor B, spontaneously hydrolysed C3, and properdin. The C3 convertases cleaves C3 to activated C3b, which will lead to formation of more C3 convertases via the amplification loop, and in turn also to the formation of the terminal complement complex (TCC)

cleaves complement factors C4 and C2 [2] and next, the fragments C4b and C2a bind to each other forming the active C3 convertase C4bC2a. In addition, MASP-1 can cleave C2 but not C4 and thus enhances the convertase formation, but is also crucial in activation of MASP-2 [3, 4]. The function of the last protease MASP-3 is still largely unknown. Similar to the MASPs there exist two smaller alternatively spliced variants lacking the serine protease domain, small MBL/ficolin/collectin associated protein (sMAP or Map19) and MBL/ficolin/collectin associated protein 1 (MAP-1 or Map44). Both molecules are proteolytic inactive, and the function of sMAP is unknown, whereas MAP-1 has been shown to be a potent regulator of the lectin pathway [5, 6]. By competing with the MASPs in the binding to the PRMs, MAP-1 inhibits lectin driven complement activation [7].

5.2.2 Activation via the Classical Pathway

The classical pathway has a similar mode of initiation as the lectin pathway: the C1 complex consisting of the pattern recognition molecule C1q and the serine proteases C1r and C1s, binds either to the Fc part of antibody:antigen complexes on the surface of pathogens or dying host cells, or directly to soluble immune complexes [8]. The conformational change in C1q caused by the ligand binding, activates the associated C1r, which can now cleave C1s. When activated, C1s will cleave first C4 and next C2, and the fragments will complex to form the C4bC2a; the same C3 convertase as generated via the lectin pathway.

5.2.3 *Activation via the Alternative Pathway*

As implied by the name, the alternative pathway varies notable from the other pathways and in addition, it carries two functions: it drives the amplification loop at the C3 level magnifying the activation mediated by the other pathways, but it can also induce an independent activation of the cascade. Spontaneous and slow hydrolysis of C3 to C3(H_2O) is continuously taking place on the surface of all cells. Under normal conditions, endogenous regulators such as Factor H inhibit any further activation on healthy host cells. On foreign or altered host cells, the protease Factor B binds to C3(H_2O) and is then cleaved by Factor D creating the alternative C3 convertase C3(H_2O)Bb, which is stabilized by properdin [9]. In addition, properdin itself can act as a PRM by direct binding to PAMPs/DAMPs and mediating formation of C3 convertases by binding fluid phase C3b [10].

5.2.4 *The Terminal Pathway*

The three pathways converge in the formation of active C3 convertases, that can efficiently cleave the most abundant complement component C3 (mean serum conc. 1.2 mg/ml) into C3a and C3b [11]. C3a acts as a potent anaphylatoxin recruiting phagocytic cells to the site of action. Via an exposed thioester, C3b is now able to bind covalently to the target cell or ligand and facilitate the formation of even more convertases. The alternative C3 convertase is responsible for the amplification loop of complement activation where C3 is rapidly cleaved and new C3 convertases are formed in parallel with the formation of the C5 convertases composed of C4bC2aC3b or C3bBb3b. This leads to deposition of C6, C7, C8 and multiple C9 that may assemble to a lytic transmembrane pore called the membrane attack complex (MAC) or terminal complement complex (TCC).

Regardless the rather strict classification described above, it is becoming more and more evident that these molecules are cross-reactive and activate each other independent of pathways and even other cascade systems such as coagulation (e.g., reviewed in [12]).

5.3 Pattern Recognition Molecules of the Lectin Pathway

Structurally, the six PRMs of the lectin pathway are alike. When transcribed and translated from their respective genes, monomeric peptides coil up in trimeric alpha helices, which further oligomerize to form the final complex molecule. The characteristic bouquet-like structure of these PRMs is achieved from disulphide bridging between cysteine residues in the N-terminal end of the peptides. This also forms the collagen like domain, responsible for the interaction with the MASPs. The heads of the bouquet structure is shaped of the C-terminal globular regions. In MBL,

Collectin-10 and Collectin-11, the carbohydrate recognition domain located at the globular region, is characterized by being C-type (calcium dependent). In the ficolins, this domain is fibrinogen-like and has a preference for recognizing acetylated carbohydrate structures such as GlcNAc [13–17]. The overall structure of the PRMs of the lectin pathway is illustrated in Fig. 5.2; exemplified as a ficolin molecule. Apart from the ligand binding domain, the PRMs have subtle differences which will be described below.

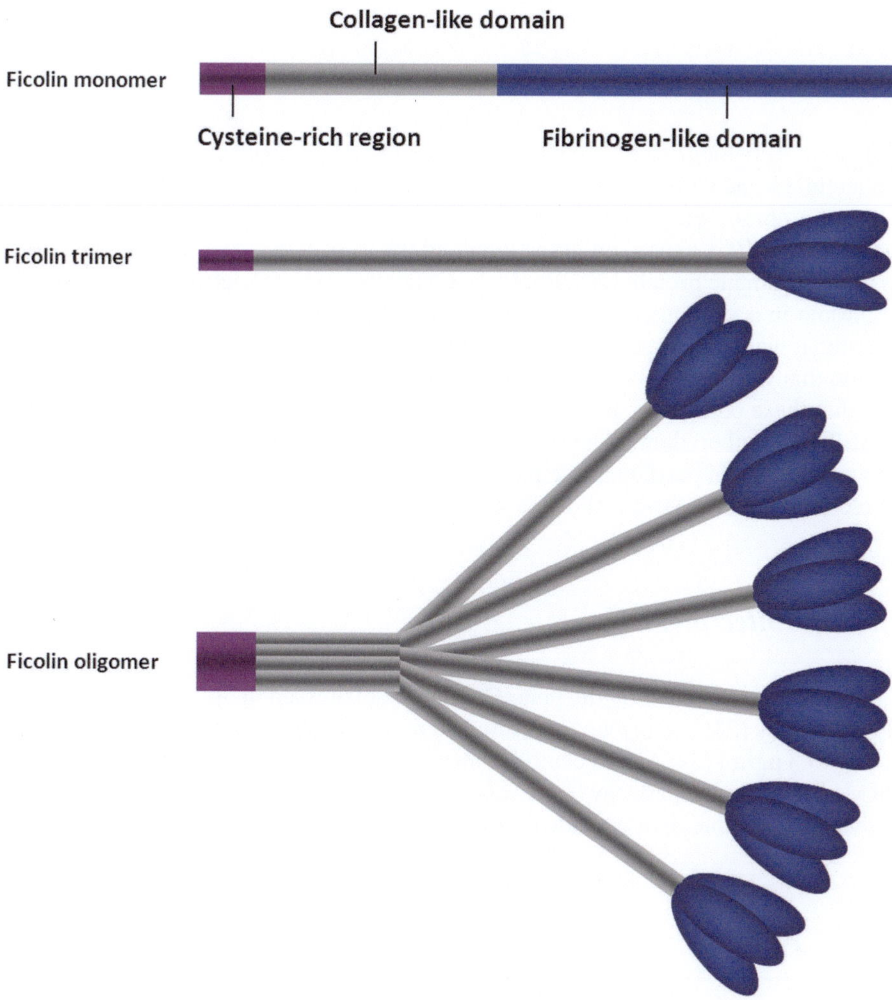

Fig. 5.2 Overall structure of the pattern recognition molecules (PRMs) of the lectin pathway. The illustration shows a ficolin molecule, but the general organization is the same for all the PRMs

5.3.1 MBL

The classical complement activating collectin MBL is encoded by the *MBL2* gene, primarily produced in the liver and present in the circulation in a mean concentration of 1,500 ng/ml (range: 5–5,000 ng/ml). This large inter individual variation in concentration is assigned to commonly occurring single nucleotide polymorphisms (SNPs) in the regulatory and structural part of the gene [18]. MBL binds to ligands such as D-mannose, *N*-acetylglucosamine (GlcNAc) and L-fucose—all carbohydrate structures found on the surface of microbes—and it activates the complement cascade via this binding [19].

5.3.2 Collectin-10

Collectin-10 belongs to the newly discovered so-called novel collectins along side Collectin-11 and Collectin-12. Collectin-12 is not described further here, since it is a transmembrane PRM and has not been shown to activate complement (yet). The gene *COLEC10* is expressed in the liver and the molecule was originally described as a cytoplasmatic protein [20]. Recently, Collectin-10 was determined to be present in plasma in a median concentration of 3.0 μg/ml (range 1.5–5.5 μg/ml) and additionally, it was found in complex with the MASPs [21]. Briefly after, it was demonstrated that Collectin-10 circulates in heterocomplexes with Collectin-11 in serum and that these complexes further associates with the MASPs with a preference for MASP-1/-3 rather than MASP-2 [22]. *In vitro* experiments in the same study showed that when purified, these native Collectin-10/-11 complexes in the presence of recombinant MASP-2 could cleave purified C4 on a ligand of mannan or DNA.

5.3.3 Collectin-11

Collectin-11 is encoded by the gene *COLEC11*, expressed in all organs but primarily the kidney [23] and present in serum in a relatively low concentration of approximately 0.3 μg/ml [24, 25]. Recently, Collectin-11 was demonstrated to belong to the PRMs of the lectin pathway, as it was reported to associate with the MASPs and activate the complement cascade on *Candida albicans* [26, 27].

5.3.4 Ficolin-1

Corresponding to their protein names, the three ficolins are encoded by the *FCN1*, *FCN2* and *FCN3* genes, respectively [28]. Ficolin-1 is primarily synthesized in monocytes and granulocytes, where it is stored in neutrophilic granules [29, 30].

The trigger mechanism by which Ficolin-1 is released is still unknown, but interestingly, once the granules are secreted, the Ficolin-1 molecules bind back to the cell surface in a calcium-dependent manner via the fibrinogen like domain. The ligand on the cells has been shown to be sialic acid [30–32]. Accordingly, Ficolin-1 is found in a relatively low mean concentration in the circulation of approximately 0.3 μg/ml in normal healthy individuals [29, 33, 34]. Additionally, Ficolin-1 has been reported to bind to different subsets of lymphocytes, namely $CD56^{dim}$ NK-cells and activated T-cells suggesting a hitherto undescribed link between innate and adaptive immunity [35]. Finally, Ficolin-1 has been shown to associate with MASP-2 *in vivo* [36], and *in vitro* experiments demonstrated complement activation potential [37], indicating functions of host defence against pathogens other than the modulating role indicated by the self recognition.

5.3.5 Ficolin-2

Ficolin-2 is produced by hepatocytes and released to the blood stream [38]. In healthy individuals it circulates in a mean concentration of approximately 5 μg/ml (range 1–12 μg/ml) [39–42]. The *FCN2* gene is highly polymorphic with numerous reported single nucleotide polymorphisms (SNPs) throughout the promoter, introns and exons. Variations located in the promoter has been shown to significantly affect the concentration of protein secreted, whereas two SNPs in exon 8 increases or decreases the binding affinity to GlcNAc, respectively [39, 42]. When it comes to recognition and binding, Ficolins-2 alternates from the other PRMs due to the vide variety of reported ligands. In the fibrinogen-like domain Ficolin-2 contains no less than four distinct binding grooves S1, S2, S3 and S4, all of which have different recognition specificity [14]. In accordance, numerous ligands have been reported, many of which are in the category of acetylated compounds: GlcNAc, GalNAc, CysNAc, acetylated low density lipoproteins, 1,3-*β*-glucan [13, 14, 43, 44]. Also, microorganisms such as capsulated *Staphylococcus aureus*, *Streptococcus pneumonia*, *Salmonella typhimurium*, *Escherichia coli*, *Pseudomonas aeruginosa*, and *Aspergillus fumigatus* [41, 45–47]. Activation of the lectin pathway of complement via Ficolin-2 has been shown to occur in association with MASP-2 as the enzymatic cleaver of C2 and C4 [48, 49]. In addition, Ficolin-2 is reported to interact with the long pentraxin-3 (PTX3) on the surface of the opportunistic fungal pathogen *Aspergillus fumigatus* and hereby boost the complement response [47].

5.3.6 Ficolin-3

The gene encoding Ficolin-3, *FCN3* is located on chromosome 1 as opposed to *FCN1* and *FCN2*, which are both located on chromosome 9. The gene is highly conserved with only a few reported polymorphisms in the coding region [50].

One of which is a deletion in exon 5, *FCN3+1637delC*, leading to premature termination of the transcript and in the homozygous state, functional deficiency of the protein [51, 52]. Ficolin-3 is predominantly produced in the lungs, bile duct epithelial cells and hepatocytes and circulates with a mean of 25 μg/ml—the highest concentration of the recognition molecules in the lectin pathway [51]. Few ligands have been described for Ficolin-3 here among the Gram positive *Aerococcus viridians* [41, 53] and latest, the Gram negative bacteria *Hafnia alvei*, a commensal of the human gastrointestinal tract [54]. *In vitro*, Ficolin-3 has been shown to bind to apoptotic cells and mediate their clearance [55]. A recently developed *in vitro* ELISA assay for analysis of Ficolin-3 complement activation function, utilizes the binding of Ficolin-3 to acetylated bovine serum albumin (acBSA) [16].

5.4 Extracorporeal Circulation and Biocompatibility

A variety of medical treatments exploit extracorporeal circulation e.g., autotransfusion, extracorporeal membrane oxygenation, plasmapheresis, haemofiltration, haemodialysis and cardiopulmonary bypass. This is a hazardous procedure because of the involvement of direct contact between blood and non-self surface.

5.4.1 Haemodialysis

In the case of kidney failure, e.g., end-stage renal disease, haemodialysis is vital for clearing the blood of urea and other waste products and accordingly keep the patient alive. However, long-term haemodialysis patients have increased morbidity due to dialysis-induced chronic inflammation [56].

Ficolin-2 has in independent studies been shown to bind to polysulfone haemodialysis tubes [57–59]. In two studies of Mares et al. proteins adsorbed to the polysulfone dialyser membrane during haemodialysis was eluted, separated by 2-dimensinal gel electrophoresis and analyzed by mass spectrometry [58, 59]. Both studies found Ficolin-2 to be massively present in the membranes alongside MASP-1, MASP-2 and C3c, clearly indicating lectin pathway activity. For confirmation, the level of Ficolin-2 and MASP-2 in plasma during haemodialysis was measured in ELISA, and it showed a steady decrease corresponding to the progressive adsorption to the polysulfone membrane. Xu et al. also investigated the profile of adhered proteins, but to reused polysulfone haemodialysis membranes [57]. Their results were similar: Ficolin-2 and C3 were the most abundant proteins found. MASP-1 and MASP-3 was also detected but not MASP-2.

5.4.2 Cardiopulmunary Bypass Surgery

Tubing for extracorporeal circulation during e.g., open heart surgery has, in general, been thoroughly tested for biocompatibility. However, some of the major complications during cardiopulmonary bypass are activation of the complement system, inflammation and postoperative haemorrhage. Heparin is administered to patients systemically during surgery to avoid blood clotting but is also widely used as coating for extracorporeal tubing, since it has been shown to reduce some of the above mentioned issues [60, 61].

Various approaches regarding coating have been used commercially to achieve the highest degree of biocompatibility in tubing for extracorporeal circulation. Bioline® circuits are layered with heparin-albumin whereas the coating of Phisio® circuits is designed to imitate the surface of host cells and utilizes a phosporylcholine polymer. A newly published study analyzing these two coatings in respect to lectin pathway molecules and activation in two randomized groups of patients undergoing elective heart surgery showed a surprising result [62]. During the operation, Ficolin-2 was significantly depleted from plasma in the patients with Bioline® tubing as compared to Phisio® and the level did not restore 24 h post-operationally. Curiously, there was no difference in the degree of *in vitro* measured complement activation between the two coatings nor the level of any of the other PRMs measured. In accordance, previous investigations of Phisio® and Bioline® circuits had shown a comparable degree of biocompatibility between the two, regarding whole body inflammation response and *in vivo* complement activity [63]. Another study by Reser et al. compared the clinical outcome of patients who had undergone coronary bypass surgery with different tubing coatings. Again, the tubing investigated was Phisio®, Bioline® and a third coating Softline®, a heparin-free synthetic polymer comprised of hydrophilic and hydrophobic areas, thought to reduce surface tension on the contact surfaces [64]. In respect to the overall survival, there was no significant difference between the tubing groups. In a third study comparing Bioline® with non-heparinized coating complement activity was not analyzed, but an overall positive effect of the Bioline® circuits was found with reduced systemic inflammation and a shorter stay at the intensive care unit [65].

5.4.3 Silicone Rubber Biomaterial

In a study by Andersen et al. various coatings for biomaterial surfaces in general was investigated for their biocompatibility in respect to complement activation [66]. Silicone rubber treated or untreated with a coating of plasma polymerized vinyl pyrrolidone (ppVP) was analyzed for complement activation potential. The results showed that both untreated silicon and a control of polystyrene were potent activators of complement measured as direct deposition of C3b to the material and fluid phase activated C3c. In contrast, ppVP treated silicone showed a markedly reduced complement activity. Further analysis revealed that while all three activation pathways were active on the untreated silicone surface and polystyrene, only the lectin pathway measured as binding of Ficolin-2 and MBL was active on ppPV treated silicone. Hence, the residual complement activity observed on ppPV coated silicone was ascribed to be mediated by Ficolin-2 and MBL.

5.5 Solid Organ Transplantation

The resolution to escape the detention of haemodialysis is renal transplantation. This, however, entails a whole new series of challenges to the recipient. But shared with the dialysis tube the donor organ has the feature of being a foreign surface to the recipient. The role of adaptive immunity and antibody mediated graft rejection is well established, however, the complement system is also a known actor. Thus, deposition of C4d is an accepted marker in antibody mediated graft rejection [67]. As it turns out, the role of the PRMs of the lectin pathway is starting to be established in transplantation in general, but particularly in kidney transplantation.

5.5.1 Kidney Transplantation

In a study from Imai et al. Ficolin-3 was found on the peritubular capillary of the kidney allograft along with IgM and C4d [68]. This could indicate a role for Ficolin-3 mediated lectin pathway activity; however no deposition of any of the MASPs were detected. Bay et al. performed a large cohort study based on pre-transplant serum samples, where the results showed that a high pre-transplant level of Ficolin-3 was significantly associated with a decreased graft survival [69]. The results have been confirmed by a similar retrospective study from Smedbråten et al. concluding high Ficolin-3 level to be an independent risk factor of kidney graft loss [70]. Though the rejected grafts were not investigated, this indirectly substantiates the possible role of Ficolin-3 and lectin pathway in kidney graft rejection.

5.6 Sampling and Preparation of Blood

Though not directly vital to the patients, an important *in vitro* biosurface is the inside of the tubes where blood samples are collected for analysis of complement components. Today, various compounds are used in blood collection tubes to isolate plasma by inhibiting the coagulation system. Citrate and EDTA are chelaters of calcium, which is essential for clotting to occur – calcium is however also essential for the complement cascade to run. Samples of this type can be analyzed by subsequent addition of calcium to the sample buffer. The human body itself produces an anti-coagulator, heparin that binds to and activates the natural inhibitor antithrombin, which in turn inhibits of the function of thrombin. Heparin does, however, also interfere with the complement system [60] and is ligand for Ficolin-2 [15]. In blood collection, hirudin, a protein derived from the saliva of leeches, is a relatively newly exploited, highly potent anti-coagulation agent. It inhibits thrombin directly, and as of yet, no interferences with other blood components have been reported [71]. Regarding the preparation of serum, blood has traditionally been drawn into a sterile, non-coated glass tube where the negatively charged surface of the glass acts as activator of the intrinsic pathway via factor XII [72]. In recent times however, the addition of coagulation activators to commercially available serum tubes is becoming widespread. The mineral silica is an oxidized form of silicon and when coated as a powder to the blood vial it will speed up the intrinsic pathway activation extensively, thus reducing the clotting time before serum can be prepared from the sample.

5.6.1 Clot Activating Silica in Serum Tubes

It was recently documented by our group and confirmed by others, that Ficolin-2 is selectively depleted from serum samples prepared in vials containing silica [73, 74]. Indirectly it was shown, that the major part of Ficolin-2 in the initial blood sample bound to the silica particles and was thus contained in the blood clot. The entire population of Ficoin-2 was not depleted, a residual entity remained in both studies. An explanation for this could be that Ficolin-2 forms head-to-head complexes (as visualized by electron micrographs in the study of Ohashi and Erickson investigating the structure of pig ficolins α [75]), thus covering the binding groove for silica. Interestingly, the binding of Ficolin-2 to silica did not result in activation of complement or consumption of complement components, in any of the studies [73, 74]. Though shown to bind to heparin as well, no results of Ficolin-2 depletion in heparin coated blood collection tubes has been reported. The findings described above, could indicate an alternative role for Ficolin-2 in the coagulation system, this however remains to be elucidated.

Though the binding to silica and depletion of Ficolin-2 in serum samples is merely of technical importance, the clinical implications could easily be imagined. The inaccurate measurements will lead to faulty conclusions e.g., regarding the role Ficolin-2 in various disease settings, and since the protein is not depleted entirely only drastically reduced, the problem may not be suspected.

A recent example is a study where the possible role of Ficolin-2 as a predictor of outcome in patients with Crohn's disease was investigated [76]. Despite indications from previous studies, Ficolin-2 was found to be a poor predictor—but interestingly the blood samples used for analysis, were collected in serum tubes containing silica. Consequently, the measured Ficolin-2 levels would be unreliable and any possible differences could have been distorted.

5.7 Final Remarks

The data reviewed here, have shown that especially Ficolin-2 binds to several otherwise biocompatible surfaces and the question remains what other surfaces could be targeted. The clinical consequences of these findings have not been investigated and the impact on the health of the patients is not known. Due to the high evolutionary conservation level of *FCN2* [77] and that no Ficolin-2 deficiency has been reported to date, it is reasonable to speculate that the *in vivo* depletion could have a significant impact on patient health and outcome at least during specific disease settings or in immune-compromised individuals. Especially considering the spectrum of opportunistic microorganisms, that Ficolin-2 has been reported to recognise, described in the introduction. Moreover, low levels of Ficolin-2 have been associated with severe respiratory diseases, e.g., chronic bronchiectasis, pulmonary *Mycobacterium tuberculosis* infection, and susceptibility to infection during allergic inflammation of the lungs [78–80]. Most recently, Ficolin-2 has been shown in two independent studies to have an inhibitory effect on hepatitis C virus by binding to the envelope protein [81, 82]. Finally, independent *in vitro* studies have shown Ficolin-2 to bind to the opportunistic fungus *Aspergillus fumigatus*, the cause of invasive pulmonary aspergillosis [47, 83, 84].

The novel collectins described in the introduction, are, as the name imply still so newly discovered and undefined, that the research regarding them are not involving biocompatibility. On the other hand, the studies available on complement responses to biosurfaces have not had the chance to take these molecules into account. Accordingly, this an open field for investigation.

Acknowledgements This work was supported by The Svend Andersen Research Foundation, The Novo Nordisk Research Foundation, The Research Foundation of the Capital Region of Denmark and The Research Foundation at Rigshospitalet.

References

1. Ricklin D, Hajishengallis G, Yang K, Lambris JD. Complement: a key system for immune surveillance and homeostasis. Nat Immunol. 2010;11:785–97.
2. Thiel S. Complement activating soluble pattern recognition molecules with collagen-like regions, mannan-binding lectin, ficolins and associated proteins. Mol Immunol. 2007;44:3875–88.
3. Chen C-B, Wallis R. Two mechanisms for mannose-binding protein modulation of the activity of its associated serine proteases. J Biol Chem. 2004;279:26058–65.
4. Heja D, Kocsis A, Dobo J, Szilagyi K, Szasz R, Zavodszky P, et al. Revised mechanism of complement lectin-pathway activation revealing the role of serine protease MASP-1 as the exclusive activator of MASP-2. Proc Natl Acad Sci USA. 2012;109:10498–503.
5. Degn SE, Hansen AG, Steffensen R, Jacobsen C, Jensenius JC, Thiel S. MAp44, a human protein associated with pattern recognition molecules of the complement system and regulating the lectin pathway of complement activation. J Immunol. 2009;183:7371–8.

6. Skjoedt M-O, Hummelshoj T, Palarasah Y, Honore C, Koch C, Skjodt K, et al. A novel mannose-binding lectin/ficolin-associated protein is highly expressed in heart and skeletal muscle tissues and inhibits complement activation. J Biol Chem. 2010;285:8234–43.
7. Skjoedt M-O, Hummelshoj T, Palarasah Y, Hein E, Munthe-Fog L, Koch C, et al. Serum concentration and interaction properties of MBL/ficolin associated protein-1. Immunobiology. 2011;216:625–32.
8. Kishore U, Reid KB. C1q: structure, function, and receptors. Immunopharmacology. 2000;49:159–70.
9. Fearon DT, Austen KF. Properdin: binding to C3b and stabilization of the C3b-dependent C3 convertase. J Exp Med. 1975;142:856–63.
10. Spitzer D, Mitchell LM, Atkinson JP, Hourcade DE. Properdin can initiate complement activation by binding specific target surfaces and providing a platform for de novo convertase assembly. J Immunol. 2007;179:2600–8.
11. Sahu A, Lambris JD. Structure and biology of complement protein C3, a connecting link between innate and acquired immunity. Immunol Rev. 2001;180:35–48.
12. Markiewski MM, Nilsson B, Ekdahl KN, Mollnes TE, Lambris JD. Complement and coagulation: strangers or partners in crime? Trends Immunol. 2007;28:184–92.
13. Krarup A, Thiel S, Hansen A, Fujita T, Jensenius JC. L-ficolin is a pattern recognition molecule specific for acetyl groups. J Biol Chem. 2004;279:47513–9.
14. Garlatti V, Belloy N, Martin L, Lacroix M, Matsushita M, Endo Y, et al. Structural insights into the innate immune recognition specificities of L- and H-ficolins. EMBO J. 2007;26:623–33.
15. Gout E, Garlatti V, Smith DF, Lacroix M, Dumestre-Pérard C, Lunardi T, et al. Carbohydrate recognition properties of human ficolins: glycan array screening reveals the sialic acid binding specificity of M-ficolin. J Biol Chem. 2010;285:6612–22.
16. Hein E, Honoré C, Skjoedt M-O, Munthe-Fog L, Hummelshøj T, Garred P. Functional analysis of Ficolin-3 mediated complement activation. PLoS One. 2010;5, e15443.
17. Hummelshoj T, Thielens NM, Madsen HO, Arlaud GJ, Sim RB, Garred P. Molecular organization of human Ficolin-2. Mol Immunol. 2007;44:401–11.
18. Garred P, Larsen F, Seyfarth J, Fujita R, Madsen HO. Mannose-binding lectin and its genetic variants. Genes Immun. 2006;7:85–94.
19. Matsushita M, Fujita T. Activation of the classical complement pathway by mannose-binding protein in association with a novel C1s-like serine protease. J Exp Med. 1992;176:1497–502.
20. Ohtani K, Suzuki Y, Eda S, Kawai T, Kase T, Yamazaki H, et al. Molecular cloning of a novel human collectin from liver (CL-L1). J Biol Chem. 1999;274:13681–9.
21. Axelgaard E, Jensen L, Dyrlund TF, Nielsen HJ, Enghild JJ, Thiel S, et al. Investigations on collectin liver 1. J Biol Chem. 2013;288:23407–20.
22. Henriksen ML, Brandt J, Andrieu J-P, Nielsen C, Jensen PH, Holmskov U, et al. Heteromeric complexes of native collectin kidney 1 and collectin liver 1 are found in the circulation with MASPs and activate the complement system. J Immunol. 2013;191:6117–27.
23. Keshi H, Sakamoto T, Kawai T, Ohtani K, Katoh T, Jang S, et al. Identification and characterization of a novel human collectin CL-K1. Microbiol Immunol. 2006;50:1001–13.
24. Selman L, Henriksen ML, Brandt J, Palarasah Y, Waters A, Beales PL, et al. An enzyme-linked immunosorbent assay (ELISA) for quantification of human collectin 11 (CL-11, CL-K1). J Immunol Methods. 2012;375:182–8.
25. Yoshizaki T, Ohtani K, Motomura W, Jang S-J, Mori K, Kitamoto N, et al. Comparison of human blood concentrations of collectin kidney 1 and mannan-binding lectin. J Biochem. 2012;151:57–64.
26. Ma YJ, Skjoedt M-O, Garred P. Collectin-11/MASP complex formation triggers activation of the lectin complement pathway–the fifth lectin pathway initiation complex. J Innate Immun. 2013;5:242–50.
27. Hansen S, Selman L, Palaniyar N, Ziegler K, Brandt J, Kliem A, et al. Collectin 11 (CL-11, CL-K1) is a MASP-1/3-associated plasma collectin with microbial-binding activity. J Immunol. 2010;185:6096–104.

28. Garred P, Honoré C, Ma YJ, Munthe-Fog L, Hummelshøj T. MBL2, FCN1, FCN2 and FCN3-The genes behind the initiation of the lectin pathway of complement. Mol Immunol. 2009;46:2737–44.
29. Honoré C, Rørvig S, Munthe-Fog L, Hummelshøj T, Madsen HO, Borregaard N, et al. The innate pattern recognition molecule Ficolin-1 is secreted by monocytes/macrophages and is circulating in human plasma. Mol Immunol. 2008;45:2782–9.
30. Rørvig S, Honore C, Larsson L-I, Ohlsson S, Pedersen CC, Jacobsen LC, et al. Ficolin-1 is present in a highly mobilizable subset of human neutrophil granules and associates with the cell surface after stimulation with fMLP. J Leukoc Biol. 2009;86:1439–49.
31. Garlatti V, Martin L, Gout E, Reiser J-B, Fujita T, Arlaud GJ, et al. Structural basis for innate immune sensing by M-ficolin and its control by a pH-dependent conformational switch. J Biol Chem. 2007;282:35814–20.
32. Honoré C, Rørvig S, Hummelshøj T, Skjoedt M-O, Borregaard N, Garred P. Tethering of Ficolin-1 to cell surfaces through recognition of sialic acid by the fibrinogen-like domain. J Leukoc Biol. 2010;88:145–58.
33. Munthe-Fog L, Hummelshoj T, Honore C, Moller ME, Skjoedt MO, Palsgaard I, Borregaard N, Madsen HO, and Garred P. Variation in FCN1 affects biosynthesis of ficolin-1 and is associated with outcome of systemic inflammation. Genes Immun. 2012;13:515–22.
34. Wittenborn T, Thiel S, Jensen L, Nielsen HJ, Jensenius JC. Characteristics and biological variations of M-ficolin, a pattern recognition molecule, in plasma. J Innate Immun. 2010;2:167–80.
35. Genster N, Ma YJ, Munthe-Fog L, Garred P. The pattern recognition molecule ficolin-1 exhibits differential binding to lymphocyte subsets, providing a novel link between innate and adaptive immunity. Mol Immunol. 2013;57:181–90.
36. Kjaer TR, Hansen AG, Sørensen UBS, Nielsen O, Thiel S, Jensenius JC. Investigations on the pattern recognition molecule M-ficolin: quantitative aspects of bacterial binding and leukocyte association. J Leukoc Biol. 2011;90:425–37.
37. Frederiksen PD, Thiel S, Larsen CB, Jensenius JC. M-ficolin, an innate immune defence molecule, binds patterns of acetyl groups and activates complement. Scand J Immunol. 2005;62:462–73.
38. Endo Y, Sato Y, Matsushita M, Fujita T. Cloning and characterization of the human lectin P35 gene and its related gene. Genomics. 1996;36:515–21.
39. Munthe-Fog L, Hummelshøj T, Hansen BE, Koch C, Madsen HO, Skjødt K, et al. The impact of FCN2 polymorphisms and haplotypes on the Ficolin-2 serum levels. Scand J Immunol. 2007;65:383–92.
40. Kilpatrick DC, Fujita T, Matsushita M. P35, an opsonic lectin of the ficolin family, in human blood from neonates, normal adults, and recurrent miscarriage patients. Immunol Lett. 1999;67:109–12.
41. Krarup A, Sørensen UBS, Matsushita M, Jensenius JC, Thiel S. Effect of capsulation of opportunistic pathogenic bacteria on binding of the pattern recognition molecules mannan-binding lectin, L-ficolin, and H-ficolin. Infect Immun. 2005;73:1052–60.
42. Hummelshoj T, Munthe-Fog L, Madsen HO, Fujita T, Matsushita M, Garred P. Polymorphisms in the FCN2 gene determine serum variation and function of Ficolin-2. Hum Mol Genet. 2005;14:1651–8.
43. Ma YG, Cho MY, Zhao M, Park JW, Matsushita M, Fujita T, et al. Human mannose-binding lectin and L-ficolin function as specific pattern recognition proteins in the lectin activation pathway of complement. J Biol Chem. 2004;279:25307–12.
44. Faro J, Chen Y, Jhaveri P, Oza P, Spear GT, Lint TF, et al. L-ficolin binding and lectin pathway activation by acetylated low-density lipoprotein. Clin Exp Immunol. 2008;151:275–83.
45. Matsushita M, Endo Y, Taira S, Sato Y, Fujita T, Ichikawa N, et al. A novel human serum lectin with collagen- and fibrinogen-like domains that functions as an opsonin. J Biol Chem. 1996;271:2448–54.
46. Lynch NJ, Roscher S, Hartung T, Morath S, Matsushita M, Maennel DN, et al. L-ficolin specifically binds to lipoteichoic acid, a cell wall constituent of Gram-positive bacteria, and activates the lectin pathway of complement. J Immunol. 2004;172:1198–202.

47. Ma YJ, Doni A, Hummelshøj T, Honoré C, Bastone A, Mantovani A, et al. Synergy between ficolin-2 and pentraxin 3 boosts innate immune recognition and complement deposition. J Biol Chem. 2009;284:28263–75.
48. Lacroix M, Dumestre-Pérard C, Schoehn G, Houen G, Cesbron J, Arlaud GJ, et al. Residue Lys57 in the collagen-like region of human L-ficolin and its counterpart Lys47 in H-ficolin play a key role in the interaction with the mannan-binding lectin-associated serine proteases and the collectin receptor calreticulin. J Immunol. 2009;182:456–65.
49. Matsushita M, Endo Y, Fujita T. Cutting edge: complement-activating complex of ficolin and mannose-binding lectin-associated serine protease. J Immunol. 2000;164:2281–4.
50. Hummelshøj T, Munthe-Fog L, Madsen HO, Garred P. Functional SNPs in the human ficolin (FCN) genes reveal distinct geographical patterns. Mol Immunol. 2008;45:2508–20.
51. Munthe-Fog L, Hummelshøj T, Ma YJ, Hansen BE, Koch C, Madsen HO, et al. Characterization of a polymorphism in the coding sequence of FCN3 resulting in a Ficolin-3 (Hakata antigen) deficiency state. Mol Immunol. 2008;45:2660–6.
52. Munthe-Fog L, Hummelshøj T, Honoré C, Madsen HO, Permin H, Garred P. Immunodeficiency associated with FCN3 mutation and ficolin-3 deficiency. N Engl J Med. 2009;360:2637–44.
53. Tsujimura M, Ishida C, Sagara Y, Miyazaki T, Murakami K, Shiraki H, et al. Detection of serum thermolabile beta-2 macroglycoprotein (Hakata antigen) by enzyme-linked immunosorbent assay using polysaccharide produced by aerococcus viridans. Clin Diagn Lab Immunol. 2001;8:454–9.
54. Swierzko A, Lukasiewicz J, Cedzynski M, Maciejewska A, Jachymek W, Niedziela T, et al. New functional ligands for ficolin-3 among lipopolysaccharides of Hafnia alvei. Glycobiology. 2012;22:267–80.
55. Honoré C, Hummelshoj T, Hansen BE, Madsen HO, Eggleton P, Garred P. The innate immune component ficolin 3 (Hakata antigen) mediates the clearance of late apoptotic cells. Arthritis Rheum. 2007;56:1598–607.
56. Goodkin DA, Bragg-Gresham JL, Koenig KG, Wolfe RA, Akiba T, Andreucci VE, et al. Association of comorbid conditions and mortality in hemodialysis patients in Europe, Japan, and the United States: the Dialysis Outcomes and Practice Patterns Study (DOPPS). J Am Soc Nephrol. 2003;14:3270–7.
57. Xu X, Yang Y, Zhu N. Characteristics and molecular mechanism of adhesion proteins on reused hemodialysis membranes. Blood Purif. 2009;27:321–9.
58. Mares J, Thongboonkerd V, Tuma Z, Moravec J, Matejovic M. Specific adsorption of some complement activation proteins to polysulfone dialysis membranes during hemodialysis. Kidney Int. 2009;76:404–13.
59. Mares J, Richtrova P, Hricinova A, Tuma Z, Moravec J, Lysak D, et al. Proteomic profiling of blood-dialyzer interactome reveals involvement of lectin complement pathway in hemodialysis-induced inflammatory response. Proteomics Clin Appl. 2010;4:829–38.
60. Mollnes TE. Complement and biocompatibility. Vox Sang. 1998;74 Suppl 2:303–7.
61. Kreisler KR, Vance RA, Cruzzavala J, Mahnken JD. Heparin-bonded cardiopulmonary bypass circuits reduce the rate of red blood cell transfusion during elective coronary artery bypass surgery. J Cardiothorac Vasc Anesth. 2005;19:608–11.
62. Hein E, Munthe-Fog L, Thiara AS, Fiane AE, Mollnes TE, Garred P. Heparin-coated cardiopulmonary bypass circuits selectively deplete the pattern recognition molecule ficolin-2 of the lectin complement pathway in vivo. Clin Exp Immunol. 2015;179:294–9.
63. Thiara AS, Andersen VY, Videm V, Mollnes TE, Svennevig K, Hoel TN, et al. Comparable biocompatibility of Phisio- and Bioline-coated cardiopulmonary bypass circuits indicated by the inflammatory response. Perfusion. 2010;25:9–16.
64. Reser D, Seifert B, Klein M, Dreizler T, Hasenclever P, Falk V, et al. Retrospective analysis of outcome data with regards to the use of Phisio®-, Bioline®- or Softline®-coated cardiopulmonary bypass circuits in cardiac surgery. Perfusion. 2012;27:530–4.
65. Palatianos GM, Foroulis CN, Vassili MI, Astras G, Triantafillou K, Papadakis E, et al. A prospective, double-blind study on the efficacy of the bioline surface-heparinized extracorporeal perfusion circuit. Ann Thorac Surg. 2003;76:129–35.

66. Andersen TE, Palarasah Y, Skjødt M-O, Ogaki R, Benter M, Alei M, et al. Decreased material-activation of the complement system using low-energy plasma polymerized poly(vinyl pyrrolidone) coatings. Biomaterials. 2011;32:4481–8.
67. Asgari E, Zhou W, Sacks S. Complement in organ transplantation. Curr Opin Organ Transplant. 2010;15:486–91.
68. Imai N, Nishi S, Alchi B, Ueno M, Fukase S, Arakawa M, et al. Immunohistochemical evidence of activated lectin pathway in kidney allografts with peritubular capillary C4d deposition. Nephrol Dial Transplant. 2006;21:2589–95.
69. Bay JT, Hein E, Sørensen SS, Hansen JM, Garred P. Pre-transplant levels of ficolin-3 are associated with kidney graft survival. Clin Immunol. 2013;146:240–7.
70. Smedbråten YV, Sagedal S, Mjøen G, Hartmann A, Fagerland MW, Rollag H, et al. High Ficolin-3 level at the time of transplantation is an independent risk factor for graft loss in kidney transplant recipients. Transplantation. 2014;00:1–6.
71. Mollnes TE, Brekke O, Fung M, Fure H, Christiansen D, Bergseth G, et al. Essential role of the C5a receptor in E coli-induced oxidative burst and phagocytosis revealed by a novel lepirudin-based human whole blood model of inflammation. Blood. 2002;100:1869–77.
72. Cochrane CG, Wuepper KD. The first component of the kinin-forming system in human and rabbit plasma. Its relationship to clotting factor XII (Hageman factor). J Exp Med. 1971;134:986–1004.
73. Hein E, Bay JT, Munthe-Fog L, Garred P. Ficolin-2 reveals different analytical and biological properties dependent on different sample handling procedures. Mol Immunol. 2013;56:406–12.
74. Brady AM, Spencer BL, Falsey AR, Nahm MH. Blood collection tubes influence serum ficolin-1 and ficolin-2 levels. Clin Vaccine Immunol. 2014;21:51–5.
75. Ohashi T, Erickson HP. The disulfide bonding pattern in ficolin multimers. J Biol Chem. 2004;279:6534–9.
76. Schaffer T, Schoepfer AM, Seibold F. Serum ficolin-2 correlates worse than fecal calprotectin and CRP with endoscopic Crohn's disease activity. J Crohns Colitis. 2014;8(9):1125–32.
77. Garred P, Honoré C, Ma YJ, Rørvig S, Cowland J, Borregaard N, et al. The genetics of ficolins. J Innate Immun. 2010;2:3–16.
78. Kilpatrick DC, Chalmers JD, MacDonald SL, Murray M, Mohammed A, Hart SP, et al. Stable bronchiectasis is associated with low serum L-ficolin concentrations. Clin Respir J. 2009;3:29–33.
79. Luo F, Sun X, Wang Y, Wang Q, Wu Y, Pan Q, et al. Ficolin-2 defends against virulent mycobacteria tuberculosis infection in vivo, and its insufficiency is associated with infection in humans. PLoS One. 2013;8, e73859. Rottenberg ME (editor).
80. Cedzynski M, Atkinson APM, St. Swierzko A, MacDonald SL, Szala A, Zeman K, et al. L-ficolin (ficolin-2) insufficiency is associated with combined allergic and infectious respiratory disease in children. Mol Immunol. 2009;47:415–9.
81. Zhao Y, Ren Y, Zhang X, Zhao P, Tao W, Zhong J, et al. Ficolin-2 inhibits hepatitis C virus infection, whereas apolipoprotein E3 mediates viral immune escape. J Immunol. 2014;193:783–96.
82. Hamed MR, Brown RJP, Zothner C, Urbanowicz RA, Mason CP, Krarup A, et al. Recombinant human L-ficolin directly neutralizes hepatitis C virus entry. J Innate Immun. 2014;6(5):676–84.
83. Bidula S, Kenawy H, Ali YM, Sexton D, Schwaeble WJ, Schelenz S. Role of ficolin-A and lectin complement pathway in the innate defense against pathogenic Aspergillus species. Infect Immun. 2013;81:1730–40.
84. Bidula S, Sexton DW, Abdolrasouli A, Shah A, Reed A, Armstrong-James D, et al. Serum opsonin, L-ficolin, is detected in human lungs of transplant patients following fungal infections and modulates inflammation and killing of A. fumigatus. J Infect Dis. 2015;1–36.

Chapter 6
Foreign Body Reaction to Subcutaneous Implants

Michail Kastellorizios, Namita Tipnis, and Diane J. Burgess

Abstract Subcutaneously implanted materials trigger the host's innate immune system, resulting in the foreign body reaction. This reaction consists of protein adsorption on the implant surface, inflammatory cell infiltration, macrophage fusion into foreign body giant cells, fibroblast activation and ultimately fibrous encapsulation. This series of events may affect the function of subcutaneous implants, such as inhibition of drug diffusion from long-acting drug delivery depots and medical device failure. The foreign body reaction is a complex phenomenon and is not yet fully understood; ongoing research studies aim to elucidate the cellular and molecular dynamics involved. Recent studies have revealed information about the specific role of macrophages and their differential activation towards pro- and anti-inflammatory states, as well as species differences in the timing of collagen deposition and fibrosis. Understanding of the diverse processes involved in the foreign body reaction has led to multiple approaches towards its negation. Delivery of tissue response modifiers, such as corticosteroids, NSAIDs, antifibrotic agents, and siRNAs, has been used to prevent or minimize fibrosis. Of these, delivery of dexamethasone throughout the implantation period is the most common method to prevent inflammation and fibrosis. More recent approaches employ surface modifications to minimize protein adsorption to 'ultra-low' levels and reduce fibrosis. However, the diverse nature of the processes involved in the foreign body reaction favor the use of corticosteroids due to their wide spectrum action compared to other approaches. To date, combination approaches, such as hydrophilic coatings that reduce protein adsorption combined with delivery of dexamethasone are the most effective.

Keywords Foreign body reaction • Biocompatibility • Implants • Subcutaneous implantation • Inflammation

---

M. Kastellorizios • N. Tipnis • D.J. Burgess (✉)
Department of Pharmaceutical Sciences, University of Connecticut,
69 N Eagleville Rd, U3092, Storrs, CT 06269, USA
e-mail: michalis108@gmail.com; Namita.Tipnis@uconn.edu; d.burgess@uconn.edu

J.D. Lambris et al. (eds.), *Immune Responses to Biosurfaces*,
Advances in Experimental Medicine and Biology 865,
DOI 10.1007/978-3-319-18603-0_6

6.1 Overview of the Foreign Body Reaction

The foreign body reaction is a result of the wound healing response, which is altered by the presence of a foreign body. In the absence of a foreign body, tissue trauma triggers a series events that comprise wound healing, i.e. inflammation, proliferation and remodeling [1–4]. These events result in wound closure and new tissue formation (regenerative or scar tissue). The presence of a foreign body interferes with the molecular cascades involved in wound healing, particularly the action of macrophages and their differentiation into foreign body giant cells. As a result, wound healing is altered into what is known as the foreign body reaction, which consists of protein adsorption on the implant surface, macrophage recruitment and giant cell formation, fibroblast activation, and fibrous encapsulation of the foreign body [5–9]. These steps, as they apply to subcutaneously implanted materials, are shown in Fig. 6.1 and described below.

6.1.1 *Protein Adsorption*

Upon implantation, materials are exposed to the subcutaneous tissue's extracellular matrix as well as edema caused by tissue trauma during implantation. This results in proteins such as albumin, fibronectin, fibrinogen, and complement being adsorbed on the surface of the implant [10–14]. The majority of the research conducted on protein adsorption on biomaterials has been focused on plasma protein adsorption. Gifford *et al.* reported protein adsorption characteristics to subcutaneously implanted biosensors [15]. It was determined that multiple proteins with molecular weights up to 15 kDa were adsorbed on the sensor surface after subcutaneous implantation in rats, with albumin fragments being the predominant species.

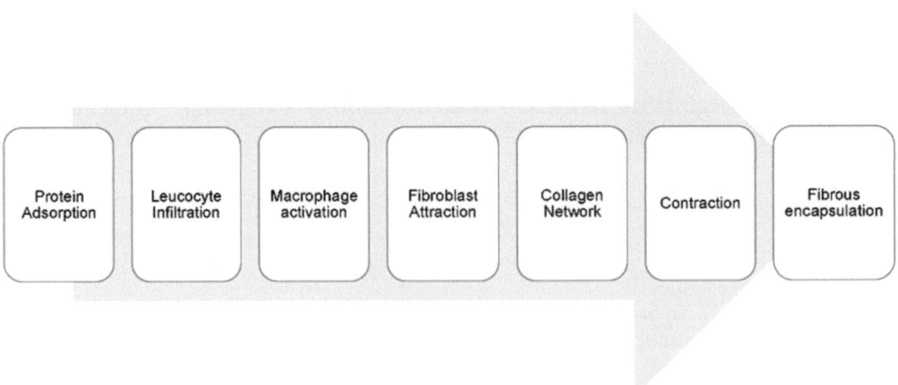

Fig. 6.1 Steps in the foreign body reaction

6.1.2 Acute Inflammation

Blood/implant interactions during initial tissue trauma lead to the activation of the complement system [16–18]. This triggers the body's innate immune system, which leads to edema and white blood cells infiltration (primarily neutrophils and monocytes) at the implantation site. Neutrophils are recruited within a few hours following implantation and their primary function is to ingest bacteria and other debris *via* phagocytosis. Neutrophils are the primary inflammatory cell type for the first 2 days following implantation, at which point they subside and macrophages, differentiated from recruited monocytes, become predominant. Macrophages remain at the site of implantation for a few days in order to ingest foreign material and recruit other cell types, such as fibroblasts, to aid in wound healing. Macrophages recognize the implant as foreign due to protein adsorption on the implant surface, and the continuous presence of the implant results in macrophage differentiation and fusion into foreign body giant cells [19–22]. Giant cells can contain up to about 100 nuclei, as multiple macrophages fuse to ingest large materials. The presence of macrophages and foreign body giant cells has been used as a marker for the foreign body reaction, even though their exact role has not been elucidated. On a molecular level, acute inflammation is characterized by increased levels of pro-inflammatory cytokines, such as IL-4, IL-8, and TNF. These cytokines are excreted by neutrophils and macrophages and play an important role in inflammation regulation. Recent advances towards deciphering the role of inflammatory cytokines are described in the next section.

Acute inflammation is diagnosed *via* physical examination (redness and swelling of the inflamed area) or *via* histological evaluation of the tissue surrounding the implant. The latter is more commonly employed in research endeavors, and the presence of neutrophils and macrophages in high levels around the implants is used to confirm inflammation.

6.1.3 Fibrosis

During the final part of the foreign body reaction, collagen fibers are deposited around the implant and ultimately contract to form a dense, acellular, fibrous capsule [6, 7]. Collagen is deposited by fibroblasts [23, 24] activated by macrophages, and the fibrous capsule that results from their action isolates the implant from the local tissue microenvironment. Isolation serves as the body's last defense against a foreign body, and results from the body being unable to digest or otherwise eliminate the foreign material from the tissue.

Fibrous encapsulation as a result of the foreign body reaction is part of the extracellular matrix (ECM) remodeling phase during wound healing. Under normal conditions, where no foreign body is present during wound healing, fibroblasts produce collagen to replace the ECM that was lost during tissue injury. The collagen fibers produced under these conditions do not form a fibrous capsule and do not have any particular orientation, which results in healthy, loose connective tissue [4].

The presence of a foreign body alters the remodeling phase of wound healing. Macrophages and foreign body giant cells attached on the surface of the foreign body secret matrix metalloproteinases and their inhibitors, which modulate the concentration of cytokines around the implant. Fibroblasts respond and move up concentration gradients of certain cytokines (such as platelet derived growth factor (PDGF)) [25–30]. Accordingly, fibroblasts orient around the foreign body and deposit collagen fibers that then encapsulate the foreign body in a fibrous membrane.

6.1.4 Recent Advances in the Elucidation of the Foreign Body Reaction Mechanisms

The study of the foreign body reaction and its underlying mechanisms has been a research focus for the past decades. A comprehensive review of the main steps of the foreign body reaction to biomaterials is presented by Anderson *et al.* (2008) with emphasis on the cellular components of the reaction [6]. More recent studies report on molecular dynamics during the foreign body reaction, as well as further insights into the role of specific cell types.

With the aid of advanced analytical techniques such as multiplex immunoassaying and ELISA, the up-regulation of pro- and anti-inflammatory molecules throughout long-term implantation periods (from 7 to 30 days) has been mapped [31, 32]. It was determined that the pro-inflammatory cytokines IL-6 and TNF (which promote macrophage activation) and IL-4 and IL-13 (which promote macrophage fusion into foreign body giant cells) are up-regulated during the first period following implantation, with IL-13 peaking at 21 days. The anti-inflammatory interleukin IL-10 is up-regulated during the later stages of the foreign body reaction [33]. In addition, genetically modified, knock-out mice that do not express the chemokine receptor CCR2 demonstrated a stronger acute inflammatory phase (indicated by neutrophil infiltration) but a reduced chronic inflammatory phase (indicated by reduced numbers of macrophages) compared to control mice [34, 35]. Similarly, the absence of the toll-like receptor TLR4 in a knock-out mouse model resulted in a thicker fibrotic band encapsulating the implants [36]. Macrophage polarization has been studied to decipher the various roles of macrophages in the foreign body reaction. It is known that macrophages can be activated to either promote inflammation (traditionally activated macrophages, M1) or to promote tissue remodeling (alternatively activated macrophages, M2). It is now known that during the foreign body reaction, macrophages are activated in a way that results in actions attributed to both M1 and M2 macrophages [37].

More recently, the foreign body reaction to subcutaneous implants of fixed size, shape and composition was compared between a small (rat) and a large (Gottingen minipig) animal model [38]. It was determined that while the steps that make up the foreign body reaction are preserved across the different species, their relative timing of occurrence is significantly different. Minipigs demonstrated earlier onset of fibrosis, with a dense collagen accumulation starting from day 7 as opposed to

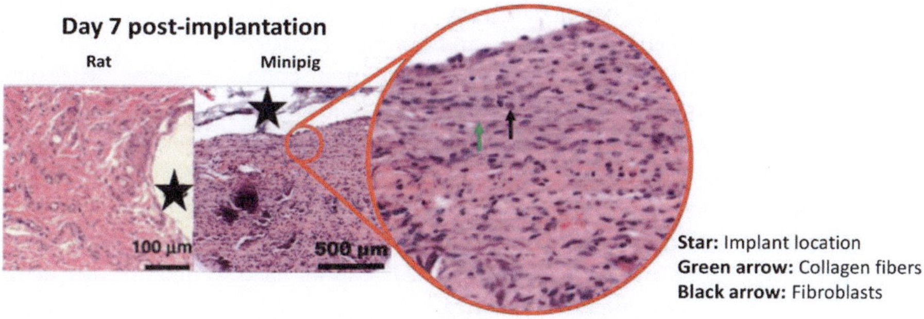

Fig. 6.2 Comparison of the foreign body reaction in rats and minipigs 7 days post-implantation. Star denotes implant location. Connective tissue is stained pink, collagen fibers light pink and inflammatory cells purple (H&E staining). *Green arrow*: collagen fibers; *black arrow*: fibroblasts [38]

day 14 in rats. A comparison of rat and minipig tissue 7 days post implantation is shown in Fig. 6.2.

The advances described above constitute significant contributions towards understanding the underlying mechanisms of the foreign body reaction, and newer research is continuously surfacing that addresses a different aspect of this complex process. However, the inter-relation of the molecular and cellular dynamics during the foreign body reaction, as well as the entire spectrum of their complex actions is yet to be fully elucidated.

6.2 Tools to Evaluate the Foreign Body Reaction

The complexity of the foreign body reaction has made the development of a single, quantifiable marker for its diagnosis and evaluation challenging. The most commonly used method to measure the foreign body reaction is *via* histological evaluation of the tissue response in animal models, usually rodents. More thorough techniques that quantify molecular and cellular markers for specific aspects of the reaction have been developed, however, histological evaluation remains the most reliable method and is used to validate newer techniques.

6.2.1 In Vivo Methods

6.2.1.1 Invasive Techniques

The most straight-forward approach to evaluate the biocompatibility of a material designed for subcutaneous implantation is to test the tissue response to the material at different time points following implantation [38–41]. For this, serial sacrifice of rodents is typically employed to obtain samples of tissue samples from the area surrounding the implants, which are then stained and examined microscopically. Hematoxylin & Eosin (H&E) staining is most commonly used, since it can stain and differentiate cells, nuclei, extracellular matrix and collagen fibers. Masson's Trichrome, which stains collagen, is also used when the focus of the study is the fibrotic phase of the foreign body reaction. When partial anti-inflammatory effects are observed, a reduction in the severity of the foreign body reaction can be evaluated by cell (neutrophil and macrophage) counting around the implant, as well as measuring the fibrotic capsule thickness. In addition, immune-staining for neutrophils (NIMP-R24) and macrophages (MAC387) can be used to quantify the acute and chronic inflammatory phases [42].

More recent approaches employ quantification of pro- and anti-inflammatory cytokines in tissue surrounding the implants [43–46]. This method is especially useful when the foreign body reaction between different materials is compared, and it can give insight on mechanistic aspects of the reaction, as described in the previous section.

6.2.1.2 Noninvasive or Minimally Invasive Techniques

The serial sacrifice of animals involved in the methods described above creates a need for minimally invasive methods to monitor the foreign body reaction during long periods of time *in vivo*. To this end, the microdialysis technique has been adapted to monitor the up-regulation of inflammatory cytokines in the interstitial fluid [33, 47, 48]. Microdialysis catheters can be implanted in the subcutaneous tissue, mimicking a subcutaneous implant. An isotonic solution is pumped through the catheter and molecules present in the tissue fluid that can pass through the catheter's pores, are collected and analyzed. This technique has been successfully applied to measure concentrations of the inflammatory cytokines MCP-1, IL-6 and IL-7. Applications in measuring tissue response to implants alongside the implanted microdialysis catheter require further development of the technique. Another approach is to use imaging techniques to monitor the foreign body reaction. For example, cellular apoptosis associated with inflammation has been monitored *via* fluorescence [42], while fibrin deposition from mast cells has been detected using a modified near infrared probe [49].

6.2.2 *In Vitro Methods*

A method to evaluate the biocompatibility of new and existing biomaterials *in vitro* is necessary, especially in early stage investigations. Currently, there is no standardized *in vitro* method to test the foreign body reaction. The cytotoxicity testing protocols described by the United States Pharmacopeia are typically followed when a new material is synthesized, however, lack of cytotoxicity does not guarantee biocompatibility.

Cell culture-based methods to test the foreign body reaction have been developed in recent years. The majority of these methods include incubation of the implant in a macrophage culture with subsequent testing of macrophage adhesion to the surface of the implant, or identification of inflammatory cytokines to determine macrophage polarization [43, 44, 46, 50]. More complex methods combine macrophages with endothelial cells or fibroblasts, in order to cover a wider area of the foreign body reaction [51]. Of these methods, macrophage adhesion on the implant surface coupled with quantification of inflammatory cytokine expression in the media has shown some ability to predict biocompatibility *in vivo* between different materials. In addition, an *in vitro* model that mimics the foreign body reaction *via* surface modification of polymeric materials has been recently reported [52]. This model, however, was not developed for the evaluation of the foreign body reaction *in vitro*.

6.3 Prevention of the Foreign Body Reaction

Recent years have seen a rise in complex subcutaneous formulations as well as subcutaneously implanted medical devices, whose function can be inhibited by the foreign body reaction. For example, inflammatory cell adhesion on biosensor surfaces can decrease biosensor stability *in vivo*. Moreover, fibrous encapsulation prevents these tissue analytes from reaching the biosensor [39, 53], and, in the case of controlled drug release formulations, the diffusion of drug to the local tissue [54]. Accordingly, prevention of the foreign body reaction to subcutaneous implants remains a major research focus. Methods studied to that effect include release of tissue response modifiers, implant surface modification to minimize the reaction, or a combination of these two approaches.

6.3.1 Release of Tissue Response Modifiers

Tissue response modifiers target cellular components of the foreign body reaction and disrupt the cascade of inflammatory events. Non-steroid anti-inflammatory drugs (NSAIDs), glucocorticoids, anti-fibrotic agents, and siRNAs have been used as tissue response modifiers.

6.3.1.1 Non-steroid Anti-inflammatory Drugs

NSAIDs are effective in inhibiting the early stages of the foreign body reaction by acting on neutrophils [55–59]. Their anti-inflammatory effect, however, is insufficient for long-term applications. For example, release of salicylic acid was shown to only reduce the severity of fibrosis (reduction in fibrotic band thickness) [60]. NSAIDs are therefore not extensively used to battle the foreign body response.

6.3.1.2 Glucocorticoids

Glucocorticoids have been used extensively to prevent the foreign body reaction due to their efficacy and wide spectrum of activities [61–63]. They target neutrophils, macrophages, mast cells, lymphocytes and fibroblasts. They increase the expression of anti-inflammatory cytokines, reduce pro-inflammatory cytokines, and reduce collagen synthesis in fibroblasts.

The most commonly used glucocorticoid is dexamethasone, due to its high potency which allows it to be administered in low doses [64, 65]. From a pharmacological perspective, dexamethasone is very effective in preventing the foreign body reaction and it has been determined that a constant presence of dexamethasone is required for a long-term anti-inflammatory effect [38, 41]. To achieve long-term release and effect, dexamethasone has been incorporated in polymer microspheres [66, 67], scaffolds [68, 69], electrospun fibers [70], and microsphere/hydrogel composites [38–40, 64, 71, 72].

The efficacy of subcutaneously delivered dexamethasone has been evaluated in small (normal and diabetic rats) [39–41, 71] and large (Gottingen minipigs)[38] animal models for implantation periods ranging from 1 to 3 months. Dexamethasone-loaded poly(lactic-co-glycolic) (PLGA) microspheres embedded in a polyvinyl alcohol hydrogel were used to coat small subcutaneous implants that mimic implantable glucose biosensors. PLGAs of molecular weight appropriate for the implantation duration were chosen to achieve gradual release of dexamethasone.

As mentioned in Sect. 6.1.4 above, Gottingen minipigs showed an earlier onset of fibrosis compared to rats. The temporal asymmetry between the two species' reaction to the subcutaneous implants necessitated modification of dexamethasone-loaded PLGA microspheres to tailor the dexamethasone release kinetics to the timing of the foreign body reaction.

6.3.1.3 Anti-fibrotic Agents

Fibrosis in vital organs (such as liver and kidney fibrosis) is not dissimilar to the chronic inflammatory phase of the foreign body reaction [73]. Accordingly, antifibrotic agents that are used in the treatment of such diseases can be used to prevent the foreign body reaction by targeting fibroblasts and inhibiting collagen production. Ganceto *et al.* reported that delivery of pirfenidone, an antifibrotic and anti-inflammatory agent, reduced fibroblast activation by macrophages as well as collagen production [74]. However, fibrosis was not completely negated, possibly due to lack of sufficient anti-inflammatory action during earlier implantation periods. A summary of antifibrotic molecules that can potentially be used to prevent the foreign body reaction is shown in Table 6.1 below.

Table 6.1 List of antifibrotic agents with potential in use to the prevent foreign body reaction, adapted from [73]

Fibrotic pathway/target	Inhibiting compound(s)
Inflammation/immunosuppression	Glucocorticoids, retinoids, colchicine, azathioprine, cyclophosphamide, thalidomide, pentoxifylline, theophylline
Collagen synthesis	Prolyl-4-hydroxylase inhibitors (e.g. HOE0 077 or phenanthrolinones)
TGFβ	Decorin, pirfenidone, relaxin, BMP-7, hepatocyte growth factor, SMAD7
CTGF	Antisense oligonucleotides, cAMP, TNF
Endothelin-1	Bosentan
Angiotensin II	ARBs, ACE inhibitors
Rho GTPases	Y-27362, fasudil
MMP2 and 9	Bay 12-9566
TIMP-1	Monoclonal antibodies specific for TIMP-1
B cell antagonists	Rituximab

6.3.1.4 Gene Silencing

The progress from protein adsorption to acute inflammation and finally fibrosis is controlled by chemotaxis induced by pro- and anti-inflammatory cytokine expression and release in the tissue. Recent early stage approaches utilize siRNA delivery to silence genes that express inflammatory cytokines. In one report, siRNA was delivered to silence the COL1A1 gene and subsequently down-regulate collagen production, which resulted in reduction of fibrotic capsule thickness [75]. In another report, siRNA that targeted the mammalian target of rapamycin (mTOR) to down-regulate collagen production showed promising results in an *in vitro* cell culture model, but had no significant anti-fibrotic effects *in vivo* [76]. Utilization of gene silencing is a promising approach to negate the foreign body reaction, however, its usefulness is uncertain since research on this is still at a very early stage. In addition, siRNAs pose stability challenges which might prove prohibitory for long-term applications where tissue response modifiers are needed to remain stable for long periods of time, up to several months, prior to release in the tissue.

6.3.2 Modification of Implant Surface

The interphase between implant and subcutaneous tissue plays a major role on the severity of the foreign body reaction. Macrophages adhered on the surface of the implant attract and orient fibroblasts around the implant, which ultimately results in fibrous encapsulation [77]. Protein adsorption on the implant surface initiates these events; accordingly, modification of the implant surface to minimize protein adsorption may lead to reduced reaction. There are different modifications that have been studied, and they all share the same principle of action: reduce hydrophobicity of the implant surface and/or modify its surface to mimic native structures.

The simplest approach is to incorporate hydrogel coatings, around the implants [7]. Natural (such as chitosan [78], alginate [79], collagen [80], dextran [81], and hyaluronan [82]) as well as synthetic (such as PVA [83–86] and polyethylene oxide, PEO [46, 87, 88]), have been used in the past to prepare hydrogel coatings for subcutaneous implants. Such coatings mask the usually hydrophobic surface of the implants and reduce protein adsorption, which has been shown to reduce but not eliminate fibrosis.

Recent studies have explored more advanced surface modifications to negate fibrosis. Of note is the use of zwitterionic hydrogels, which have been determined to prevent fibrosis and promote wound healing and neovascularization for a period of 3 months [89]. This is a significant development since long-term prevention of the foreign body reaction without delivery of tissue response modifiers was not achieved previously. Another notable development is the use of porous materials to control macrophage polarization [90]. It has been shown that macrophages that infiltrate pores of 34 μm activate primarily to M1 macrophages (anti-inflammatory), as opposed to macrophages outside the material pores that activate primarily to M2 macrophages (pro-inflammatory). Lastly, decellularized dermal tissue has been used to improve tissue/implant interactions. A recent study conducted in primates using human acellular dermal matrices demonstrated minimal fibrosis 2 months after implantation [91].

6.4 Conclusions

The foreign body reaction is a major obstacle towards long-term functionality of materials and devices implanted subcutaneously, with implantable biosensors probably being the most affected. Typically, implanted or semi-implanted biosensors cannot be used for more than a few days at a time due to tissue reaction that starts from the trauma caused during implantation. This is a serious setback since advancements in electrochemical and microelectronic components have led to miniaturized, fully implantable biosensors that cannot function in the body due to the foreign body reaction. Accordingly, understanding the mechanisms involved in as well as preventing and/or suppressing the foreign body reaction is of urgency.

The foreign body reaction is not a single occurrence, but a cascade of interlocked events. The molecular and cellular mechanisms that make up the reaction are of both a specific and non-specific nature. For example, non-specific protein adsorption occurs during the first few minutes after implantation whereas specific recruitment of macrophages during the subsequent days is tightly controlled by chemotaxis. The diversity of the components that make up the foreign body reaction have made the development of a single test for its evaluation so far impossible. Currently, research-stage methods to evaluate the foreign body reaction in vitro utilize incubation of the material with macrophages and other relevant cell lines. However, in vivo testing in animal models is necessary, with histological evaluation of the tissue reaction at multiple time points the most prevalent method.

To overcome the obstacles posed by the foreign body reaction, prevention or suppression approaches are under investigation. Most commonly, the anti-inflammatory drug, dexamethasone, is delivered throughout the implantation period to suppress the foreign body reaction. Dexamethasone has been incorporated in composite coatings that have been shown to prolong the functionality of implantable biosensors in vivo. Other methods to suppress the reaction are under investigation, such as delivery of antifibrotic agents and gene silencing. Dexamethasone remains the most effective approach to date due to its high potency and wide range of efficacy that covers most of the spectrum of the foreign body reaction. It is unclear whether preventing the initial tissue and blood protein adsorption on the implant surface can prevent the subsequent steps that would lead to fibrous encapsulation. However, novel materials with modified surfaces have been developed that produce minimal reaction that can potentially be suppressed more easily in a prevention/suppression combination approach.

It is expected that strategies to suppress the foreign body reaction developed over the past two decades will be integrated with subcutaneous implants such as biosensors and these devices will soon be tested in clinical settings. In addition, further elucidation of the foreign body reaction and novel materials to prevent as well as new methods to suppress it will likely continue to prevail in research studies on this subject.

Acknowledgements The authors would like to acknowledge the US Army Medical Research and Materiel Command, Telemedicine & Advanced Technology Research Center (Awards W81XWH-07-1-0668 and W81XWH-09-1-0711) for financial support.

References

1. Silver IA. The mechanics of wound healing. Equine Vet J. 1979;11(2):93–6. PubMed.
2. Carrico TJ, Mehrhof Jr AI, Cohen IK. Biology of wound healing. Surg Clin North Am. 1984;64(4):721–33. PubMed.
3. Hammerle CH, Giannobile WV, Working Group 1 of the European Workshop on P. Biology of soft tissue wound healing and regeneration--consensus report of Group 1 of the 10th European Workshop on Periodontology. J Clin Periodontol. 2014;41 Suppl 15:S1–5. PubMed.
4. Safferling K, Sutterlin T, Westphal K, Ernst C, Breuhahn K, James M, et al. Wound healing revised: a novel reepithelialization mechanism revealed by in vitro and in silico models. J Cell Biol. 2013;203(4):691–709. Pubmed Central PMCID: 3840932.
5. Namdari S, Melnic C, Huffman GR. Foreign body reaction to acellular dermal matrix allograft in biologic glenoid resurfacing. Clin Orthop Relat Res. 2013;471(8):2455–8. Pubmed Central PMCID: 3705065.
6. Anderson JM, Rodriguez A, Chang DT. Foreign body reaction to biomaterials. Semin Immunol. 2008;20(2):86–100. Pubmed Central PMCID: 2327202.

7. Morais JM, Papadimitrakopoulos F, Burgess DJ. Biomaterials/tissue interactions: possible solutions to overcome foreign body response. AAPS J. 2010;12(2):188–96. Pubmed Central PMCID: 2844517.
8. Kyriakides TR, Bornstein P. Matricellular proteins as modulators of wound healing and the foreign body response. Thromb Haemost. 2003;90(6):986–92. PubMed.
9. Nomoto K. Mechanism of the response of the living body to foreign material. Allergy. 1990;39(5):437–41. PubMed.
10. Zaveri TD, Lewis JS, Dolgova NV, Clare-Salzler MJ, Keselowsky BG. Integrin-directed modulation of macrophage responses to biomaterials. Biomaterials. 2014;35(11):3504–15. Pubmed Central PMCID: 3970928.
11. Cei S, Karapetsa D, Aleo E, Graziani F. Protein Adsorption on a Laser-Modified Titanium Implant Surface. Implant Dent. 2015;24(2):134–41. PubMed.
12. Nune KC, Misra RD. Pre-adsorption of protein on electrochemically grooved nanostructured stainless steel implant and relationship to cellular activity. J Biomed Nanotechnol. 2014;10(7):1320–35. PubMed.
13. Wang LC, Chen XG, Xu QC, Liu CS, le Yu J, Zhou YM. Plasma protein adsorption pattern and tissue-implant reaction of poly (vinyl alcohol)/carboxymethyl-chitosan blend films. J Biomater Sci Polym Ed. 2008;19(1):113–29. PubMed.
14. Bagnall RD, Arundel PA. A method for the prediction of protein adsorption on implant surfaces. J Biomed Mater Res. 1983;17(3):459–66. PubMed.
15. Gifford R, Kehoe JJ, Barnes SL, Kornilayev BA, Alterman MA, Wilson GS. Protein interactions with subcutaneously implanted biosensors. Biomaterials. 2006;27(12):2587–98. PubMed.
16. Franz S, Rammelt S, Scharnweber D, Simon JC. Immune responses to implants—a review of the implications for the design of immunomodulatory biomaterials. Biomaterials. 2011;32(28):6692–709. PubMed.
17. Feng XL, Yi CX, Zhang YM, Wang YM, Wang YR, Peng C, et al. Immunological effects of polyacrylamide hydrogel injection in human body. Zhonghua zheng xing wai ke za zhi. 2004;20(5):349–50. PubMed.
18. Schmidt S, Haase G, Csomor E, Lutticken R, Peltroche-Llacsahuanga H. Inhibitor of complement, Compstatin, prevents polymer-mediated Mac-1 up-regulation of human neutrophils independent of biomaterial type tested. J Biomed Mater Res A. 2003;66(3):491–9. PubMed.
19. Yang J, Jao B, McNally AK, Anderson JM. In vivo quantitative and qualitative assessment of foreign body giant cell formation on biomaterials in mice deficient in natural killer lymphocyte subsets, mast cells, or the interleukin-4 receptoralpha and in severe combined immunodeficient mice. J Biomed Mater Res A. 2014;102(6):2017–23. PubMed.
20. Anderson JM, Defife K, McNally A, Collier T, Jenney C. Monocyte, macrophage and foreign body giant cell interactions with molecularly engineered surfaces. J Mater Sci Mater Med. 1999;10(10/11):579–88. PubMed.
21. Kao WJ, McNally AK, Hiltner A, Anderson JM. Role for interleukin-4 in foreign-body giant cell formation on a poly (etherurethane urea) in vivo. J Biomed Mater Res. 1995;29(10): 1267–75. PubMed.
22. Smetana Jr K. Multinucleate foreign-body giant cell formation. Exp Mol Pathol. 1987;46(3):258–65. PubMed.
23. Brokopp CE, Schoenauer R, Richards P, Bauer S, Lohmann C, Emmert MY, et al. Fibroblast activation protein is induced by inflammation and degrades type I collagen in thin-cap fibroatheromata. Eur Heart J. 2011;32(21):2713–22. Pubmed Central PMCID: 3205479.
24. Laato M, Kahari VM, Niinikoski J, Vuorio E. Epidermal growth factor increases collagen production in granulation tissue by stimulation of fibroblast proliferation and not by activation of procollagen genes. Biochem J. 1987;247(2):385–8. Pubmed Central PMCID: 1148420.
25. Virakul S, Dalm VA, Paridaens D, van den Bosch WA, Hirankarn N, van Hagen PM, et al. The tyrosine kinase inhibitor dasatinib effectively blocks PDGF-induced orbital fibroblast activation. Graefes Arch Clin Exp Ophthalmol. 2014;252(7):1101–9. PubMed.

26. van Steensel L, Hooijkaas H, Paridaens D, van den Bosch WA, Kuijpers RW, Drexhage HA, et al. PDGF enhances orbital fibroblast responses to TSHR stimulating autoantibodies in Graves' ophthalmopathy patients. J Clin Endocrinol Metab. 2012;97(6):944–53. PubMed.
27. Wang X, Waldeck H, Kao WJ. The effects of TGF-alpha, IL-1beta and PDGF on fibroblast adhesion to ECM-derived matrix and KGF gene expression. Biomaterials. 2010;31(9):2542–8. Pubmed Central PMCID: 2813970.
28. Lachapelle F, Avellana-Adalid V, Nait-Oumesmar B, Baron-Van EA. Fibroblast growth factor-2 (FGF-2) and platelet-derived growth factor AB (PDGF AB) promote adult SVZ-derived oligodendrogenesis in vivo. Mol Cell Neurosci. 2002;20(3):390–403. PubMed.
29. Kim WJ, Mohan RR, Mohan RR, Wilson SE. Effect of PDGF, IL-1alpha, and BMP2/4 on corneal fibroblast chemotaxis: expression of the platelet-derived growth factor system in the cornea. Invest Ophthalmol Vis Sci. 1999;40(7):1364–72. PubMed.
30. Xu J, Clark RA. Extracellular matrix alters PDGF regulation of fibroblast integrins. J Cell Biol. 1996;132(1-2):239–49. Pubmed Central PMCID: 2120701.
31. Rodriguez A, Meyerson H, Anderson JM. Quantitative in vivo cytokine analysis at synthetic biomaterial implant sites. J Biomed Mater Res A. 2009;89(1):152–9. Pubmed Central PMCID: 3864694.
32. Ward WK, Li AG, Siddiqui Y, Federiuk IF, Wang XJ. Increased expression of Interleukin-13 and connective tissue growth factor, and their potential roles during foreign body encapsulation of subcutaneous implants. J Biomater Sci Polym Ed. 2008;19(8):1065–72. PubMed.
33. von Grote EC, Venkatakrishnan V, Duo J, Stenken JA. Long-term subcutaneous microdialysis sampling and qRT-PCR of MCP-1, IL-6 and IL-10 in freely-moving rats. Mol BioSyst. 2011;7(1):150–61. Pubmed Central PMCID: 3864033.
34. Castro PR, Marques SM, Viana CT, Campos PP, Ferreira MA, Barcelos LS, et al. Deletion of the chemokine receptor CCR2 attenuates foreign body reaction to implants in mice. Microvasc Res. 2014;95:37–45. PubMed.
35. Sawyer AJ, Tian W, Saucier-Sawyer JK, Rizk PJ, Saltzman WM, Bellamkonda RV, et al. The effect of inflammatory cell-derived MCP-1 loss on neuronal survival during chronic neuroinflammation. Biomaterials. 2014;35(25):6698–706. Pubmed Central PMCID: 4128094.
36. Auquit-Auckbur I, Caillot F, Arnoult C, Menard JF, Drouot L, Courville P, et al. Role of toll-like receptor 4 in the inflammation reaction surrounding silicone prosthesis. Acta Biomater. 2011;7(5):2047–52. PubMed.
37. van Putten SM, Ploeger DT, Popa ER, Bank RA. Macrophage phenotypes in the collagen-induced foreign body reaction in rats. Acta Biomater. 2013;9(5):6502–10. PubMed.
38. Kastellorizios M, Papadimitrakopoulos F, Burgess DJ. Prevention of foreign body reaction in a pre-clinical large animal model. J Control Release. 2015;202:101–7. PubMed.
39. Wang Y, Papadimitrakopoulos F, Burgess DJ. Polymeric "smart" coatings to prevent foreign body response to implantable biosensors. J Control Release. 2013;169(3):341–7. PubMed.
40. Bhardwaj U, Sura R, Papadimitrakopoulos F, Burgess DJ. PLGA/PVA hydrogel composites for long-term inflammation control following s.c. implantation. Int J Pharm. 2010;384 (1-2):78–86. PubMed.
41. Bhardwaj U, Sura R, Papadimitrakopoulos F, Burgess DJ. Controlling acute inflammation with fast releasing dexamethasone-PLGA microsphere/PVA hydrogel composites for implantable devices. J Diabetes Sci Technol. 2007;1(1):8–17. Pubmed Central PMCID: 2769608.
42. Bartsch I, Willbold E, Yarmolenko S, Witte F. In vivo fluorescence imaging of apoptosis during foreign body response. Biomaterials. 2012;33(29):6926–32. PubMed.
43. Hiromoto S, Inoue M, Taguchi T, Yamane M, Ohtsu N. In vitro and in vivo biocompatibility and corrosion behaviour of a bioabsorbable magnesium alloy coated with octacalcium phosphate and hydroxyapatite. Acta Biomater. 2015;11:520–30. PubMed.
44. Rajam AM, Jithendral P, Mandal AB, Rose C. Evaluation of in vitro macrophage response and in vivo host response to growth factors incorporated chitosan nanoparticle impregnated collagen-chitosan scaffold. J Biomed Nanotechnol. 2014;10(3):508–18. PubMed.

45. Blakney AK, Swartzlander MD, Bryant SJ. The effects of substrate stiffness on the in vitro activation of macrophages and in vivo host response to poly (ethylene glycol)-based hydrogels. J Biomed Mater Res A. 2012;100(6):1375–86. Pubmed Central PMCID: 3339197.
46. Lynn AD, Kyriakides TR, Bryant SJ. Characterization of the in vitro macrophage response and in vivo host response to poly (ethylene glycol)-based hydrogels. J Biomed Mater Res A. 2010;93(3):941–53. PubMed.
47. Sides CR, Stenken JA. Microdialysis sampling techniques applied to studies of the foreign body reaction. Eur J Pharm Sci. 2014;57:74–86. Pubmed Central PMCID: 4004681.
48. Keeler GD, Durdik JM, Stenken JA. Comparison of microdialysis sampling perfusion fluid components on the foreign body reaction in rat subcutaneous tissue. Eur J Pharm Sci. 2014;57:60–7. Pubmed Central PMCID: 4004663.
49. Tsai YT, Zhou J, Weng H, Tang EN, Baker DW, Tang L. Optical imaging of fibrin deposition to elucidate participation of mast cells in foreign body responses. Biomaterials. 2014;35(7):2089–96. Pubmed Central PMCID: 3934503.
50. Wolf MT, Vodovotz Y, Tottey S, Brown BN, Badylak SF. Predicting In Vivo Responses to Biomaterials via Combined In Vitro and In Silico Analysis. Tissue Eng Part C Methods. 2015;21(2):148–59. PubMed.
51. McDonald SM, Matheson LA, McBane JE, Kuraitis D, Suuronen E, Santerre JP, et al. Use of monocyte/endothelial cell co-cultures (in vitro) and a subcutaneous implant mouse model (in vivo) to evaluate a degradable polar hydrophobic ionic polyurethane. J Cell Biochem. 2011;112(12):3762–72. PubMed.
52. Damanik FF, Rothuizen TC, van Blitterswijk C, Rotmans JI, Moroni L. Towards an in vitro model mimicking the foreign body response: tailoring the surface properties of biomaterials to modulate extracellular matrix. Sci Rep. 2014;4:6325. Pubmed Central PMCID: 4168285.
53. Wisniewski N, Klitzman B, Miller B, Reichert WM. Decreased analyte transport through implanted membranes: differentiation of biofouling from tissue effects. J Biomed Mater Res. 2001;57(4):513–21. PubMed.
54. Voskerician G, Shive MS, Shawgo RS, von Recum H, Anderson JM, Cima MJ, et al. Biocompatibility and biofouling of MEMS drug delivery devices. Biomaterials. 2003;24(11):1959–67. PubMed.
55. Wang QS, Cui YL, Gao LN, Guo Y, Li RX, Zhang XZ. Reduction of the pro-inflammatory response by tetrandrine-loading poly (L-lactic acid) films in vitro and in vivo. J Biomed Mater Res A. 2014;102(11):4098–107. PubMed.
56. Ziyan L, Yongmei Z, Nan Z, Ning T, Baolin L. Evaluation of the anti-inflammatory activity of luteolin in experimental animal models. Planta Med. 2007;73(3):221–6. PubMed.
57. Suleyman H, Gul HI, Gul M, Alkan M, Gocer F. Anti-inflammatory activity of bis (3-aryl-3-oxo-propyl) methylamine hydrochloride in rat. Biol Pharm Bull. 2007;30(1):63–7. PubMed.
58. Pelzer LE, Guardia T, Osvaldo Juarez A, Guerreiro E. Acute and chronic antiinflammatory effects of plant flavonoids. Farmaco. 1998;53(6):421–4. PubMed.
59. Swingle KF, Shideman FE. Phases of the inflammatory response to subcutaneous implantation of a cotton pellet and their modification by certain anti-inflammatory agents. J Pharmacol Exp Ther. 1972;183(1):226–34. PubMed.
60. Bryers JD, Jarvis RA, Lebo J, Prudencio A, Kyriakides TR, Uhrich K. Biodegradation of poly (anhydride-esters) into non-steroidal anti-inflammatory drugs and their effect on Pseudomonas aeruginosa biofilms in vitro and on the foreign-body response in vivo. Biomaterials. 2006;27(29):5039–48. PubMed.
61. Abraham SM, Lawrence T, Kleiman A, Warden P, Medghalchi M, Tuckermann J, et al. Antiinflammatory effects of dexamethasone are partly dependent on induction of dual specificity phosphatase 1. J Exp Med. 2006;203(8):1883–9. Pubmed Central PMCID: 2118371.
62. Hart PH, Whitty GA, Burgess DR, Croatto M, Hamilton JA. Augmentation of glucocorticoid action on human monocytes by interleukin-4. Lymphokine Res. 1990;9(2):147–53. PubMed.
63. Kruse NJ, Rowe DW, Fujimoto WY, Bornstein P. Inhibitory effects of glucocorticoids on collagen synthesis by mouse sponge granulomas and granuloma fibroblasts in culture. Biochim Biophys Acta. 1978;540(1):101–16. PubMed.

64. Bhardwaj U, Papadimitrakopoulos F, Burgess DJ. A review of the development of a vehicle for localized and controlled drug delivery for implantable biosensors. J Diabetes Sci Technol. 2008;2(6):1016–29. Pubmed Central PMCID: 2769817.
65. Norton LW, Koschwanez HE, Wisniewski NA, Klitzman B, Reichert WM. Vascular endothelial growth factor and dexamethasone release from nonfouling sensor coatings affect the foreign body response. J Biomed Mater Res A. 2007;81(4):858–69. Pubmed Central PMCID: 4070388.
66. Wang Y, Gu B, Burgess DJ. Microspheres prepared with PLGA blends for delivery of dexamethasone for implantable medical devices. Pharm Res. 2014;31(2):373–81. PubMed.
67. Hickey T, Kreutzer D, Burgess DJ, Moussy F. Dexamethasone/PLGA microspheres for continuous delivery of an anti-inflammatory drug for implantable medical devices. Biomaterials. 2002;23(7):1649–56. PubMed.
68. Vallejo-Heligon SG, Klitzman B, Reichert WM. Characterization of porous, dexamethasone-releasing polyurethane coatings for glucose sensors. Acta Biomater. 2014;10(11):4629–38. Pubmed Central PMCID: 4186909.
69. Acarregui A, Herran E, Igartua M, Blanco FJ, Pedraz JL, Orive G, et al. Multifunctional hydrogel-based scaffold for improving the functionality of encapsulated therapeutic cells and reducing inflammatory response. Acta Biomater. 2014;10(10):4206–16. PubMed.
70. Vacanti NM, Cheng H, Hill PS, Guerreiro JD, Dang TT, Ma M, et al. Localized delivery of dexamethasone from electrospun fibers reduces the foreign body response. Biomacromolecules. 2012;13(10):3031–8. Pubmed Central PMCID: 3466020.
71. Wang Y, Vaddiraju S, Qiang L, Xu X, Papadimitrakopoulos F, Burgess DJ. Effect of dexamethasone-loaded poly (lactic-co-glycolic acid) microsphere/poly (vinyl alcohol) hydrogel composite coatings on the basic characteristics of implantable glucose sensors. J Diabetes Sci Technol. 2012;6(6):1445–53. Pubmed Central PMCID: 3570887.
72. Galeska I, Kim TK, Patil SD, Bhardwaj U, Chatttopadhyay D, Papadimitrakopoulos F, et al. Controlled release of dexamethasone from PLGA microspheres embedded within polyacid-containing PVA hydrogels. AAPS J. 2005;7(1):E231–40. Pubmed Central PMCID: 2751512.
73. Love RJ, Jones KS. Biomaterials, fibrosis, and the use of drug delivery systems in future antifibrotic strategies. Crit Rev Biomed Eng. 2009;37(3):259–81. PubMed.
74. Gancedo M, Ruiz-Corro L, Salazar-Montes A, Rincon AR, Armendariz-Borunda J. Pirfenidone prevents capsular contracture after mammary implantation. Aesthet Plast Surg. 2008;32(1): 32–40. PubMed.
75. Rujitanaroj PO, Jao B, Yang J, Wang F, Anderson JM, Wang J, et al. Controlling fibrous capsule formation through long-term down-regulation of collagen type I (COL1A1) expression by nanofiber-mediated siRNA gene silencing. Acta Biomater. 2013;9(1):4513–24. Pubmed Central PMCID: 3523808.
76. Takahashi H, Wang Y, Grainger DW. Device-based local delivery of siRNA against mammalian target of rapamycin (mTOR) in a murine subcutaneous implant model to inhibit fibrous encapsulation. J Control Release. 2010;147(3):400–7. Pubmed Central PMCID: 2975752.
77. Moore LB, Sawyer AJ, Charokopos A, Skokos EA, Kyriakides TR. Loss of monocyte chemoattractant protein-1 alters macrophage polarization and reduces NFkappaB activation in the foreign body response. Acta Biomater. 2015;11:37–47. Pubmed Central PMCID: 4278755.
78. Uchegbu IF, Schatzlein AG, Tetley L, Gray AI, Sludden J, Siddique S, et al. Polymeric chitosan-based vesicles for drug delivery. J Pharm Pharmacol. 1998;50(5):453–8. PubMed.
79. de Vos P, Hoogmoed CG, Busscher HJ. Chemistry and biocompatibility of alginate-PLL capsules for immunoprotection of mammalian cells. J Biomed Mater Res. 2002;60(2):252–9. PubMed.
80. Sano A, Hojo T, Maeda M, Fujioka K. Protein release from collagen matrices. Adv Drug Deliv Rev. 1998;31(3):247–66. PubMed.
81. Draye JP, Delaey B, Van de Voorde A, Van Den Bulcke A, De Reu B, Schacht E. In vitro and in vivo biocompatibility of dextran dialdehyde cross-linked gelatin hydrogel films. Biomaterials. 1998;19(18):1677–87. PubMed.

82. Vercruysse KP, Prestwich GD. Hyaluronate derivatives in drug delivery. Crit Rev Ther Drug Carrier Syst. 1998;15(5):513–55. PubMed.
83. Quinn CA, Connor RE, Heller A. Biocompatible, glucose-permeable hydrogel for in situ coating of implantable biosensors. Biomaterials. 1997;18(24):1665–70. PubMed.
84. Vaddiraju S, Wang Y, Qiang L, Burgess DJ, Papadimitrakopoulos F. Microsphere erosion in outer hydrogel membranes creating macroscopic porosity to counter biofouling-induced sensor degradation. Anal Chem. 2012;84(20):8837–45. Pubmed Central PMCID: 3791326.
85. Vaddiraju S, Singh H, Burgess DJ, Jain FC, Papadimitrakopoulos F. Enhanced glucose sensor linearity using poly (vinyl alcohol) hydrogels. J Diabetes Sci Technol. 2009;3(4):863–74. Pubmed Central PMCID: 2769944.
86. Patil SD, Papadimitrakopoulos F, Burgess DJ. Dexamethasone-loaded poly (lactic-co-glycolic) acid microspheres/poly (vinyl alcohol) hydrogel composite coatings for inflammation control. Diabetes Technol Ther. 2004;6(6):887–97. PubMed.
87. Espadas-Torre C, Meyerhoff ME. Thrombogenic properties of untreated and poly (ethylene oxide)-modified polymeric matrices useful for preparing intraarterial ion-selective electrodes. Anal Chem. 1995;67(18):3108–14. PubMed.
88. Swartzlander MD, Lynn AD, Blakney AK, Kyriakides TR, Bryant SJ. Understanding the host response to cell-laden poly (ethylene glycol)-based hydrogels. Biomaterials. 2013;34(4): 952–64. Pubmed Central PMCID: 3683297.
89. Zhang L, Cao Z, Bai T, Carr L, Ella-Menye JR, Irvin C, et al. Zwitterionic hydrogels implanted in mice resist the foreign-body reaction. Nat Biotechnol. 2013;31(6):553–6. PubMed.
90. Sussman EM, Halpin MC, Muster J, Moon RT, Ratner BD. Porous implants modulate healing and induce shifts in local macrophage polarization in the foreign body reaction. Ann Biomed Eng. 2014;42(7):1508–16. PubMed.
91. Sandor M, Singh D, Silverman RP, Xu H, De Deyne PG. Comparative host response of 2 human acellular dermal matrices in a primate implant model. Eplasty. 2014;14:e7. Pubmed Central PMCID: 3914385.

Chapter 7
Molecular Characterization of Macrophage-Biomaterial Interactions

Laura Beth Moore and Themis R. Kyriakides

Abstract Implantation of biomaterials in vascularized tissues elicits the sequential engagement of molecular and cellular elements that constitute the foreign body response. Initial events include the non-specific adsorption of proteins to the biomaterial surface that render it adhesive for cells such as neutrophils and macrophages. The latter undergo unique activation and in some cases undergo cell-cell fusion to form foreign body giant cells that contribute to implant damage and fibrotic encapsulation. In this review, we discuss the molecular events that contribute to macrophage activation and fusion with a focus on the role of the inflammasome, signaling pathways such as JAK/STAT and NF-κB, and the putative involvement of micro RNAs in the regulation of these processes.

Keywords Biomaterials • Foreign body response • Foreign body giant cells • Inflammation • Inflammasome • Macrophage

7.1 Introduction

The foreign body response (FBR) is initiated following injury due to biomaterial implantation. An influx of proteins from the blood and interstitial fluid creates a random and temporary proteinaceous coating on the biomaterial surface. Proteins from serum as well as interstitial proteins attach and undergo changes in structure including denaturation on the surface of the implant and aid in subsequent cell

L.B. Moore
Department of Genetics, Yale University, New Haven, CT USA

Vascular Biology and Therapeutics Program, Yale University, New Haven, CT USA

T.R. Kyriakides, Ph.D. (✉)
Vascular Biology and Therapeutics Program, Yale University, New Haven, CT USA

Biomedical Engineering, Yale University, New Haven, CT USA

Department of Pathology, Yale University School of Medicine, New Haven, CT 06519, USA
e-mail: themis.kyriakides@yale.edu

J.D. Lambris et al. (eds.), *Immune Responses to Biosurfaces*,
Advances in Experimental Medicine and Biology 865,
DOI 10.1007/978-3-319-18603-0_7

adhesion [1–3]. Additionally, platelets release chemoattractant signals that stimulate cell migration and fibrinogen that contributes to the formation of a provisional matrix [2]. Following protein interactions, inflammatory cells such as neutrophils and monocytes migrate to the implant, interact with surface-adsorbed proteins and undergo activation. Recruited macrophages, as well as resident macrophages, are induced to secrete their own chemoattractant signals and cytokines, which contribute to the development of the FBR [1]. Recent evidence has also implicated activation of the inflammasome in this process, which leads to the production of pro-inflammatory cytokines such IL-1β, and eventual fibrotic encapsulation of the implant. In contrast to the typical wound environment, macrophages undergo unique activation and a subset undergoes fusion to create foreign body giant cells (FBGC) [4]. FBGCs are considered a hallmark of the FBR and can cause direct degradation of the implanted biomaterial, which often leads to its malfunction [1, 4, 5]. Specifically, it has been shown that FBGCs secrete reactive oxygen species, degradative enzymes, and create an acidic environment at the biomaterial interface [4]. In fact, direct erosion of implants by FBGCs has been demonstrated by scanning electron microscopy [5, 6]. Subsequently, pro-fibrotic signals at the implant site induce the formation of a collagenous and largely avascular capsule, which envelopes the biomaterial within 2–4 weeks after implantation [2, 4]. Confinement in the capsule prevents true integration of the implant with the surrounding tissue, which is responsible for the loss of function for numerous biomaterials including sensors [1, 7, 8]. Figure 7.1 provides an overview of the participation of macrophages in the FBR.

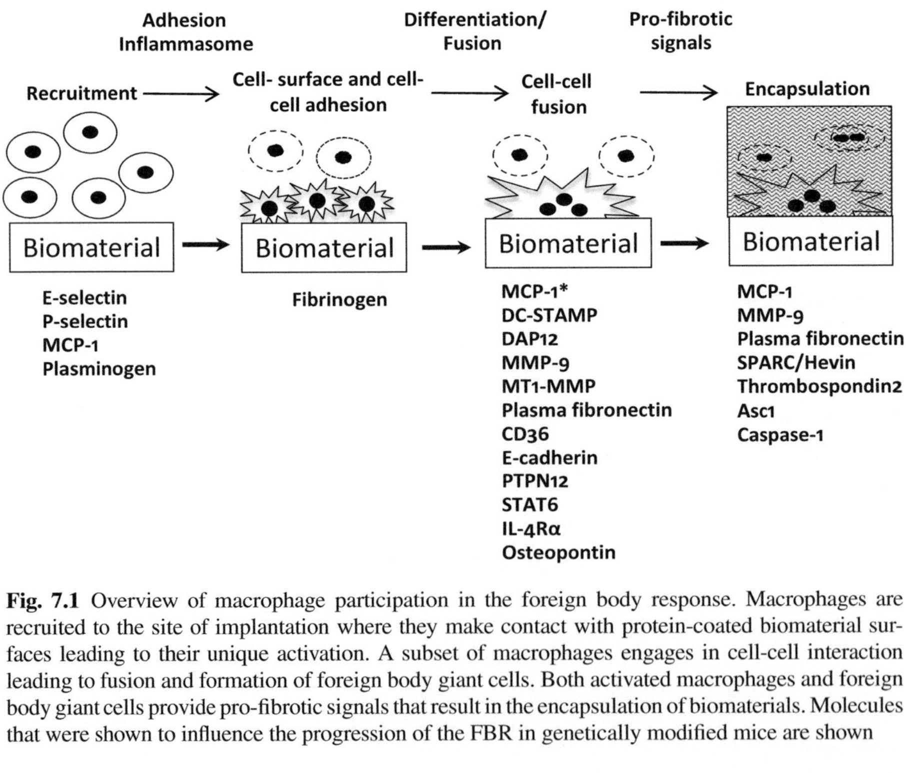

Fig. 7.1 Overview of macrophage participation in the foreign body response. Macrophages are recruited to the site of implantation where they make contact with protein-coated biomaterial surfaces leading to their unique activation. A subset of macrophages engages in cell-cell interaction leading to fusion and formation of foreign body giant cells. Both activated macrophages and foreign body giant cells provide pro-fibrotic signals that result in the encapsulation of biomaterials. Molecules that were shown to influence the progression of the FBR in genetically modified mice are shown

7.2 Biomaterials and the Inflammasome

Various stimuli including silica, uric acid, ATP, and disruption of cellular integrity induce the assembly of a protein complex consisting of nucleotide-binding domain and leucine-rich repeat-containing –type (Nlrp), apoptosis-associated speck-like protein containing CARD (Asc), and Caspase-1, known as the inflammasome [9, 10]. Caspase-1 is normally inactive but undergoes activation during complex formation. The primary role of the inflammasome is to convert pro- IL-1β to IL-1β, which is then secreted into the extracellular milieu [10–12]. Induction of pro- IL-1β production is mediated by a separate signal involving Toll-like receptor (TLR) 4. Therefore, engagement of TLR4 and Nlrp leads to the coordinated production, activation, and secretion of IL-1β. The stimuli mentioned above can induce inflammasome formation and caspase-1 activation by causing K+ efflux, generation of reactive oxygen species, or lysosomal destabilization. Macrophage-biomaterial interactions have been shown to induce IL-1β release in a number of in vitro and in vivo studies suggesting the involvement of the inflammasome. However, studies demonstrating a direct link between macrophage-biomaterial interactions and inflammasome activation are limited. St Pierre et al., showed in a series of in vitro studies that uptake of titanium microparticles by macrophages induced the release of IL-1β and utilized siRNAs against inflammasome components to inhibit this process [13]. Similarly, Maitra et al showed that isolated polyethylene-based implant-derived particles and alkane polymers could induce pro- IL-1β production and IL-1β release in macrophages [14]. In this study, inhibition of inflammasome components was not pursued but the authors did demonstrate caspase-1 activation. In addition, numerous studies have shown that nanoparticles and microparticles can activate the inflammasome. More recently, Bueter el al demonstrated activation of the inflammasome by chitosan, which based on studies with selective inhibitors, was dependent on K+ efflux, reactive oxygen species, and lysosomal destabilization [15]. This and the studies described above involve biomaterial-based foreign bodies that are readily taken up by single cells. However, many in vitro studies show release of IL-1β by macrophages in contact with biomaterials suggesting the involvement of mechanisms that do not involve uptake. Consistent with this suggestion, investigators showed that the inflammasome is activated in the context of cell-biomaterial interactions [16]. Specifically, addition of 150 μm poly-methyl methacrylate (PMMA) microspheres to macrophages induced inflammasome activation and IL-1β secretion. Similarly, injection of the same microspheres in a mouse intraperitoneal model resulted in increased levels of IL-1β. Utilizing the same model, investigators showed lack of IL-1β production in mice deficient for Caspase-1, Nlrp3, or Asc. Interestingly, in long term subcutaneous (SC) implant studies (4 weeks), mice deficient in either Caspase-1 or Asc displayed reduced implant encapsulation. In contrast, encapsulation was normal in Nlrp3 KO mice indicating the participation of a separate Nlrp receptor in the progression of the FBR. Therefore, more research is needed to identify the inflammasome components that mediate implant fibrosis. Figure 7.2 illustrates putative modes of inflammasome activation by biomaterials.

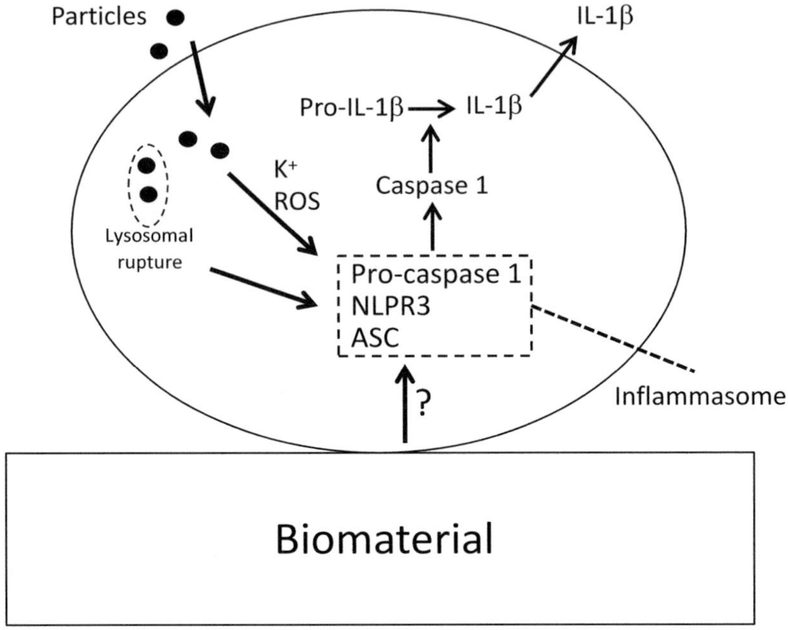

Fig. 7.2 Biomaterial-induced inflammasome activation. Nano- and micro-sized particles can induce assembly of the inflammasome complex leading to conversion pro-caspase 1 to caspase 1, which then converts pro- IL-1β to IL-1β. Secretion of IL-1β contributes to the inflammatory response. Small size particles induce activation following uptake by macrophages and lysosomal rupture. In addition, certain particles can induce activation via the generation of reactive oxygen species and/or K^+ efflux. Large biomaterials, too big to be taken up by cells, can also induce inflammasome activation and IL-1β secretion via a process that depends on membrane dynamics. However, the exact mechanism of activation has not been elucidated

7.3 Macrophages and FBGCs in the FBR

Homologous cell fusion is a highly orchestrated process that occurs in numerous cell types under both physiological and pathological conditions, including trophoblasts in placental development, myoblasts in skeletal muscle formation, and cells of the monocytic lineage in osteoclast and FBGC formation [17–21]. The latter can be induced in vitro by treatment of macrophages plated on fusion-permissive surfaces, such as naked polystyrene, with IL-4. In vitro fusion studies involving macrophages from IL-4Rα knockout mice have demonstrated the importance of IL-4 in the fusion process [22]. Interestingly, FBGC formation was recently shown to be normal in IL-4Rα knockout mice suggesting the existence of additional fusogenic signals in vivo [23]. FBGCs can be damaging to biomaterials and devices and contribute to inflammation at the implant site [1, 4, 19, 21, 24, 25]. In addition, macrophages and FBGCs provide signals that contribute to the formation of the collagenous capsule. For example, they secrete pro-migratory molecules and TGF-β, which leads to recruitment of fibroblasts that deposit extracellular matrix and encapsulate the implant [1, 4, 26]. Foreign body capsules can reach thickness of 50–200 μm and completely envelope implants in a largely avascular space that consists of dense and highly organized collagen fibers [1, 26].

7.4 Macrophage Activation in the FBR

Macrophages have been recently categorized based on the expression of specific molecules that reflect their activation state [27]. Most commonly, they are subdivided into two activation/polarization states: classically activated (M1) or alternatively activated (M2) but these states should be considered as broad characterizations. M1 macrophages are thought to be involved in pro-inflammatory signaling whereas M2 are classified as anti-inflammatory cells that contribute to tissue repair [27–29]. Both of these states can be induced in vitro by treatment of cells with IFN-γ/LPS (M1) or IL-4 (M2). Because FBGC formation can be induced by IL-4 these cells often categorized as M2 [29–31]. However, recent studies support the idea of a M1/M2 activation continuum rather than distinct states [27, 29, 32–35]. In fact, in vivo analysis of traditional M1 and M2 activation markers in an interperitoneal (IP) implantation model by qRT-PCR, immunohistochemistry, and enzyme linked immunosorbent assays (ELISAs) demonstrated a unique polarization state that was highlighted by both M1 and M2 markers [36]. Specifically, FBGC expressed both M1 (iNOS, IL-1β, TNF) and M2 (Arg1, CD36, IL-10) markers [36]. Likewise, analysis of gene expression following IP implantation of boiled egg white demonstrated induction of both M1 (TNF, IL-6) and M2 (IL-4, IL-10) markers during the ensuing FBR [32]. Finally, analysis of subcutaneous polyvinyl alcohol (PVA) sponge implants demonstrated overlapping M1-M2 macrophage phenotypes during the FBR, with cells expressing TNF, IL-6, Arg1, TGFβ, and Ym1 [33]. These studies highlight the unique plasticity and activation state of macrophages during the FBR [29, 34].

7.5 Molecular Pathways

7.5.1 JAK/STAT Pathway

Fusion of macrophages is the consequence of a multistep mechanism induced by IL-4 and followed by the acquisition of a fusion competent state, chemotaxis, and subsequent cytoskeletal rearrangements during and after fusion [17, 19, 37, 38]. Progression to fusion results in the increased expression of cell surface and secreted molecules including DNAX-activating protein of molecular mass 12 kDa (DAP12), dendritic cell-specific transmembrane protein (DC-STAMP), matrix metalloproteinase 9

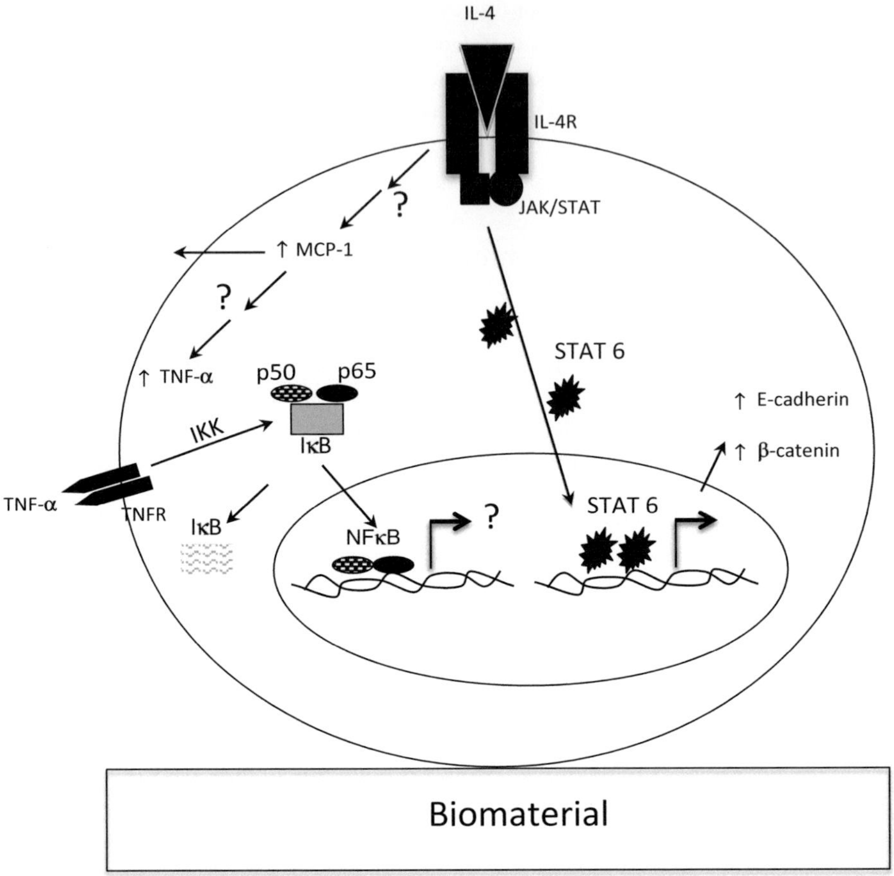

Fig. 7.3 JAK/STAT and NFkB pathways in biomaterial-adherent macrophages. Exposure of biomaterial-adherent macrophage to IL-4 induces activation of the JAK/STAT pathway leading to phosphorylation of STAT6 that translocates to the nucleus where it promotes the expression of genes including E-cadherin and β-catenin. A separate more complex and not completely defined IL-4-induced pathway causes upregulation of MCP-1 and TNF, which are secreted and bind their respective receptors. TNF activates the canonical NF-κB pathway by phosphorylation of IκB by IKK and its subsequent ubiquitination and degradation. Degradation of IκB leads to liberated p50/p65 complex that translocates to the nucleus and induces transcription. Both pathways are essential for macrophage fusion in vitro and the process depends on IL-4 and interaction with specific surfaces

(MMP9), monocyte chemotactic protein-1 (MCP-1), and epithelial-cadherin (ECad) [2, 4, 17, 39, 40]. As mentioned above, induction of a fusion-competent state in vitro can be achieved by addition of IL-4 and involves at least two separate pathways including JAK/STAT [37]. As shown in Fig. 7.3, IL-4 induces the JAK1/3 and STAT6 signaling cascade, leading to upregulation of ECad and β-catenin that localize to the cell periphery where they are thought to facilitate cell-cell interactions [37, 41].

7.5.2 MCP1 and Rac1-Dependent Cytoskeletal Remodeling

IL-4 induces DAP12-dependent signaling through the ITAM motif and TREM2 receptor, with downstream SYK signaling increasing DC-STAMP expression [42]. Moreover, machinery reminiscent of macrophage phagocytosis, including MCP-1 mediated Rac-1-dependent cytoskeleton rearrangements and phosphatidyl serine (PtdSer) exposure and subsequent recognition by CD36, have also been linked to macrophage fusion [17, 43–45]. The reason the two pathways are considered separate involves observations in MCP1-KO mice and their macrophages that are defective in fusion but display normal upregulation of Ecad and β-catenin in response to IL-4. In contrast, they display reduced Rac-1 activation and Rac-1-mediated cytoskeletal remodeling as well as TNF and MMP-9 expression [2, 37]. However, fusion in MCP-1 macrophages can be restored by addition of exogenous MMP-9, which leads to changes in the subcellular localization of E-cad from the cell periphery to the cytoplasm. These observations suggest that the secretion of MMP-9 and the possible cleavage of Ecad represent a point where the two pathways display functional overlap [37]. While several molecular mediators of fusion have been implicated in the FBR, the exact signaling pathways regulating this process remain unknown and it is anticipated that their identification would lead to the development of novel strategies to attenuate FBGC formation and the FBR [4].

7.5.3 TNF and NF-κB Pathway

In vitro and in vivo studies have demonstrated the importance of TNF in the FBR [4, 32, 33, 37]. In fact, analysis of levels of TNF during macrophage-biomaterial interactions can be helpful in evaluating the biocompatibility of new materials [4]. For example, TNF has been used as a marker of inflammation and indicator of severity of the FBR in studies ranging from the effect of topographical alterations, lymphocyte enhancement of FBGC activation, and the biocompatibility of novel materials such as poly(ethyleneglycol)-based hydrogels [46–48]. The importance of TNF is apparent when one considers the fact that the fusion-deficient phenotype of MCP-1 macrophages can be rescued via exogenous TNF treatment [37]. TNF induction in macrophages during the FBR is intriguing because it is a potent inducer of the NF-κB pathway. In the absence of activation, the canonical NF-κB components p50 and p65 (RelA) are held in the cytoplasm by the inhibitory IκB. When present, TNF induces activation of the IKK complex, promoting the phosphorylation and ubiquitination of IκB, which is subsequently degraded allowing the release and translocation of the p50-p65 heterodimer complexes to the nucleus where they induce transcription of target genes (Fig. 7.3) [49]. Alternatively, the non-canonical NF-κB pathway involves NIK-dependent induction of IKK which phosphorylates the p100 precursor thereby releasing the p52-RelB complex [49]. Canonical activation of NF-κB has been noted *in vivo* following implantation of titanium and copper implants in rats and propylene mesh in mice [50, 51]. Additionally, induction of the non-canonical NF-κB pathway has been demonstrated as essential to RANKL mediated osteoclast fusion [52]. Although the canonical NF-κB pathway has been shown to be important in IL-4-induced macrophage fusion and the non-canonical pathway for osteoclast fusion, it has been suggested that cross talk between pathways does occur, potentially allowing for compensation [53]. Nevertheless, It has been recently established that the canonical NF-κB pathway is required for macrophage fusion during the FBR both in vitro and in vivo [36]. Specifically, induction and nuclear translocation of NF-κB components p50 and RelA were shown at day 3 following IL-4 stimulation. NF-κB induction occurred in temporal manner consistent with TNF expression and was minimal in fusion-deficient MCP-1 KO mice. Additionally, inhibition of canonical NF-κB pathway by treatment with the pharmacological inhibitor Bay11, resulted in decreased fusion. More importantly, induction and nuclear translocation of p50/RelA was observed in vivo in implant-adherent macrophages undergoing fusion at day 4 following implantation in an IP model [36]. These observations suggest that TNF contributes to FBGC formation and the FBR, in part, by activating the canonical NF-κB pathway. However, the downstream effects of this pathway and the genes that are regulated by p50/p65 in this process have not been identified.

7.6 FBGC Formation and FBR Phenotypes in Genetically Modified Mice

With the advent of genetically modified mice, investigators have utilized models of biomaterial implantation in order to elucidate the contribution of specific molecules in the FBR. Despite the lack of standardized approaches in these studies and the variable approaches used, such as multiple implantation locations and time points, numerous biomaterials, and different modes of analysis, the cumulative body of acquired knowledge is informative. For example, it was shown in short term studies that mice deficient in either plasminogen or fibrinogen displayed reduction in cell recruitment and/or cell attachment to biomaterials [54]. In addition, mice lacking components critical for monocyte/macrophage recruitment such as E- and P-selectin displayed reduced accumulation of inflammatory cells in an IP implantation model and this was associated with a reduced fibrotic response [55]. Similarly, mice lacking MCP-1 displayed reduced macrophage accumulation and FBGC formation and significant attenuation of capsule thickness in an IP implant model [37]. Interestingly, the same mice with SC implants displayed reduced FBGC formation despite normal macrophage recruitment and capsule thickness [2]. Several knockout mice or cells isolated from them displayed altered FBGC formation including MMP-9, DC-STAMP, DAP12, IL-4Rα, MT1-MMP, plasma fibronectin, osteopontin, PTPN12, STAT6, and CD36 [22, 40, 42, 44, 56–61]. As mentioned in Section 2 above, Helming et al demonstrated compromised fusion of IL4Rα-KO macrophages in vitro [22]. Consistent with the findings of Helming et al, anti-IL-4 antibodies were shown to block FBGC formation in a cage implant model [62]. In contrast, Yang et al showed normal FBGC formation in IL4Rα-KO mice in a SC implant model [23]. Therefore, the requirement for IL-4 signaling and perhaps other signals in FBGC formation in vitro and in vivo remains to be elucidated. Moreover, the complex phenotype of biomaterial-adherent macrophages, featuring characteristics of both M1 and M2 activation, suggests the contribution of additional signaling molecules.

Alterations in capsule formation have also been detected in genetically modified mice, including those lacking the angiogenesis inhibitor thrombospondin-2 (TSP2), which formed capsules with increased vessel density and aberrant collagen fibers [63]. SPARC-KO mice displayed reduced collagen capsule thickness, and double deletion of SPARC and its homologue have resulted in increased vessel density [64, 65]. Plasminogen activator inhibitor-1 KO mice displayed reduced fibrosis in a PVA sponge implant model [66]. More recently, Zaveri et al demonstrated a surprising role for macrophage integrin Mac1 in influencing capsule thickness [67]. Obviously, these molecules represent significant variation in function (enzymes, receptors, cell adhesion proteins, cytokines, extracellular matrix proteins) and subcellular localization (cytoplasmic, membrane-bound, secreted), which highlights the complexity of the processes they regulate. Moreover, several of these molecules have been shown to be induced significantly during FBGC formation and progression of the FBR suggesting that regulation of gene expression plays a significant role in these processes.

7.7 MicroRNAs and FBGC Formation

MicroRNAs (miRs) are small noncoding RNAs that regulate gene expression via post-transcriptional modification of transcripts [68]. See Eulalio et al and Krol et al for detailed reviews on the generation and mode of action of miRs [70, 71]. MiRs have confirmed participation in essentially all cellular processes examined to date, including cellular development, metabolism, apoptosis, proliferation, and differentiation [71–73]. If one considers that a single miR may influence post-transcriptional control of hundreds of targets, and that an mRNA transcript will be influenced by many miRs, it is not surprising that miRs regulate over half of the human genome [72–74]. Though little work has been done regarding the role of miRs in cellular fusion, evidence exists that they are involved in monocyte/macrophage differentiation and processes involved in cell fusion such as cytoskeletal remodeling and the NF-κB pathway [75–78]. In addition, miRs haven been shown to regulate several molecules that have been implicated in FBGC formation. For example, miRs regulate the hematopoietic stem cell lineage, including the differentiation of monocytes into macrophages, which strongly suggests their importance in determining macrophage phenotype [72]. Furthermore, miR-21 has recently been linked to the process of differentiation of monocyte-derived dendritic cells, a process that is dependent on IL-4 and granulocyte- macrophage colony stimulating factor (GM-CSF) [79].

MiR-705 has been shown to regulate MMP9 expression in the uterine matrix, and miRs 143/145 regulate cytoskeletal remodeling during phenotype switch in smooth muscle cells [80, 81]. More importantly, miRs have been implicated in homotypic cell fusion including that of myoblasts, osteoclasts, and most recently FBGC. Sugatini et al. reported that miR-223 and miR-21 regulated RANKL-mediated osteoclastogenesis [82, 83]. Unlike miR-223 where the targeted transcript(s) involved in the regulation of osteoclast formation are not known [84–86], miR-21 was shown to downregulate PDCD4 to promote fusion [83]. In addition, miR-7b has been demonstrated to directly target DC-STAMP during osteoclastogenesis, thus inhibiting NFATC1 and c-FOS to attenuate fusion [87]. As mentioned above, myoblast fusion during developmental myogenesis as well as following injury is also regulated by microRNAs. Specifically, reduction of myoblast fusion was observed to be regulated by miR-1192 targeting HMGB1 as well as miR-206 and miR-1 downregulating CX43 gap junctions during myogenesis [88, 89]. The participation of miRs in FBGC formation is largely unexplored with a single study showing that miR-7a-1 can regulate DC-STAMP during IL-4 induced macrophage fusion [90]. Finally, deletion of dicer, which is a central molecule in miR processing, resulted in a significant increase of IL-4 dependent fusion [90].

7.8 Conclusion

Investigation of the molecules and signaling pathways that regulate FBGC formation and the FBR offers the dual promise of facilitating the development of strategies to improve the function and longevity of biomaterials as well as enhance our fundamental understanding of key cellular processes. Identification of required molecules should lead to the rational design of biomaterials with the capacity to modulate their expression and/or function in a beneficial manner. Current approaches are predominantly focused on surface modifications that have had limited success in curbing the FBR and in general apply to a small subset of biomaterials. In contrast, molecular approaches could be applied in numerous applications. Equally important, the elucidation of the molecules and pathways that regulate FBGC formation should provide insights relevant to other types of cell fusion including osteoclast and myoblast formation. Similarly, a more in-depth understanding of the processes that regulate the encapsulation of biomaterials could enhance our ability to combat other fibrotic diseases.

References

1. Ratner BD. Reducing capsular thickness and enhancing angiogenesis around implant drug release systems. J Control Release. 2002;78(1-3):211–8.
2. Kyriakides TR, et al. The CC chemokine ligand, CCL2/MCP1, participates in macrophage fusion and foreign body giant cell formation. Am J Pathol. 2004;165(6):2157–66.

3. Wilson CJ, et al. Mediation of biomaterial-cell interactions by adsorbed proteins: a review. Tissue Eng. 2005;11(1-2):1–18.
4. Anderson JM, Rodriguez A, Chang DT. Foreign body reaction to biomaterials. Semin Immunol. 2008;20(2):86–100.
5. Dadsetan M, et al. In vivo biocompatibility and biodegradation of poly(ethylene carbonate). J Control Release. 2003;93(3):259–70.
6. Zhao Q, et al. Foreign-body giant cells and polyurethane biostability: in vivo correlation of cell adhesion and surface cracking. J Biomed Mater Res. 1991;25(2):177–83.
7. Klueh U, Dorsky DI, Kreutzer DL. Enhancement of implantable glucose sensor function in vivo using gene transfer-induced neovascularization. Biomaterials. 2005;26(10):1155–63.
8. Park KM, Gerecht S. Harnessing developmental processes for vascular engineering and regeneration. Development. 2014;141(14):2760–9.
9. Henao-Mejia J, et al. Inflammasomes: far beyond inflammation. Nat Immunol. 2012; 13(4):321–4.
10. Bryant C, Fitzgerald KA. Molecular mechanisms involved in inflammasome activation. Trends Cell Biol. 2009;19(9):455–64.
11. Strowig T, et al. Inflammasomes in health and disease. Nature. 2012;481(7381):278–86.
12. Latz E, Xiao TS, Stutz A. Activation and regulation of the inflammasomes. Nat Rev Immunol. 2013;13(6):397–411.
13. St Pierre CA, et al. Periprosthetic osteolysis: characterizing the innate immune response to titanium wear-particles. J Orthop Res. 2010;28(11):1418–24.
14. Maitra R, et al. Endosomal damage and TLR2 mediated inflammasome activation by alkane particles in the generation of aseptic osteolysis. Mol Immunol. 2009;47(2-3):175–84.
15. Bueter CL, et al. Spectrum and mechanisms of inflammasome activation by chitosan. J Immunol. 2014;192(12):5943–51.
16. Malik AF, et al. Inflammasome components Asc and caspase-1 mediate biomaterial-induced inflammation and foreign body response. Proc Natl Acad Sci U S A. 2011;108(50): 20095–100.
17. Helming L, Gordon S. Molecular mediators of macrophage fusion. Trends Cell Biol. 2009;19(10):514–22.
18. Brodbeck WG, Anderson JM. Giant cell formation and function. Curr Opin Hematol. 2009; 16(1):53–7.
19. Helming L, Gordon S. The molecular basis of macrophage fusion. Immunobiology. 2007;212(9-10):785–93.
20. Anderson JM. Multinucleated giant cells. Curr Opin Hematol. 2000;7(1):40–7.
21. McNally AK, Anderson JM. Macrophage fusion and multinucleated giant cells of inflammation. Adv Exp Med Biol. 2011;713:97–111.
22. Helming L, Gordon S. Macrophage fusion induced by IL-4 alternative activation is a multistage process involving multiple target molecules. Eur J Immunol. 2007;37(1):33–42.
23. Yang J, et al. In vivo quantitative and qualitative assessment of foreign body giant cell formation on biomaterials in mice deficient in natural killer lymphocyte subsets, mast cells, or the interleukin-4 receptoralpha and in severe combined immunodeficient mice. J Biomed Mater Res A. 2014;102(6):2017–23.
24. Singer AJ, Clark RA. Cutaneous wound healing. N Engl J Med. 1999;341(10):738–46.
25. Kyriakides TR, Bornstein P. Matricellular proteins as modulators of wound healing and the foreign body response. Thromb Haemost. 2003;90(6):986–92.
26. Morris AH, Kyriakides TR. Matricellular proteins and biomaterials. Matrix Biol. 2014;37:183–91.
27. Martinez FO, Gordon S. The M1 and M2 paradigm of macrophage activation: time for reassessment. F1000Prime Rep. 2014;6:13.
28. Laskin DL. Macrophages and inflammatory mediators in chemical toxicity: a battle of forces. Chem Res Toxicol. 2009;22(8):1376–85.
29. Brown BN, et al. Macrophage polarization: an opportunity for improved outcomes in biomaterials and regenerative medicine. Biomaterials. 2012;33(15):3792–802.

30. Sussman EM, et al. Porous implants modulate healing and induce shifts in local macrophage polarization in the foreign body reaction. Ann Biomed Eng. 2014;42(7):1508–16.
31. Wolf MT, et al. Predicting In Vivo Responses to Biomaterials via Combined In Vitro and In Silico Analysis. Tissue Eng Part C Methods. 2014;21(2):148–59.
32. Mooney JE, et al. Transcriptional switching in macrophages associated with the peritoneal foreign body response. Immunol Cell Biol. 2014;92(6):518–26.
33. Daley JM, et al. The phenotype of murine wound macrophages. J Leukoc Biol. 2010; 87(1):59–67.
34. Mosser DM, Edwards JP. Exploring the full spectrum of macrophage activation. Nat Rev Immunol. 2008;8(12):958–69.
35. Mantovani A, et al. Macrophage plasticity and polarization in tissue repair and remodelling. J Pathol. 2013;229(2):176–85.
36. Moore LB, et al. Loss of monocyte chemoattractant protein-1 alters macrophage polarization and reduces NFkappaB activation in the foreign body response. Acta Biomater. 2015; 11:37–47.
37. Skokos EA, et al. Lack of TNF-alpha-induced MMP-9 production and abnormal E-cadherin redistribution associated with compromised fusion in MCP-1-null macrophages. Am J Pathol. 2011;178(5):2311–21.
38. McNally AK, Anderson JM. Interleukin-4 induces foreign body giant cells from human monocytes/macrophages. Differential lymphokine regulation of macrophage fusion leads to morphological variants of multinucleated giant cells. Am J Pathol. 1995;147(5):1487–99.
39. Yagi M, et al. Role of DC-STAMP in cellular fusion of osteoclasts and macrophage giant cells. J Bone Miner Metab. 2006;24(5):355–8.
40. MacLauchlan S, et al. Macrophage fusion, giant cell formation, and the foreign body response require matrix metalloproteinase 9. J Leukoc Biol. 2009;85(4):617–26.
41. Moreno JL, et al. IL-4 promotes the formation of multinucleated giant cells from macrophage precursors by a STAT6-dependent, homotypic mechanism: contribution of E-cadherin. J Leukoc Biol. 2007;82(6):1542–53.
42. Helming L, et al. Essential role of DAP12 signaling in macrophage programming into a fusion-competent state. Sci Signal. 2008;1(43):ra11.
43. Jay SM, et al. Foreign body giant cell formation is preceded by lamellipodia formation and can be attenuated by inhibition of Rac1 activation. Am J Pathol. 2007;171(2):632–40.
44. Helming L, Winter J, Gordon S. The scavenger receptor CD36 plays a role in cytokine-induced macrophage fusion. J Cell Sci. 2009;122(Pt 4):453–9.
45. Lemaire I, Falzoni S, Adinolfi E. Purinergic signaling in giant cell formation. Front Biosci (Elite Ed). 2012;4:41–55.
46. Chen S, et al. Characterization of topographical effects on macrophage behavior in a foreign body response model. Biomaterials. 2010;31(13):3479–91.
47. Chang DT, Colton E, Anderson JM. Paracrine and juxtacrine lymphocyte enhancement of adherent macrophage and foreign body giant cell activation. J Biomed Mater Res A. 2009; 89(2):490–8.
48. Blakney AK, Swartzlander MD, Bryant SJ. The effects of substrate stiffness on the in vitro activation of macrophages and in vivo host response to poly(ethylene glycol)-based hydrogels. J Biomed Mater Res A. 2012;100(6):1375–86.
49. Hayden MS, Ghosh S. Shared principles in NF-kappaB signaling. Cell. 2008;132(3):344–62.
50. Suska F, et al. In vivo cytokine secretion and NF-kappaB activation around titanium and copper implants. Biomaterials. 2005;26(5):519–27.
51. Fet N, et al. In vivo characterisation of the inflammatory reaction following mesh implantation in transgenic mice models. Langenbecks Arch Surg. 2014;399(5):579–88.
52. Cui W, et al. The intracellular domain of CD44 promotes the fusion of macrophages. Blood. 2006;107(2):796–805.
53. Yu M, et al. NF-kappaB signaling participates in both RANKL- and IL-4-induced macrophage fusion: receptor cross-talk leads to alterations in NF-kappaB pathways. J Immunol. 2011; 187(4):1797–806.

54. Busuttil SJ, et al. A central role for plasminogen in the inflammatory response to biomaterials. J Thromb Haemost. 2004;2(10):1798–805.
55. Tang L, Jiang W, Welty SE. The participation of P- and E-selectins on biomaterial-mediated tissue responses. J Biomed Mater Res. 2002;62(4):471–7.
56. Miyamoto H, et al. Osteoclast stimulatory transmembrane protein and dendritic cell-specific transmembrane protein cooperatively modulate cell-cell fusion to form osteoclasts and foreign body giant cells. J Bone Miner Res. 2012;27(6):1289–97.
57. Gonzalo P, et al. MT1-MMP is required for myeloid cell fusion via regulation of Rac1 signaling. Dev Cell. 2010;18(1):77–89.
58. Keselowsky BG, et al. Role of plasma fibronectin in the foreign body response to biomaterials. Biomaterials. 2007;28(25):3626–31.
59. Rhee I, et al. Macrophage fusion is controlled by the cytoplasmic protein tyrosine phosphatase PTP-PEST/PTPN12. Mol Cell Biol. 2013;33(12):2458–69.
60. Miyamoto H, et al. An essential role for STAT6-STAT1 protein signaling in promoting macrophage cell-cell fusion. J Biol Chem. 2012;287(39):32479–84.
61. Tsai AT, et al. The role of osteopontin in foreign body giant cell formation. Biomaterials. 2005;26(29):5835–43.
62. Kao WJ, et al. Role for interleukin-4 in foreign-body giant cell formation on a poly(etherurethane urea) in vivo. J Biomed Mater Res. 1995;29(10):1267–75.
63. Kyriakides TR, et al. Mice that lack the angiogenesis inhibitor, thrombospondin 2, mount an altered foreign body reaction characterized by increased vascularity. Proc Natl Acad Sci U S A. 1999;96(8):4449–54.
64. Puolakkainen P, et al. Compromised production of extracellular matrix in mice lacking secreted protein, acidic and rich in cysteine (SPARC) leads to a reduced foreign body reaction to implanted biomaterials. Am J Pathol. 2003;162(2):627–35.
65. Barker TH, et al. Matricellular homologs in the foreign body response: hevin suppresses inflammation, but hevin and SPARC together diminish angiogenesis. Am J Pathol. 2005;166(3):923–33.
66. Chuang-Tsai S, et al. Reduction in fibrotic tissue formation in mice genetically deficient in plasminogen activator inhibitor-1. Am J Pathol. 2003;163(2):445–52.
67. Zaveri TD, et al. Integrin-directed modulation of macrophage responses to biomaterials. Biomaterials. 2014;35(11):3504–15.
68. Bartel DP. MicroRNAs: genomics, biogenesis, mechanism, and function. Cell. 2004;116(2):281–97.
69. Perera RJ, Ray A. MicroRNAs in the search for understanding human diseases. BioDrugs. 2007;21(2):97–104.
70. Eulalio A, Huntzinger E, Izaurralde E. Getting to the root of miRNA-mediated gene silencing. Cell. 2008;132(1):9–14.
71. Krol J, Loedige I, Filipowicz W. The widespread regulation of microRNA biogenesis, function and decay. Nat Rev Genet. 2010;11(9):597–610.
72. Anglicheau D, Muthukumar T, Suthanthiran M. MicroRNAs: small RNAs with big effects. Transplantation. 2010;90(2):105–12.
73. Suarez Y, Sessa WC. MicroRNAs as novel regulators of angiogenesis. Circ Res. 2009;104(4):442–54.
74. Selbach M, et al. Widespread changes in protein synthesis induced by microRNAs. Nature. 2008;455(7209):58–63.
75. Cheng HS, et al. Noncoding RNAs regulate NF-kappaB signaling to modulate blood vessel inflammation. Front Genet. 2014;5:422.
76. Pellegrino L, et al. miR-23b regulates cytoskeletal remodeling, motility and metastasis by directly targeting multiple transcripts. Nucleic Acids Res. 2013;41(10):5400–12.
77. Fontana L, et al. MicroRNAs 17-5p-20a-106a control monocytopoiesis through AML1 targeting and M-CSF receptor upregulation. Nat Cell Biol. 2007;9(7):775–87.
78. Montagner S, Deho L, Monticelli S. MicroRNAs in hematopoietic development. BMC Immunol. 2014;15:14.

79. Cekaite L, Clancy T, Sioud M. Increased miR-21 expression during human monocyte differentiation into DCs. Front Biosci (Elite Ed). 2010;2:818–28.
80. Nothnick WB. Regulation of uterine matrix metalloproteinase-9 and the role of microRNAs. Semin Reprod Med. 2008;26(6):494–9.
81. Xin M, et al. MicroRNAs miR-143 and miR-145 modulate cytoskeletal dynamics and responsiveness of smooth muscle cells to injury. Genes Dev. 2009;23(18):2166–78.
82. Sugatani T, Hruska KA. MicroRNA-223 is a key factor in osteoclast differentiation. J Cell Biochem. 2007;101(4):996–9.
83. Sugatani T, Vacher J, Hruska KA. A microRNA expression signature of osteoclastogenesis. Blood. 2011;117(13):3648–57.
84. Li YT, et al. Brief report: amelioration of collagen-induced arthritis in mice by lentivirus-mediated silencing of microRNA-223. Arthritis Rheum. 2012;64(10):3240–5.
85. Shibuya H, et al. Overexpression of microRNA-223 in rheumatoid arthritis synovium controls osteoclast differentiation. Mod Rheumatol. 2013;23(4):674–85.
86. Chen SY. MicroRNA-223: a double-edged sword in rheumatoid arthritis. Rheumatol Int. 2014;34(2):285–6.
87. Dou C, et al. MiR-7b directly targets DC-STAMP causing suppression of NFATc1 and c-Fos signaling during osteoclast fusion and differentiation. Biochim Biophys Acta. 2014;1839(11):1084–96.
88. Dormoy-Raclet V, et al. HuR and miR-1192 regulate myogenesis by modulating the translation of HMGB1 mRNA. Nat Commun. 2013;4:2388.
89. Anderson C, Catoe H, Werner R. MIR-206 regulates connexin43 expression during skeletal muscle development. Nucleic Acids Res. 2006;34(20):5863–71.
90. Sissons JR, et al. Cutting edge: microRNA regulation of macrophage fusion into multinucleated giant cells. J Immunol. 2012;189(1):23–7.

Chapter 8
Heparan Sulfate Proteoglycan Metabolism and the Fate of Grafted Tissues

Jeffrey L. Platt, Lucile E. Wrenshall, Geoffrey B. Johnson, and Marilia Cascalho

Abstract Tissue and organ transplants between genetically distinct individuals are always or nearly always rejected. The universality and speed of transplant rejection distinguishes this immune response from all others. Although this distinction is incompletely understood, some efforts to shed light on transplant rejection have revealed broader insights, including a relationship between activation of complement in grafted tissues, the metabolism of heparan sulfate proteoglycan and the nature of immune and inflammatory responses that ensue. Complement activation on cell surfaces, especially on endothelial cell surfaces, causes the shedding heparan sulfate, an acidic saccharide, from the cell surface and neighboring extracellular matrix. Solubilized in this way, heparan sulfate can activate leukocytes via toll like receptor-4, triggering inflammatory responses and activating dendritic cells, which migrate to regional lymphoid organs where they spark and to some extent govern cellular immune responses. In this way local ischemia, tissue injury and infection, exert systemic impact on immunity. Whether or in what circumstances this series of

J.L. Platt, M.D. (✉) • M. Cascalho, M.D., Ph.D.
Transplantation Biology, Department of Surgery, University of Michigan,
A520B Medical Sciences Research Building I, 1150W. Medical Center Drive,
Ann Arbor, MI 48109-5656, USA

Department of Microbiology & Immunology, University of Michigan, Ann Arbor, MI, USA
e-mail: plattjl@umich.edu; marilia@umich.edu

L.E. Wrenshall, M.D., Ph.D.
Department of Surgery, Wright State University, Dayton, OH, USA

Department of Neuroscience, Cell Biology and Physiology,
Wright State University, Dayton, OH, USA
e-mail: Lucile.wrenshall@wright.edu

G.B. Johnson, M.D., Ph.D.
Department of Radiology, Mayo Clinic, Rochester, MN, USA

Department of Immunology, Mayo Clinic, Rochester, MN, USA
e-mail: johnson.geoffrey@mayo.edu

J.D. Lambris et al. (eds.), *Immune Responses to Biosurfaces*,
Advances in Experimental Medicine and Biology 865,
DOI 10.1007/978-3-319-18603-0_8

events explains the distinct characteristics of the immune response to transplants is still unclear but the events offer insight into the inception of immunity under the sub-optimal conditions accompanying infection and mechanisms by which infection and tissue injury engender systemic inflammation.

Keywords Complement • Endothelial cells • Heparan sulfate • Antigen presenting cell • Dendritic cell • Toll-like receptor • Systemic inflammatory response syndrome • Sepsis • Accommodation

8.1 The Thinking Process

In June 1999, Tom and Ray Magliozzi delivered the commencement address at the Massachusetts Institute of Technology (MIT). The Magliozzi brothers were alumni of MIT and had a popular radio program, Car Talk, in which they entertained audiences with humorous stories and with advice on automobile repair and human nature. According to one report of the address [1], the brothers proposed a theory that happiness is an inverse function of phylogeny and declared their mantra to be: "*non impediti ratione cogitatonis*" or "unencumbered by the thought process." One unifying conclusion was that rational thinking is inimical to happiness but another conclusion, more pertinent for the fields of immunology and transplantation, might be that *dumb luck* often solves the most important problems; but, just as often, rational thinking prevents us from seeing that the problems are solved. Below we describe and reinterpret some conclusions we drew from fortuitous observations made as we investigated immunity and transplantation. We make no attempt to discuss the broader literature on these subjects since we cannot know which observations of others were truly fortuitous and because we believe the evolution of thinking and not the thinking process *per se* has more lasting value than the details.

8.2 The Immune Response to Transplantation

Three decades ago we began to explore the immune response to transplantation. The question that seemed most urgent at that time (and still today) was why transplantation evokes immunity that is universal, rapid and powerful. Conventional immune reactions, typified by initial exposure to *Mycobacterium bovis*, attenuated and optimized in dosage as Bacillus Calmette–Guérin (BCG) vaccine, are detected in approximately 50 % of those first exposed approximately 4–6 weeks after exposure and detection required re-administration of antigen in the form of a skin test. In contrast, immune responses to transplantation occur in nearly 100 % of recipients, can be detected within a few days and in the absence of immunosuppression destroy the grafted tissue or organ [2–5]. Our original approach to understanding what might distinguish the immune response to transplantation was to explore the numbers of leukocytes of varying phenotypes that populated rejecting grafts [6] and delayed type hypersensitivity reactions [7]. The phenotype of leukocytes in DTH differed somewhat from the phenotype of leukocytes in rejection but the kinetics and other characteristics differed more [7]. Thus, this thought process brought an end to what had been a productive line of research and led to research aimed linking phenotype with functions.

The functions of the phenotypic markers initially studied, CD2, CD3, CD4, and CD8, BA-1, among others, were not then understood but since some markers were also expressed in development [8, 9], it seemed that understanding the processes governing the evolution of phenotypes in development would shed light on the function of the markers in mature tissue. It seemed further that changes in the phenotype and function of cells might be governed by glycosaminoglycans, the unique carbohydrate substitutions on proteoglycans, the metabolism of which had been found to drive cell-cell and cell-matrix interactions in development [10]. The lines of reasoning that brought us to investigate glycosaminoglycans and proteoglycans were entirely wrong, but the investigation nevertheless would bring some understanding of processes that can determine the fate of transplants

8.3 Proteoglycans in Ontogeny and Rejection of Kidneys

Proteoglycans consist of a core protein conjugated with glycosaminoglycan chains. Glycosaminoglycan chains are O-linked linear copolymers consisting of interdigitating hexuronic acid and hexosamine residues modified by N- and O-linked sulfate esters. The expression of a given core protein determines which glycosaminoglycan chain will be added to the core protein, where on or in the cell the proteoglycan will be situated and a few biological properties. However, it is the glycosaminoglycan chains that confer the predominant biological properties of proteoglycans we shall consider. Only a few of many outstanding reviews of the structure, biosynthesis and biological properties of proteoglycans are provided as references [11–13].

To understand the connection between the phenotype and function of cells, we explored the metabolism of proteoglycans in kidney organogenesis and the impact of perturbing that metabolism [14–16]. The kidney was selected for study because morphogenesis of that organ involves complex stereotypic cell-cell and cell-matrix interactions the disruption of which might cause dramatic and reproducible change in morphology and biochemistry. Disrupting chondroitin sulfate proteoglycan synthesis had clearly and also predictable changes. But, adding heparan sulfate had the greatest impact; it shut down development of branching structures without apparently impacting on maturation of epithelial element of glomeruli, which we found to be associated with degradation of heparan sulfate. We took these results to indicate that heparan sulfate controls the earliest events in nephron formation (induction of nephrogenic mesenchyme); but, it might also have reflected the inhibition of heparan sulfate depolymerization by heparanase or the elution of heparan sulfate binding peptides. We would later return to these possibilities in entirely different systems, ischemic tissues and rejecting organs [17–20]. However, we first turned our attention to the question of whether inflammation and immunity might change the metabolism of heparan sulfate proteoglycan.

8.4 Metabolism of Heparan Sulfate Proteoglycan in Inflammation and Immunity

Having found that metabolism of heparan sulfate and chondroitin sulfate proteoglycans are linked to organogenesis of kidney [14, 21], we wondered whether metabolism of these also changes in inflammation and immunity. The query was focused on endothelial cells, which we considered the principal target of cell-mediated and humoral immunity [17, 22, 23]. Mature endothelial cells that we used produced far more heparan sulfate than chondroitin sulfate and hence we direct our work at the metabolism of heparan sulfate proteoglycan. But, there was another, far better reason for focusing on endothelial cell heparan sulfate proteoglycan. Heparan sulfate proteoglycan on cell surfaces and extracellular matrices exerted or at least supported all of the key physiologic functions of endothelial cells that inflammation, immunity and disease disrupt (Fig. 8.1). Heparan sulfate maintains the integrity of the endothelial lining, providing a key barrier to diffusion of proteins and migration of cells. Heparan sulfate maintains the fluidity of blood by tethering and activating antithrombin III and tissue factor pathway inhibitor. Heparan sulfate also regulates activation of complement, in part by its action of antithrombin III and in part by tethering factors H and properdin and helps to limit oxidant injury by tethering superoxide dismutase. Heparan sulfate also potentially regulates inflammation and immunity by attaching chemokines and many cytokines to endothelial surfaces. Therefore, we reasoned that if one had to name a molecule the metabolism of which would transform the biology of tissues and organs, one could find no better candidate to name than heparan sulfate.

This reasoning led us to investigate whether and how inflammation modifies heparan sulfate metabolism in endothelial cells. In this one case, perhaps owing to dumb luck, we were apparently *non impediti ratione cogitatonis*. Thus, activation of complement on endothelial cells, as it might occur in ischemia-reperfusion injury or graft rejection, caused the quantitative shedding of heparan sulfate from the cells [22]. Interaction of neutrophils [24] and activated T cells also caused shedding of heparan sulfate [25]. Shedding of heparan sulfate caused by complement occurred within a few minutes, shedding caused by neutrophils proceeded over 20–60 min and was less complete, shedding caused by activated T cells took place over about 1 h and the loss represented <50 % of heparan sulfate. It thus seemed that acute inflammation and immunity might, as an early manifestation, disrupt the barrier, anticoagulant, and anti-inflammatory functions of blood vessels and in this way set the stage for the dramatic changes in endothelial cell physiology and activation that

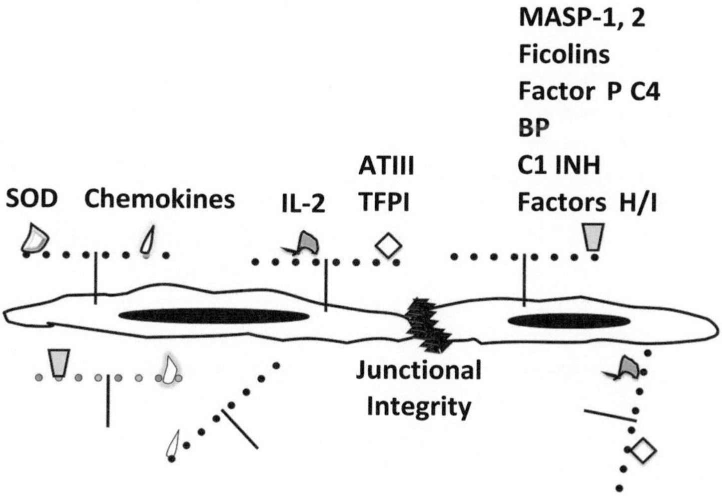

Fig. 8.1 Heparan sulfate proteoglycan and the functions of endothelium. Heparan sulfate proteoglycans, consisting of a core protein conjugated with heparan sulfate glycosaminoglycan chains (strings of pearls) contribute to many function of normal blood vessels. These functions include (a) providing negative cell surface-charge that regulates complement, coagulation and cellular interactions; (b) maintaining the junctional integrity of the endothelial barrier to efflux of cells, solutes and plasma; (c) tethering and in some cases activating proteins that regulate oxidants such as superoxide dismutase (SOD); adherence, migration and activation of leukocytes, stem cells (chemokines) and lymphocytes (IL-2); coagulation [antithrombin III (ATIII) and tissue factor pathway inhibitor (TFPI)] and complement [MBL associated serine protease (MASP-1) and (MASP-2), ficolins, Factor P, C4 binding protein (C4 BP), C1 inhibitor (C1 INH), factor H and factor I]. Shedding of heparan sulfate deprives endothelial cells of these functions leading to cellular injury, extravasation of blood cells and plasma from blood vessels, activation of complement, coagulation and hemostasis and inflammation

would be seen over the ensuing hours [26–30]. It seemed also that the burgeoning interest in endothelial cell biology could not be explored in full unless changes in heparan sulfate proteoglycan were taken into account.

What seemed most interesting then and now, from a practical perspective, was not the loss of heparan sulfate *per se*, as important as that might be, but rather the mechanism of the loss (which one might wish to prevent therapeutically). Both proteases and an endoglycosidase, heparanase (endo-β-glucuronidase), were known to degrade heparan sulfate proteoglycans associated with normal and malignant cells (although an impact on blood vessels and blood vessel functions in inflammation and immunity had not been described). Proteases might cause release of nearly full-sized proteoglycans; heparanase might release small fragments of individual glycosaminoglycan chains; both enzymes, acting together would release partly degraded fragments of proteoglycans and glycosaminoglycans. Thus, the size of heparan sulfate proteoglycans and fragments thereof could provide clues to the enzyme activities responsible for the shedding of the molecules. Determining the size of shed molecules was especially important for discovering how complement had acted because the sera used as a source of complement might contain abundant amounts of platelet heparanase [31], the activity of which would confound probing this subject. Using endothelial cells in which heparan sulfate had been biosynthetically labeled we traced the fate and size of the label, [35S]sulfate, after exposure of the cells to complement, neutrophils and activated (and resting) T cells. In each of these setting the preponderance of heparan sulfate initially released from the labeled cells was found in nearly full-size proteoglycans [32]. Thus, the earliest step, especially after complement was activated [20, 33], involved the action of proteases, the inhibition of which would preserve endothelial cell heparan sulfate, at least under the conditions used in our experiments (Fig. 8.2).

That is not to say that heparanase and/or oxidants are not important in the overall sequence of events. Activated T cells had been found to produce heparanase and use it to penetrate matrices and inhibition of heparanase appears to halt migration of T cells [34, 35]. Likewise, oxidants produced by inflammatory cells and endothelial cells might also cleave heparan sulfate [36]. However, degradation by heparanase and oxidants is much slower and requires not only prior activation of immune

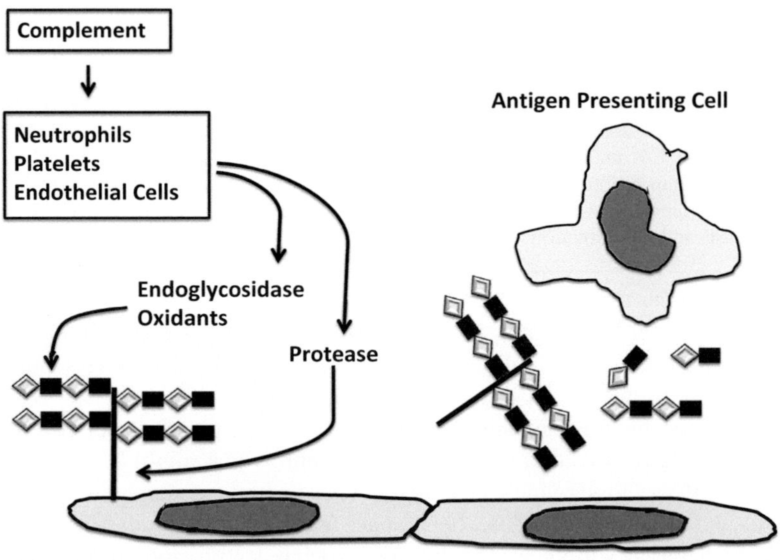

Fig. 8.2 Complement activation and the inciting of inflammation and immunity. Complement activation on endothelial cells incites inflammation and immunity via several mechanisms. Illustrated in the figure is one mechanism involving degradation of heparan sulfate proteoglycan. Complement activation causes neutrophils, platelets and endothelial cells to secrete proteases and heparanase, an endoglycosidese that specifically cleaves heparan sulfate. Heparan sulfate proteoglycans and partially degraded heparan sulfate chains so released activate inflammatory cells and dendritic cells (antigen presenting cells) which, via secretion of cytokines, such as IL-1, increased antigen processing and presentation and co-stimulation and migration to regional lymph nodes, activate naïve T cells

inflammatory cells but persistence of those cells in the vicinity of the proteoglycans to be degraded. Thus, inflammation or immunity involving complement probably does not depend on heparanase and oxidants, at least at the inception.

8.5 Heparan Sulfate and the Immune Response

If shedding of heparan sulfate was an early event in tissue injury, infection or transplantation, could it impact in some way on the activation of T cells? Heparin, which is structurally similar to but more sulfated than heparan sulfate had been shown previously to inhibit autoimmune disease and allograft rejection [37, 38]. But, heparin was commonly used as an anticoagulant in transplant recipients and seemed to have no appreciable impact on rejection in that setting. Besides, rejection of a graft would be preceded by T cell differentiation and expansion and migration and we were interested in understanding the potential impact of heparan sulfate shed at the earliest time, when cells of the immune system would either be ignited to respond or held in check (tolerance).

The initial testing of the impact of heparan sulfate on T cell activation was conducted in mixed leukocyte cultures, which, in mouse, are prepared by mixing splenocytes from different strains, or cultures of splenocytes and mitogens of various types [39]. In these systems, we observed that heparan sulfate amplified T cell proliferative responses and development of effector functions, especially under suboptimal conditions we believed would model the condition in which T cell activation normally occurs [39, 40].

However, the most interesting finding was not that heparan sulfate stimulated T cells; indeed, we observed that heparan sulfate had no appreciable direct impact on T cells. Rather, the most interesting finding was that heparan sulfate modified the function of antigen presenting cells (APC), the leukocytes, such as dendritic cells, that actually take up and present antigen to T cells. This finding made more sense than any impact heparan sulfate might have on T cells because it is the APC in transplants or infected tissues that take up antigen at sites of tissue injury, and hence the site of heparan sulfate shedding, and carry the antigen to lymphoid organs where naïve T cells reside.

A further observation was also of much interest. If the presence of heparan sulfate was limited to the first day of a 5-day mixed leukocyte culture, it had the most profoundly stimulatory impact, while the presence of heparan sulfate only during the last several days of a 5-day mixed culture had a profoundly inhibitory impact on the proliferative response. The early impact of heparan sulfate thus seemed to model rather well a circumstance in which an APC takes up antigen and becomes activated and then migrates to a microenvironment, the lymphoid tissue, which lacks a surfeit of shed of heparan sulfate.

We also found that the apparently disparate actions of heparan sulfate on APC were owed, at least in part, to secretion of IL-1 and IL-6 early after stimulation and PGE2 later and were associated with activation and nuclear translocation of

NFκB [41, 42]. We also found that the panoply of changes in cellular functions was relatively specific for heparan sulfate. Heparin, which has structural similarity to heparan sulfate but contains twice as many sulfate esters and is mainly confined to cellular granules, had far less stimulatory effect than heparan sulfate (and since heparin has some sequences modified like heparan sulfate, the heparin sequences might well have had no impact). Chondroitin sulfate, which has similar charge density but differ disaccharide units but is expressed outside cells, caused little change except at the highest concentrations used. And, only heparan sulfate incited production of PGE2 [42]. In its action on APC, heparan sulfate seemed to trigger a number of signaling intermediates in APC, the constellation of which could not be ascribed to any one receptor or cell surface perturbation and we concluded that heparan sulfate probably delivered signals via two or more discrete surface events, but the net effect would enhance T cell activation. The production of PGE2 on the other hand seemed unrelated to the overall propensity of heparan sulfate to promote immune response to sub-optimal stimuli (which we figured then as now represent the condition when immunity to foreign organisms and toxins most needs stimulation). Instead, the production of PGE2 days after APC were activated might limit the expansion of T cells responding to antigen or avoid ongoing activation of antigen specific responses.

8.6 Orchestrating T Cell Responses *In Vivo*

Studying T cells in mixed cultures of splenocytes has revealed much of what is known about the specificity of alloimmune responses. However, neither alloimmune nor conventional immune responses arise by a mixing of splenocytes. Rather, they arise when a small number of dendritic cells, take up antigen and receive activating signals such as lipopolysaccharide (LPS). Activated dendritic cells migrate from tissues to regional lymph nodes where the dendritic cells are brought together with a large number naïve T cells in lymphoid tissues. Dendritic cells are sometimes referred to as "professional antigen presenting cells" because unlike the various cells that might be used to probe T cell specificity and biochemical processes of antigen presentation, dendritic cells have the unique abilities to engulf antigen in various forms, migrate from the source of antigen to key positions in lymphoid organs, and present antigen and the key accessory signals needed to activate naïve T cells or in the absence of the accessory signals to generate energy [43–45]. Hence, to know whether and how heparan sulfate might actually impact on T cell activation (or suppression), it would be necessary to probe these events using dendritic cells.

As a first step, we asked whether heparan sulfate changes the differentiation and function of dendritic cells of the mouse [46]. Immature bone marrow derived dendritic cells were incubated with small amounts of heparan sulfate or with control substances (including heparan sulfate that had been depolymerized by treatment with HNO2, to assure absence of contaminating substances) and then "activation" was evaluated by assaying expression of proteins typically found on mature dendritic cells. Immature dendritic cells expressed MHC class II at intermediate levels and low levels of CD40, CD54, and CD86 at low levels. Dendritic cells exposed to heparan sulfate expressed high levels of these proteins. Exposure to heparan sulfate also caused functional changes in the dendritic cells—(a) the cells secreted appreciable amounts of TNF, IL-1β and IL-6; (b) the uptake and processing of antigen ceased while the MHC class II molecules became "fixed" at the surface; and (c) the number of dendritic cells needed to evoke an alloimmune response decreased by an order of magnitude. Thus, immature dendritic cells exposed to heparan sulfate behaved like mature, activated dendritic cells poised to induce cellular immune responses [44, 47]. In contrast, the cells kept under control conditions continued to appear and behaved as immature dendritic cells, which induce immunological tolerance in some systems [48–53].

It seemed as though heparan sulfate acted initially as an agonist to cause dendritic cells to mature and in this way to promote cellular immunity but that the promoting of cellular immunity was circumscribed and indeed ultimately suppressed by the later production of PGE2. Thus action of heparan sulfate on antigen presenting cells could explain both the expansion and also the eventual contraction of a cellular immune response to foreign antigen [20]. The suppression of T cell proliferation caused by PGE2 clearly differed from the condition of energy generated by immature dendritic cells.

8.7 What About Complement, PGE2 and Control of the Immune Response?

Our finding that heparan sulfate causes APC to produce PGE2 and our thinking that PGE2 might help circumscribe cellular immune responses was eclipsed by another observation. We discovered that IL-2 can be tethered to cells by heparan sulfate and that IL-2 so tethered can induce apoptosis of newly activated T cells, which express receptors for IL-2 [54]. In fact, the tethered form of IL-2 and not IL-2 in solution appeared to account for the impact of IL-2 in the generation and control of immune responses to model antigens delivered *in vivo* [55]. We imagined that interaction or lack thereof between complement and endothelial or parenchymal cells might govern immune responses—promoting responses when complement causes heparan sulfate to be shed and suppressing response when complement does not (Fig. 8.3) [56].

Although we did not forget entirely about PGE2 [20], our thinking about complement, heparan sulfate and IL-2, we did not pursue an apparent disparity concerning the involvement of complement and heparan sulfate in activation versus suppression of cellular immunity. If activation of complement causes heparan sulfate to be shed, and by now we had compelling evidence for that in vivo, and if the shed heparan sulfate activated APC which migrated to lymph nodes to activate

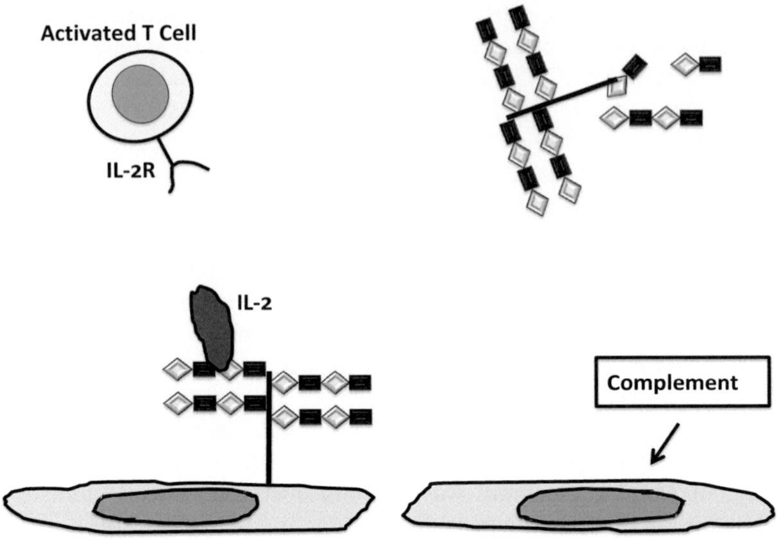

Fig. 8.3 Impact of heparan sulfate proteoglycan and shedding of heparan sulfate proteoglycan and glycosaminoglycan on control of cellular immunity. *Left*: Intact heparan sulfate chains bind IL-2, which can cause activated T cells to undergo apoptosis, contributing to immunological tolerance. *Right*: Loss of heparan sulfate caused by activation of complement deprives endothelium of IL-2, allowing activated to cells to attach, transmigrate and exert effector functions

T cells, then the production of PGE2 would subvert the expansion and function of effector T cells before they arrived in infected or transplanted tissues. We should have considered how PGE2 might impact of T cell activation in the microenvironment of lymph nodes. For, in the microenvironment of lymph nodes, PGE2 might conceivably promote foreign antigen-specific immunity and hinder auto-antigen-specific immunity, at least as we imagine these potentially occur. Decades ago, PGE2 was shown to profoundly suppress random migration of helper T cells [57]. The investigation of PGE2 focused on the impact on T cells of known specificity and function. In a lymph node, the migrating dendritic cells encounter T cells of diverse specificities and activation of those recognizing peptide-MHC complexes depends on the duration of specific interaction with TCR. Since, as mentioned above, the activated dendritic cells have peptide-MHC complexes relatively fixed, the duration of TCR engagement with the complexes will depend on the number of MHC bearing a given peptide and the propensity of T cells to migrate away. In this setting, as opposed to the conditions in a mixed leukocyte culture, PGE2 would favor the full activation of T cells. Further, if PGE2 or other factors failed to slow the separation of newly activated T cells from activated dendritic cells, then T cells bearing TCR that recognize self peptide-MHC complexes might gain access to the activated dendritic cells, leading to autoimmunity.

8.8 How Heparan Sulfate Activates Dendritic Cells

While we clearly missed the opportunity to explore what might be a pivotal involvement of heparan sulfate metabolism in sculpting, via PGE2, the T cell response to antigen, we did not forget our earlier question about how heparan sulfate might interact with leukocytes in the first place.

We had found, as mentioned above, that heparan sulfate triggers many signaling pathways in APC [41], the constellation of which seemed incompatible with utilization of a single type of receptor. However, since the stimulated cells produced cytokines, it was possible some of the pathways were activated by an autocrine loop. Indeed, we had found that complement activates endothelial cells through such an autocrine loop—the membrane attack complex triggers transcriptional activation and secretion of IL-1α which acts on the endothelial cell to evoke the broad range of changes [26, 27, 58]. However, the pathways induced by heparan sulfate involved activation of protein kinase cascades and NFκB [41]. These pathways happened to be the same pathways utilized by LPS and for that reason we used LPS as a positive control when we first tested how heparan sulfate if at all would activate dendritic cells [46]. LPS proved an excellent control because it evoked responses quite similar to heparan sulfate.

In the late 1990s, toll-like receptor-4 (TLR-4) was reported to be the cellular protein through which LPS delivered signals to cells [59]. We immediately tested whether heparan sulfate might deliver signals through TLR-4. Using wild type mice and mice with mutations that encoded defective or absent TLR-4 or CD14 as sources of dendritic cells, we found that dendritic cells from the mutant strains of mice were inured to exposure to heparan sulfate while dendritic cells from wild type mice became activated as described above [60]. These results indicated quite clearly that heparan sulfate was utilizing TLR-4 in the same way as LPS, although how exactly either agonist utilized TLR-4 was not then clear. The results also suggested to us that since heparan sulfate proteoglycan undergoes degradation during the repair and remodeling of injured tissues, TLR might serve as monitors for the overall well being of tissues, and not just for infection [60, 61].

8.9 From Inflammation to the Immune Response to Transplantation?

Our investigation of heparan sulfate metabolism began with the question of whether processes such as ischemia and complement activation that damage endothelial cells could account in part for the unique characteristics of the cellular immune response to transplantation. Now, having found that transplantation (and complement activation) causes shedding of heparan sulfate and that the shed macromolecules activate dendritic cells, enabling T cell activation under suboptimal conditions, we were poised finally to test the overall model in transplantation. Defective signaling of all TLR, owing to deficiency of MyD88 in the transplant and in the recipient, had been reported to prevent development of rejection in tissues from male mice transplanted into female mice [62], a minor transplantation antigen mismatch. Using mice with aberrant or absent TLR4, we tested the concept both for minor and major (MHC) antigen mismatches and the results could not be more clear. Absence of TLR-4 function or protein had absolutely no impact on the kinetics of rejection. What did matter however was the genetic background of the strains of mice used. What explains such a result? Our earliest work showing that complement induces shedding of heparan sulfate from endothelial cells also showed that complement induces transcriptional activation of IL-1α, and that IL-1α acting as an autocrine agonist activates endothelial cells. Although we did not think about it at the time, the conditions that identified the seminal importance of IL-1α, including the replacing of medium bathing complement-treated cells and specifically blocking IL-1α, had proved that shed heparan sulfate was not essential for activation of endothelial cells. And, IL-1α was quite sufficient for activating macrophages and dendritic cells. Thus, if shedding of heparan sulfate and action of TLR-4 was important for ischemia and immune-mediated injury, and for the genesis of immunity in truly suboptimal conditions (when PGE2 is needed), it was not at all essential for the generation of alloimmunity, as the conditions in which transplantation immunity arises are far from suboptimal.

8.10 Whither Endogenous Agonists

Our work also led to an equally clear and less appealing conclusion. Immunologists *impediti ratione cogitatonis* were not at all ready to accept the possibility that something other than LPS or other exogenous (pathogen-derived) agonists could deliver signals through TLR. It took us nearly 4 years and layer upon layer of proof that heparan sulfate was not contaminated by bacterial products, to bring our findings to publication. During much of that time, TLR were said to be the receptors for PAMPS, pathogen associated molecular patterns [63, 64]. However, the idea that TLR could recognize endogenous agonists [61], apart from any contamination, eventually gained acceptance and the agonists came to be known as DAMPS, damage associated molecular patters.

But, the term "damage associated molecular patterns" may cloud more vital and universal functions for TLR and the metabolism of proteoglycans and some other macromolecules. As we knew from the work of others [10, 65] and confirmed [14, 15] at the outset of our work, development and possibly repair and regeneration, of tissues and organs depends absolutely on the degradation of proteoglycans (and perhaps other macromolecules from which 'DAMP' derive). Blocking degradation blocks development. This degradation then is not a reflection of 'damage' but is something essential for life in multicellular organisms [61]. Perhaps, then, developmental biologists might consider changing the acronym use to refer to the agonists of TLR or toll receptors, the invertebrate homologues of TLR, from DAMPS to "DAMPS," the later referring, of course, to "development associated molecular patterns."

With the more expansive definition of DAMPS in mind, we explored the potential impact of TLR on development and maturation of mice [66]. Mice deficient of TLR-4 or a co-receptor, CD14, looked very much like wild type mice and hence did not appear, at least to us, to have the gross development defects one might expect to see if TLR regulated the development of mice as *toll* regulates the development of insects. However, as wild type mice aged, they exhibited dramatic changes in weight, bone structure, and physique, becoming heavy, obese and osteoporotic and developing measurably weaker bones. In contrast, TLR-4-deficient or -defective mice and CD14-deficient mice remained lean and strong-boned and exhibited no osteoporosis as they aged. We thus referred to the phenotype of mice lacking TLR functions as the "Adonis phenotype." Of particular interest, then, was the further observation that Adonis mice were no more active than wild type mice. Thus, despite the undoubted importance of exercise for overall health and well being, it did not explain the Adonis phenotype.

8.11 Heparan Sulfate in SIRS and Sepsis

Our interest in heparan sulfate as a potential agonist for TLR-4 led us to investigate an entirely different condition in which signaling by TLR generates biological changes. Besides their involvement in recruiting adaptive immune responses, TLR and particularly TLR-4 were best known as triggers for the sepsis syndrome and for the systemic inflammatory response syndrome or SIRS. SIRS was defined by the abrupt onset of fever, leukocytosis, shock and sometimes death in the absence of detectable infection; in the presence of infection, these findings would be called sepsis. SIRS occurs in such conditions as pancreatitis, multi-organ trauma, acute liver failure among others and the resemblance to sepsis is so close that investigators have asked repeatedly whether these conditions might cause by some means the entry of endotoxin into the system circulation. The answer has generally been no. Using the same strains of wild type and TLR-4 deficient or defective mice, we found that systemic administration of heparan sulfate had the same biological impact as LPS in wild type mice—it stimulated production of TNF and IL-6 and ultimately death—and like LPS, it had no appreciable effect on the mutant strains of mice [67]. Of particular note for those still skeptical about heparan sulfate was that a protein that specifically blocked the inflammatory impact of LPS did not impair the action of heparan sulfate.

We next asked whether release of endogenous heparan sulfate would engender SIRS. Serine proteases cleave heparan sulfate core proteins near the transmembrane domain to generate proteoglycans of the same size as those released from endothelial cells and one such protease, elastase, is released in pancreatitis. Hence we administered elastase to wild type and mice with defective TLR-4 signaling [67].

Once again, the results were clear—SIRS occurred in wild type mice while no appreciably changes occurred mutant strains. The administration of elastase did cause release of heparan sulfate in both strains and especially in spleen where large numbers of inflammatory cells are found.

These studies led us to propose a working model for the events that lead to the systemic inflammation and death in SIRS and potentially in sepsis [68, 69]. Since mice and humans with defective receptors for LPS have substantially increased risk of death from sepsis the expression and function of TLR-4 is clearly adaptive, probably facilitating the local containment and walling off of infectious agents [20]. Yet, when containment and walling off fail, and TLR beyond the site of infection are stimulated, systemic manifestations ensue—this concept probably represents the consensus model. But, we are struck by the observations in multiple clinical trials that administration of antibodies or other agents that block LPS does not improve the pathophysiology or outcome of sepsis and neither does it make sepsis worse [70]. To some, this trial and the many other failed attempts to improve the outcome of sepsis by blocking LPS indicate that still better blocking agents are needed. To us, these observations suggest the possibility that despite 150 years of research on LPS, that substance might not actually cause the life threatening manifestations of sepsis in patients with infection. And, if that is so, then we should at least consider the possibility that it is endogenous agonists for TLR, such as heparan sulfate, at not exogenous substances, such as LPS or other PAMPS, that cause the pathophysiology of sepsis (and SIRS). This model would finally unify the pathogenesis of SIRS and sepsis and possibly encourage someone to invent an acronym more poetic than DAMPS.

8.12 Concluding Remarks

Today heparan sulfate is known to have more functions in endothelial cell biology, and graft rejection, and graft acceptance than editorial space would permit us to discuss. However, we would be remiss if we failed at least to mention that the presence of heparan sulfate in tissue and organ grafts probably plays a key part in protecting grafts from injury, as another glance at Figs. 8.1 and 8.3 might suggest, and in reestablishing cell and tissue function after ischemia and reperfusion. Because, in the absence of immunosuppression, immunity poses an absolute barrier to transplantation of foreign tissues and organs, the subject of the immune response to transplantation overshadows nearly every other biological consideration. However, recent work in developmental and 'regenerative' biology suggests that restoring the integrity of tissue architecture and facilitating engraftment will pose challenges at least as great as those posed by transplant immunity. And, to the extent that a tissue or organ can be made to resist injury from ischemia or immunity, the challenge of regeneration and engraftment will be more easily met. Toward that objective, we investigate with enthusiasm, and hopefully *non impediti ratione cogitatonis*, the condition of "accommodation," which we discovered unexpectedly in transplants that apparently resisted immune and inflammatory injury that should have caused their destruction [23, 71–73]. Accommodation is now appreciated to occur not only in transplants, but also in tumors, infections and in cells exposed to environmental toxins [71, 74, 75]. Whether by way of happenstance or mechanism, we have observed *de novo* expression of heparan sulfate, previously shed from grafts, in accommodated organs [76].

Acknowledgements We thank Charles A. C. Platt for insightful discussions about saccharides. The NIH supported the work summarized here.

References

1. Dabek F. 'Car Talk' Brothers Address Graduates. The Tech. 1999;119:1, 16.
2. Platt JL, Rubinstein P. Mechanisms and characteristics of allograft rejection. In: Sabiston Jr DC, Lyerly HK, editors. Textbook of surgery: the biological basis of modern surgical practice. 15th ed. Philadelphia: W.B. Saunders; 1997. p. 400–8.
3. Platt JL, Cascalho M. Transplantation immunology. In: Mulholland MW, Lillemoe KD, Doherty GM, Maier RV, Simeone DM, Upchurch GRJ, editors. Greenfield's surgery: scientific principles and practice. 5th ed. Philadelphia: Lippincott Williams & Wilkins; 2011. p. 497–514.
4. Platt JL, Cascalho M, West LJ. Lessons from cardiac transplantation in infancy. Pediatr Transplant. 2009;13:814–9.
5. Lynch RJ, Silva IA, Chen BJ, Punch JD, Cascalho M, Platt JL. Cryptic B cell response to renal transplantation. Am J Transplant. 2013;13(7):1713–23.
6. Platt JL, LeBien TW, Michael AF. Interstitial mononuclear cell populations in renal graft rejection: identification by monoclonal antibodies in tissue sections. J Exp Med. 1982;155:17–30.
7. Platt JL, Grant BW, Eddy AA, Michael AF. Immune cell populations in cutaneous delayed-type hypersensitivity. J Exp Med. 1983;158:1227–42.
8. Platt JL, LeBien TW, Michael AF. Stages of renal ontogenesis identified by monoclonal antibodies reactive with lymphohemopoietic differentiation antigens. J Exp Med. 1983;157: 155–72.
9. Platt JL, Burke BA, Michael AF. Cellular antigens in nephroblastoma: identification with monoclonal antibodies which recognize hemopoietic cells. Clin Immunol Immunopathol. 1987;43:110–6.
10. Bernfield M, Banerjee SD. The turnover of basal lamina glycosaminoglycan correlates with epithelial morphogenesis. Dev Biol. 1982;90(2):291–305.
11. Iozzo RV. Matrix proteoglycans: from molecular design to cellular function. Annu Rev Biochem. 1998;67:609–52.
12. Sarrazin S, Lamanna WC, Esko JD. Heparan sulfate proteoglycans. Cold Spring Harb Perspect Biol. 2011;3(7):a004952 (1–33).
13. Mikami T, Kitagawa H. Biosynthesis and function of chondroitin sulfate. Biochim Biophys Acta. 2013;1830(10):4719–33.
14. Platt JL, Brown DM, Granlund K, Oegema TR, Klein DJ. Proteoglycan metabolism associated with mouse metanephric development: morphologic and biochemical effects of b-D-xyloside. Dev Biol. 1987;123:293–306.
15. Klein DJ, Brown DM, Moran A, Oegema TR, Platt JL. Chondroitin sulfate proteoglycan synthesis and reutilization of b-D-xyloside initiated chondroitin/dermatan sulfate glycosaminoglycans in fetal kidney branching morphogenesis. Dev Biol. 1989;133:515–28.

16. Klein DJ, Brown DM, Oegema TR, Brenchely PE, Anderson JC, Dickinson MAJ, et al. Glomerular basement membrane proteoglycans are derived from a large precursor. J Cell Biol. 1988;106:963–70.
17. Ihrcke NS, Wrenshall LE, Lindman BJ, Platt JL. Role of heparan sulfate in immune system-blood vessel interactions. Immunol Today. 1993;14:500–5.
18. Selvan RS, Ihrcke NS, Platt JL. Heparan sulfate in immune responses. Ann N Y Acad Sci. 1996;797:127–39.
19. Platt JL, Saadi S. The role of complement in transplantation. Mol Immunol. 1999;36:965–71.
20. Saadi S, Wrenshall LE, Platt JL. Regional manifestations and control of the immune system. FASEB J. 2002;16(8):849–56.
21. Platt JL, Trescony P, Lindman BJ, Oegema TR. Heparin and heparan sulfate delimit nephron formation in fetal metanephric kidneys. Dev Biol. 1990;139:338–48.
22. Platt JL, Vercellotti GM, Lindman BJ, Oegema Jr TR, Bach FH, Dalmasso AP. Release of heparan sulfate from endothelial cells. Implications for pathogenesis of hyperacute rejection. J Exp Med. 1990;171(4):1363–8.
23. Platt JL, Vercellotti GM, Dalmasso AP, Matas AJ, Bolman RM, Najarian JS, et al. Transplantation of discordant xenografts: a review of progress. Immunol Today. 1990; 11:450–6.
24. Key NS, Platt JL, Vercellotti GM. Vascular endothelial cell proteoglycans are susceptible to cleavage by neutrophils. Arterioscler Thromb. 1992;12:836–42.
25. Geller RL, Ihrcke NS, Platt JL. Release of endothelial cell-associated heparan sulfate proteoglycan by activated T cells. Transplantation. 1994;57:770–4.
26. Saadi S, Holzknecht RA, Patte CP, Stern DM, Platt JL. Complement-mediated regulation of tissue factor activity in endothelium. J Exp Med. 1995;182:1807–14.
27. Bustos M, Coffman TM, Saadi S, Platt JL. Modulation of eicosanoid metabolism in endothelial cells in a xenograft model: role of cyclooxygenase-2. J Clin Invest. 1997;100:1150–8.
28. Malyguine AM, Saadi S, Holzknecht RA, Patte CR, Sud N, Platt JL, et al. Induction of procoagulant function in porcine endothelial cells by human NK cells. J Immunol. 1997;159: 4659–64.
29. Saadi S, Holzknecht RA, Patte CP, Platt JL. Endothelial cell activation by pore forming structures: pivotal role for IL-1a. Circulation. 2000;101:1867–73.
30. Bustos M, Saadi S, Platt JL. Platelet-mediated activation of endothelial cells: implications for the pathogenesis of transplant rejection. Transplantation. 2001;72:509–15.
31. Ihrcke NS, Parker W, Reissner KJ, Platt JL. Regulation of platelet heparanase during inflammation: role of pH and proteinases. J Cell Physiol. 1998;175:255–67.
32. Ihrcke NS, Platt JL. Shedding of heparan sulfate proteoglycan by stimulated endothelial cells: evidence for proteolysis of cell surface molecules. J Cell Physiol. 1996;168:625–37.
33. Saadi S, Platt JL. Endothelial cell responses to complement activation. In: Volanakis JE, Frank MM, editors. The human complement system in health and disease. New York: Marcel Dekker, Inc.; 1998. p. 335–53.
34. Naparstek Y, Cohen IR, Fuks Z, Vlodavsky I. Activated T-lymphocytes produce a matrix-degrading heparan sulfate endoglycosidase. Nature. 1984;310:241–4.
35. Fridman R, Lider O, Naparstek Y, Fuks Z, Vlodavsky I, Cohen IR. Soluble antigen induces T lymphocytes to secrete an endoglycosidase that degrades the heparan sulfate moiety of subendothelial extracellular matrix. J Cell Physiol. 1987;130:85–92.
36. Raats CJ, Bakker MA, van den Born J, Berden JH. Hydroxyl radicals depolymerize glomerular heparan sulfate in vitro and in experimental nephrotic syndrome. J Biol Chem. 1997;272(42): 26734–41.
37. Lider O, Baharav E, Mekori YA, Miller T, Naparstek Y, Vlodavsky I, et al. Suppression of experimental autoimmune diseases and prolongation of allograft survival by treatment of animals with low doses of heparins. J Clin Invest. 1989;83:752–6.
38. Lider O, Mekori YA, Miller T, Bar-Tana R, Vlodavsky I, Baharav E, et al. Inhibition of T lymphocyte heparanase by heparin prevents T cell migration and T cell-mediated immunity. Eur J Immunol. 1990;20:493–9.

39. Wrenshall LE, Cerra FB, Carlson A, Bach FH, Platt JL. Regulation of murine splenocyte responses by heparan sulfate. J Immunol. 1991;147:455–9.
40. Wrenshall LE, Carlson A, Cerra FB, Platt JL. Modulation of cytolytic T cell responses by heparan sulfate. Transplantation. 1994;57:1087–94.
41. Wrenshall LE, Cerra FB, Singh RK, Platt JL. Heparan sulfate initiates signals in murine macrophages leading to divergent biological outcomes. J Immunol. 1995;154:871–80.
42. Wrenshall LE, Stevens RB, Cerra FB, Platt JL. Modulation of macrophage and B cell function by glycosaminoglycans. J Leukoc Biol. 1999;66(3):391–400.
43. Steinman RM. Dendritic cells: understanding immunogenicity. Eur J Immunol. 2007;37 Suppl 1:S53–60.
44. Gilboa E. DC-based cancer vaccines. J Clin Invest. 2007;117(5):1195–203.
45. Unanue ER. Antigen presentation in the autoimmune diabetes of the NOD mouse. Annu Rev Immunol. 2014;32:579–608.
46. Kodaira Y, Nair SK, Wrenshall LE, Gilboa E, Platt JL. Phenotypic and functional maturation of dendritic cells modulated by heparan sulfate. J Immunol. 2000;165:1599–604.
47. van Montfoort N, Camps MG, Khan S, Filippov DV, Weterings JJ, Griffith JM, et al. Antigen storage compartments in mature dendritic cells facilitate prolonged cytotoxic T lymphocyte cross-priming capacity. Proc Natl Acad Sci U S A. 2009;106(16):6730–5.
48. Steinman RM, Nussenzweig MC. Avoiding horror autotoxicus: the importance of dendritic cells in peripheral T cell tolerance. Proc Natl Acad Sci U S A. 2002;99(1):351–8.
49. Hawiger D, Inaba K, Dorsett Y, Guo M, Mahnke K, Rivera M, et al. Dendritic cells induce peripheral T cell unresponsiveness under steady state conditions in vivo. J Exp Med. 2001;194(6):769–79.
50. Thomson AW, Robbins PD. Tolerogenic dendritic cells for autoimmune disease and transplantation. Ann Rheum Dis. 2008;67 Suppl 3:iii90–6.
51. Frick JS, Grunebach F, Autenrieth IB. Immunomodulation by semi-mature dendritic cells: a novel role of Toll-like receptors and interleukin-6. Int J Med Microbiol. 2010;300(1):19–24.
52. Johnson DJ, Ohashi PS. Molecular programming of steady-state dendritic cells: impact on autoimmunity and tumor immune surveillance. Ann N Y Acad Sci. 2013;1284:46–51.
53. Morelli AE, Thomson AW. Orchestration of transplantation tolerance by regulatory dendritic cell therapy or in-situ targeting of dendritic cells. Curr Opin Organ Transplant. 2014;19(4):348–56.
54. Wrenshall LE, Platt JL. Regulation of T cell homeostasis by heparan sulfate-bound IL-2. J Immunol. 1999;163:3793–800.
55. Wrenshall LE, Platt JL, Elliot TS, Wight TN, Miller JD. Propagation and control of T cell responses by heparan sulfate-bound IL-2. J Immunol. 2003;170:5470–4.
56. Platt JL, Johnson GB, Kodaira Y, Wrenshall LE. Tolerance and the microenvironment. Transplantation. 2001;72:S23–4.
57. Jordan ML, Hoffman RA, Debe EF, West MA, Simmons RL. Prostaglandin E2 mediates subset-specific effects on the functional responses of allosensitized T lymphocyte clones. Transplantation. 1987;43(1):117–23.
58. Brunn GJ, Saadi S, Platt JL. Differential regulation of endothelial cell activation by complement and interleukin 1alpha. Circ Res. 2006;98:793–800.
59. Poltorak A, He X, Smirnova I, Liu MY, Huffel CV, Du X, et al. Defective LPS signaling in C3H/HeJ and C57BL/10ScCr mice: mutations in Tlr4 gene. Science. 1998;282:2085–8.
60. Johnson GB, Brunn GJ, Kodaira Y, Platt JL. Receptor-mediated monitoring of tissue well-being via detection of soluble heparan sulfate by toll-like receptor 4. J Immunol. 2002; 168(10):5233–9.
61. Johnson GB, Brunn GJ, Tang AH, Platt JL. Evolutionary clues to the functions of the Toll-like family as surveillance receptors. Trends Immunol. 2003;24(1):19–24.
62. Goldstein DR, Tesar BM, Akira S, Lakkis FG. Critical role of the Toll-like receptor signal adaptor protein MyD88 in acute allograft rejection. J Clin Invest. 2003;111(10):1571–8.
63. Janeway Jr CA, Medzhitov R. Lipoproteins take their toll on the host. Curr Biol. 1999;9(23):R879–82.

64. Medzhitov R. Toll-like receptors and innate immunity. Nat Rev Immunol. 2001;1:135.
65. Rapraeger A, Bernfield M. Cell surface proteoglycan of mammary epithelial cells: protease releases a heparin sulfate-rich extodomain from a putative membrane-associated domain. J Biol Chem. 1985;260:4103–9.
66. Johnson GB, Riggs BL, Platt JL. A genetic basis for the "Adonis" phenotype of low adiposity and strong bones. FASEB J. 2004;18(11):1282–4.
67. Johnson GB, Brunn GJ, Platt JL. An endogenous pathway to systemic inflammatory response syndrome (SIRS)-like responses through toll-like receptor 4. J Immunol (Cutting Edge). 2004;172(1):20–4.
68. Brunn GJ, Wijdicks MF, Platt JL. Toward a modern concept of sepsis: new answers to ancient questions. Discov Med. 2006;6(31):11–7.
69. Brunn GJ, Platt JL. The etiology of sepsis: turned inside out. Trends Mol Med. 2006;12(1):10–6.
70. Opal SM, Laterre PF, Francois B, LaRosa SP, Angus DC, Mira JP, et al. Effect of eritoran, an antagonist of MD2-TLR4, on mortality in patients with severe sepsis: the ACCESS randomized trial. JAMA. 2013;309(11):1154–62.
71. Koch CA, Khalpey ZI, Platt JL. Accommodation: preventing injury in transplantation and disease. J Immunol. 2004;172(9):5143–8.
72. Lynch RJ, Platt JL. Accommodation in organ transplantation. Curr Opin Organ Transplant. 2008;13:165–70.
73. Cascalho MI, Chen BJ, Kain M, Platt JL. The paradoxical functions of B cells and antibodies in transplantation. J Immunol. 2013;190(3):875–9.
74. Koch CA, Khalpey ZI, Platt JL. Humoral immunity in xenotransplantation: B-cell tolerance and accommodation. Curr Opin Organ Transplant. 2004;9(2):170–5.
75. Koch CA, Kanazawa A, Nishitai R, Knudsen BE, Ogata K, Plummer TB, et al. Intrinsic resistance of hepatocytes to complement-mediated injury. J Immunol. 2005;174:7302–9.
76. Williams JM, Holzknecht ZE, Plummer TB, Lin SS, Brunn GJ, Platt JL. Acute vascular rejection and accommodation: divergent outcomes of the humoral response to organ transplantation. Transplantation. 2004;78(10):1471–8.

Part III
Transplantation

Chapter 9
Xenotransplantation of Cells, Tissues, Organs and the German Research Foundation Transregio Collaborative Research Centre 127

Bruno Reichart, Sonja Guethoff, Paolo Brenner, Thomas Poettinger, Eckhard Wolf, Barbara Ludwig, Alexander Kind, Tanja Mayr, and Jan-Michael Abicht

Abstract Human organ transplantation is the therapy of choice for end-stage organ failure. However, the demand for organs far exceeds the donation rate, and many patients die while waiting for a donor. Clinical xenotransplantation using discordant species, particularly pigs, offers a possible solution to this critical shortfall. Xenotransplantation can also increase the availability of cells, such as neurons, and tissues such as cornea, insulin producing pancreatic islets and heart valves. However, the immunological barriers and biochemical disparities between pigs and primates (human) lead to rejection reactions despite the use of common immunosuppressive drugs. These result in graft vessel destruction, haemorrhage, oedema, thrombus formation, and transplant loss. Our consortium is pursuing a broad range of strategies to overcome these obstacles. These include genetic modification of the donor animals to knock out genes responsible for xenoreactive surface epitopes and to express multiple xenoprotective molecules such as the human complement regulators CD46, 55, 59, thrombomodulin and others. We are using (new) drugs including complement

B. Reichart (✉) • S. Guethoff • T. Mayr
Walter Brendel Centre of Experimental Medicine, Ludwig-Maximilians-Universität (LMU), Munich, Germany
e-mail: Bruno.Reichart@med.uni-muenchen.de

P. Brenner • T. Poettinger
Department of Cardiac Surgery, LMU, Munich, Germany

E. Wolf
Chair for Molecular Animal Breeding and Biotechnology, Gene Center, LMU, Munich, Germany

B. Ludwig
Department of Medicine III, Technical University, Dresden, Germany

A. Kind
Livestock Biotechnology, Technical University, Munich, Germany

J.-M. Abicht
Department of Anaesthesiology, LMU, Munich, Germany

J.D. Lambris et al. (eds.), *Immune Responses to Biosurfaces*,
Advances in Experimental Medicine and Biology 865,
DOI 10.1007/978-3-319-18603-0_9

inhibitors (e.g. to inhibit C3 binding), anti-CD20, 40, 40L, and also employing physical protection methods such as macro-encapsulation of pancreatic islets. Regarding safety, a major objective is to assure that possible infections are not transmitted to recipients. While the aims are ambitious, recent successes in preclinical studies suggest that xenotransplantation is soon to become a clinical reality.

Keywords Xenogeneic heart transplantation • Xenogeneic islet transplantation • Complement and xenogeneic transplantation • Genetically modified donors • Safety after xenogeneic transplantation

9.1 Introduction

9.1.1 The Shortage of Human Organs, Discordant (Porcine) Xenogeneic Transplants as an Alternative

Human (allogeneic) organ transplantations have been very successful over the last six decades and remain the therapy of choice for end-stage organ failure whenever conventional medical or surgical therapies are of no avail. Heart transplantation as an example, not only offers an approximately 40 % chance to survive 20 post-operative years [1] (own experience) but also a good quality of life. However, the demand for organs far surpasses the donation rate and a substantial proportion of patients die while still on the waiting list, 17.4 % in Germany [2].

Patients on the list for deceased kidney donations wait almost 42 months, although the advantages of transplantation over dialysis are evident after only 1 year [3].

The shortage of organs is undoubtedly grave, in Germany fewer than ten donors per million residents were reported in 2014, and several alternatives have been suggested. One possible solution would be clinical xenotransplantation using non-human primates as concordant donors and triple drug immunosuppression as applied in human allotransplants [4]. However, ethical and logistical considerations preclude this. Apes are endangered species and their use is out of the question, other non-human primates are too small and their growth too slow.

In contrast, discordant species, notably pigs, offer an abundant new source of organs for various reasons:

- Similarities in size, anatomy, nutrition and physiology to man
- Short generation intervals (12 months) and high fertility (10–14 offspring per litter)
- Well-established and economic housing and breeding conditions with high hygienic standards
- Availability of advanced reproductive biotechnologies and genetic engineering techniques

- Minor concerns regarding the slaughtering of pigs, at least in Western countries, because they are raised for meat production on an industrial scale

Nevertheless, despite these obvious advantages, serious ethical concerns do exist in society regarding the use of pigs as donors and these have to be allayed [5].

9.1.2 The Need for Discordant Cellular Transplantations

There are clinical needs for a huge variety of cell types, some of which are already being investigated as transplants, such as neurons [6] und corneas [7, 8]. At the moment there is a particular focus on pancreatic insulin producing islets.

An epidemic of obesity in Western populations has dramatically increased the threat of diabetes mellitus, with the number of patients expected to almost double within the next two decades [9]. Although anti-diabetic therapy is successful for most patients, hypoglycaemia is a life-threatening complication in 5–10 % of cases. At present, allogeneic pancreatic islet cell transplantation offers a solution for type 1 diabetes and greatly improves life quality with relatively low operative risk. According to the American Collaborative Islet registry (CITR), the 5-year graft survival rate approximates 50 %, with successful second and third interventions possible [10–12]. These results compare favourably with other difficult transplantation procedures, such as lungs, where the 5-year survival rate is around 50 % [1].

In Germany, there is only the Dresden group successfully carrying out islet transplantation. Results are very encouraging, within the observation time patients exhibited stable glycaemic control with reduction of HbA1c and required less insulin [13]. However the shortage of suitable donors and the extraordinarily high number of islet cells required for each patient severely restricts the availability of such treatment.

Discordant xenogeneic islet transplantation would therefore offer a practical solution. This is supported by the impressive results of porcine islet transplantation into diabetic primate models, using islets from wild-type pigs and immunosuppression of the recipient [14], encapsulated islets from wild-type pigs [15], or islets from genetically engineered donor pigs [16].

9.1.3 The Need for Biological Valve Prostheses for Younger Patients

Another focus of our consortium is on biological heart valve prostheses, of which between 120,000 and 240,000 devices are implanted worldwide each year. The new transcatheter techniques for aortic valve replacement have further increased their number. These biological valves, mounted on small catheters and rapidly expanding balloons are usually introduced via the femoral artery without the need for a thoracotomy or the heart-lung-machine [17].

Biological valves (porcine, bovine) do not need anticoagulation if the patient is in sinus-rhythm. They do however have restricted durability, and the rate of degeneration correlates with the youth of the patient at the time of intervention [18]. In contrast mechanical devices have unlimited durability, but necessitate life-long anticoagulation.

Extending the durability of biological valves, especially in younger patients, has posed a tremendous challenge since their introduction more than 40 years ago. Newly developed prostheses have been made from decellularised heart valve matrices that are revitalised in vivo by cells from the recipient, forming a functional epithelium and live interstitium. Decellularised heart valve matrix from wild-type pigs does however attract inflammatory cells and mediates platelet activation mostly due to preformed antibodies [19]. Pigs genetically modified to overcome these immune mechanisms may thus provide a superior source of such materials.

9.2 Discordant Xenotransplantation to Solve the Shortage of Human Cells, Organs and Tissues

Xenotransplantation would undoubtedly provide substantial advantages for human regenerative medicine. The main problems arise from disparities between swine and primates resulting from approximately 90 million years of evolutionary divergence, which can affect important protein-protein and other biochemical interactions. Considerable immunological and physiological incompatibilities must therefore be overcome before xenogeneic grafts can be clinically effective. Fortunately, our understanding of these barriers is increasing rapidly and rational strategies are being developed to overcome them. Humoral rejection from preformed antibodies and the blood coagulation system present the immediate obstacles, in the longer term the greatest challenge comes from the adaptive immune response.

The possible transfer of infectious agents to a graft recipient—also a major risk in allo-transplantation—might be exacerbated with organs, tissues, cells from non-human species. On the other hand, xenotransplantation offers the opportunity to systematically examine donors for infectious agents before transplantation [20]. To control the infectious burden, animals must be raised in a clean environment (designated pathogen free, DPF) and xenograft recipients must be monitored post-operatively.

9.2.1 Responses to Xenogeneic Cells and Vascularised Organs, the Role of Complement

An unmodified porcine organ transplanted into a human or primate recipient is confronted with a series of rejection responses. The first is hyperacute rejection, followed by acute humoral xenograft rejection, also known as acute vascular or delayed xenograft reaction. Both hyperacute and acute humoral xenograft rejection are ultimately the result of preformed and acquired antibodies binding to cell surface antigens on the graft endothelium. The hyperacute reaction is directed primarily against porcine α1,3-galactosyl-galactose (α-Gal) epitopes [21], and the acute humoral reaction against non-Gal epitopes [22]. Antibody binding initiates (mostly) classical complement activation, followed by graft vessel destruction, interstitial haemorrhage and oedema.

Hyperacute rejection leads to graft failure within minutes or hours, and acute humoral rejection within days or at most 3 weeks.

In addition to antibody-mediated activation of the xenograft endothelium, incompatibilities between human (or primate) blood coagulation components and the porcine vessel wall also contribute to the formation of microthrombi, ischemic injury and ultimately consumptive coagulopathy (reduced plasma fibrinogen, thrombocytopenia, increased D-dimers and INR) [23–29]. Porcine thrombomodulin is implicated as responsible, since it binds weakly to primate thrombin, leading to levels of activated protein C insufficient to interrupt coagulation [30].

If left unprotected, tissues like porcine insulin producing islets succumb to early graft loss known as immediate blood mediated inflammatory reaction (IBMIR), which is also driven by preformed antibodies, complement and excessive coagulation [31, 32].

As with allogeneic procedures, cellular reactions should be expected following xenogeneic transplantation.

9.2.2 Strategies to Overcome Immunologic Xenogeneic Reactions

9.2.2.1 Micro- and Macro-Encapsulated Porcine Islets

The need for immunosuppressive drugs can be obviated by encapsulation, in which all islets are surrounded with a porous biopolymer composed mainly of alginate (micro-encapsulation) [33, 34]. This contains pores large enough for small molecules such as water, glucose, oxygen and most importantly insulin to permeate, but excludes immune competent cells and larger molecules such as antibodies. However, it is not known how long these capsules can maintain function in vivo, since inferior bio-compatibility may lead to occlusion of pores. Recent studies have shown that the encapsulation material, especially alginate, activates both the complement and coagulation cascades resulting in the formation of tight fibrotic capsules around the microsystems [35, 36].

Nevertheless, the New Zealand company Living Cell Technologies, a pioneer in this field, has so far carried out transplants with 14 diabetic patients chosen because they suffered frequent episodes of unaware hypoglycaemia [34]. A dose-finding and safety study showed improvement in some treated patients, for example through reduced glycated haemoglobin levels.

Macro-encapsulation might offer an alternative to obviate the problems seen in microencapsulation. The device that is proposed here, is the size of a cardiac

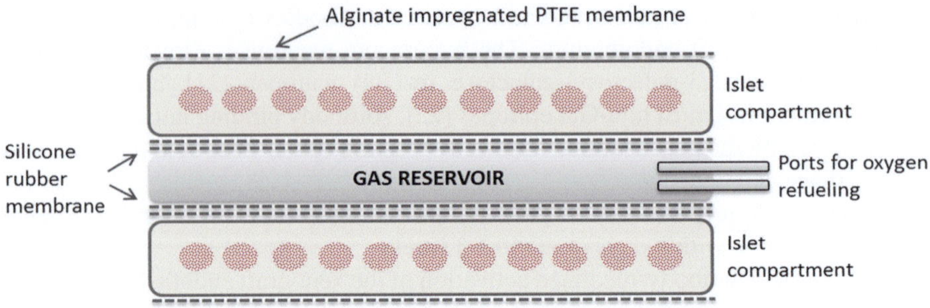

Fig. 9.1 Macroencapsulation system for islets. Cross-section of the device (Beta-O2, Israel). The islets remain immobilised within the alginate compartment which is covered by a PTFE membrane. Alginate and PTFE are porous allowing the permeability of small molecules such as glucose and insulin, but not big sized antibodies and immune competent cells. Oxygen is supplied to the islets via two ports connected to a gas reservoir integrated in the device

pacemaker (or an ice-hockey puck) and is implanted to the outer side of the peritoneum (Fig. 9.1) [37]. The islets remain immobilised within the alginate, which is totally covered by a PTFE membrane. A gas mixture is delivered to the islets via two ports, assuring for optimal oxygen supply to the islets, a feature developed by the company Beta-O_2, of Israel. Avoiding hypoxia is of critical importance to prevent early and long term graft loss.

So far six diabetic pigs have been successfully treated with the system. Within the mean observation time of 85 days, the animals remained normoglycaemic at the upper limit of the norm and glucose challenge tests led to a measurable increase of porcine C-peptide [38].

In a proof of principle study, a 63-year-old diabetic patient received the device (loaded with human islets) for 10 months without application of immunosuppression, within that time the transplanted islets remained living and reacted to metabolic challenges [37].

9.2.2.2 Genetically Modified Porcine Islets, Heart Valves, Vascularised Organs (Hearts)

Genetic modifications of the tissues must ameliorate the various xenogeneic reactions of the recipient.

Unmodified porcine cell surfaces, including the endothelium, present α-Gal sugar epitopes that are absent in old world monkeys (e.g. baboons) and humans. Animals (humans) lacking α-Gal develop antibodies to α-Gal during early life [39] and these initiate hyperacute rejection.

Inactivation of the enzyme responsible for α-Gal synthesis, α-1,3-galactosyl transferase (GGTA1), by genetic knock out (KO) of the GGTA1 gene and generation of homozygous pigs was therefore a major breakthrough [40]. Since then several independent herds have been established and α-Gal-KO organs have been tested in numerous pig-to-baboon transplantation studies resulting in maximum survival of 3 months for kidneys [41] and 8 months for beating but not working hearts (Fig. 9.2a) [27].

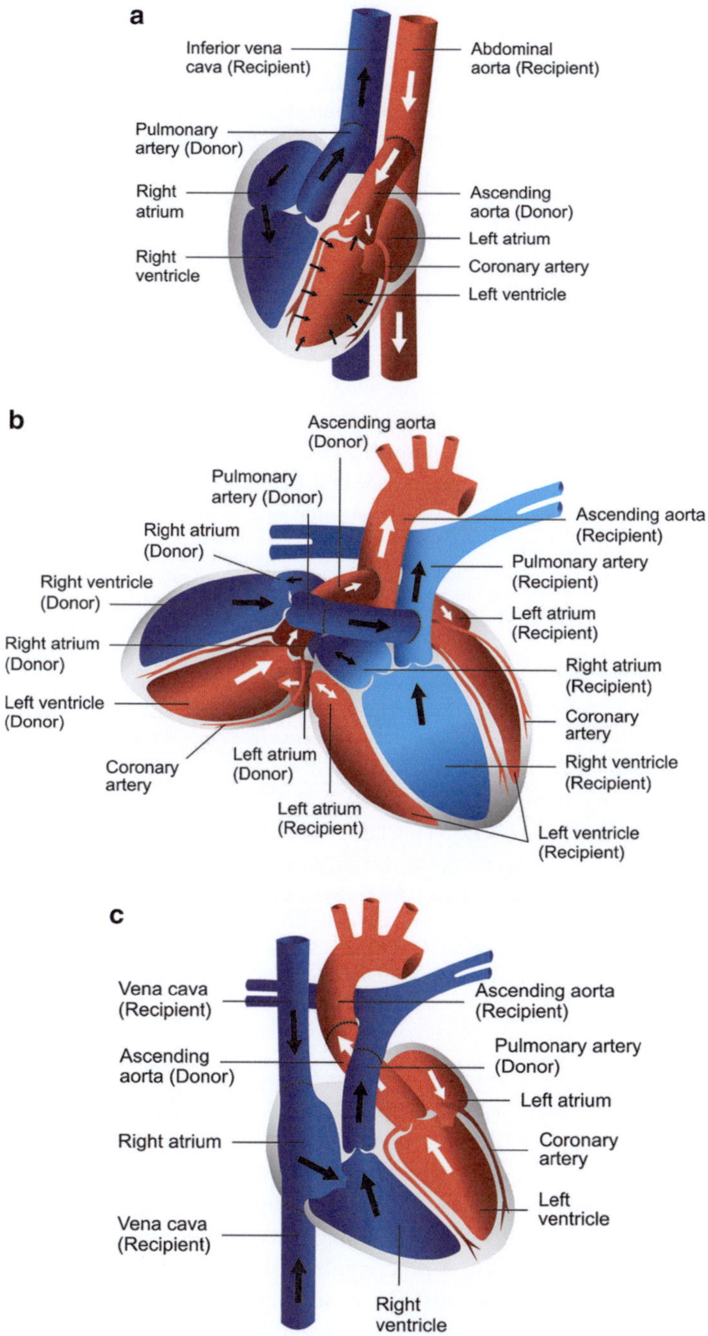

Fig. 9.2 Techniques of heart transplantation; (**a**) Heterotopic experimental abdominal transplantation; the graft is connected to the abdominal vessels of the recipient and perfused via the coronary arteries. No heart-lung-machine is necessary for this non-working model; (**b**) Heterotopic thoracic transplantation; connection of both the donor and recipient atriums (and of course the aortas and pulmonary artery trunks), whereby all ventricles are filled. The donor organ is working; (**c**) Orthotopic heart transplantation; the clinical state of the art technique; (**b**) and (**c**) necessitate the use of the heart-lung-machine, incurring some technical challenges and inducing a non-specific (complement driven) inflammatory reaction. Illustrations by Nina Bantschow

Table 9.1 The "genetic toolbox" central to our strategies to minimise or abolish hyperacute and delayed humoral rejection. These genetic modifications of donor pigs prolong graft survival by different mechanisms and are combined in various multi-genetically modified pigs

Genetic modification	Mechanisms
α-Gal-KO	Deletion of Gal antigen expression
α-Gal/CMAH-KO	Deletion of Gal antigen and sialic acid Neu5Gc expression
h-CD46 or h-CD55 or h-CD59	Downregulation of the human (h) complement system
h-TM	Binds human thrombin, cofactor in activating Protein C
CD39	Inhibition of ADP-induced platelet activation and aggregation
HO-1	Anti-apoptotic, cytoprotective and anti-inflammatory
A20	Anti-inflammatory and anti-apoptotic
LEA 29Y	Blockade of a T-cell co-stimulation pathway
HLA-E	Graft protection against human natural killer cells.
Multi-genetically modified pigs	
α-Gal-KO + h-CD46[a,b]	Multi-genetically modified animals are correspondingly more complex. The genetic modifications include the gene knockouts and transgenes listed above. Obtaining these animals with appropriate levels of transgene expression requires multiple experimental iterations to optimize transgene structure and identify the best founder animals, a process that can take several years.
α-Gal-KO + h-CD46 + h-TM[a,b]	
α-Gal-KO + h-CD46 + HO-1[a,b]	
α-Gal-KO + h-CD55 + HO-1[a,b]	
α-Gal-KO + h-CD46 + LEA 29Y[a,b]	
α-Gal-KO + h-CD46 + HLA-E[a,b]	
α-Gal-KO + h-CD46 + h-CD55 + h-CD59, A20, HO-1[b,c]	
α-Gal/CMAH-KO + h-CD46 + h-CD55 + h-CD59, A20, HO-1[c]	
α-Gal/CMAH-KO + h-CD46 + h-CD55 + h-CD59, A20, HO-1 + CD39 + hTM[c]	

[a]Produced by E. Wolf (Ludwig-Maximilians-Universität Munich, Oberschleißheim, Germany) using Gal-KO/CD46 background provided by D. Ayares (Revivicor, Blacksburg, USA)
[b]Genetically modified pig hearts available for our current cardiac transplant experiments in baboons
[c]Produced by A. Schnieke (Technical University Munich, Weihenstephan, Germany) and by H. Niemann (HO-1, A20, h-CD59, Friedrich Löffler Institute, Mariensee, Germany)

Once hyperacute rejection is overcome, acute humoral rejection presents the next obstacle. Hearts from α-Gal-KO pigs transplanted into baboons were found to exhibit widespread thrombotic microangiopathy, ischemia, focal haemorrhage and necrosis as a consequence of progressive humoral rejection and disordered thromboregulation, as mentioned above. The underlying mechanisms are not completely understood, but are thought to involve changes to the porcine endothelium following transplantation that lead to a procoagulant state. Antibodies to antigens other than α-Gal (non-Gal) epitopes also seem to play a major role in acute humoral xenogeneic rejection [22]. However, the number and diversity of non-Gal antigens preclude their removal by gene targeting. The preferred strategy is thus to prevent complement-mediated destruction of the xenograft. Various transgenic pigs expressing human complement regulators (CD46, CD55, CD59; [42–45]; reviewed in [46, 47]) on the vascular endothelium have been generated, combined with α-Gal-KO animals, and tested in pig-to-baboon transplantation experiments (Table 9.1).

Transgenes that modulate endothelial activation such as heme oxygenase 1 (HO-1), are also thought to be beneficial [48]; the expression of human HLA-E protects against the cytotoxic activity of natural killer cells [49].

Expression of human thrombomodulin on porcine cells, especially on the endothelium, has been favourable in preclinical experiments (see also next chapter), ameliorating or even avoiding signs of thrombotic microangiopathy [50–53].

Additional immunosuppressive treatment (tacrolimus, mycophenolate mofetil, cortisone, ATG each alone or in combination) will clearly be necessary to overcome the consistent delayed humoral and cell-mediated rejection mechanisms. Total thoracic lymph node irradiation with 6–7 Gy [54] or co-stimulation blockade with CD40 or CD40L antibodies must be added [55]. A useful genetic approach is to express a T-cell co-stimulation blocker such as the improved form of CTLA4-Ig, LEA29Y, on transplanted cells or organs. This has been successfully demonstrated by researchers within our consortium, who transplanted LEA29Y expressing pig islets into diabetic humanised immune-deficient mice [56].

Since xenogeneic interventions can be planned in advance, the bone marrow (and therefore antibody production) is suppressed prior to transplantation using anti-CD20 to destroy B-cells, or even Bortezomib in combination with cortisone to obviate plasma cells, cyclophosphamide for myeloablation [57]. Extracorporeal immune-adsorption (induces complement activation) may be used to remove pre-existing antibodies.

9.3 Preclinical Results

In a recent review, Ekser and colleagues [46] listed the longest survival times reported for xenografted porcine cells and organs. Micro-encapsulated pancreatic islets from wild-type animals survived more than 800 days [15], non-encapsulated islets transgenic for human-complement regulator protein CD46 survived almost 400 days [16]. Hearts placed heterotopically within the thorax (Fig. 9.2b) beat for up to 50 days [58] (personal experience), and those placed orthotopically (Fig. 9.2c) for 57 days [59]. Transplanted CTLA4-Ig transgenic neuronal cells to treat Parkinson's disease, and wild-type decellularised corneas were also successful for hundreds of days [7, 60]. Whole kidneys from CD55 transgenic animals survived for 90 days [61], although porcine renal erythropoietin is not recognised by primate recipients and must be replaced. Liver and lung transplants do however stand in contrast to these successes, surviving only 8 and 5 days [62, 63].

A team led by Muhammad Mansoor Mohiuddin (NIH) has achieved a major breakthrough in the field of xenogeneic heart transplantation. In his most recent group, two out of five genetically modified (α-Gal-KO, CD46, h-thrombomodulin) pig hearts, have been beating for more than one and two post-operative years in baboons [51, 53, 64]. The immunosuppressive regimen is simple and obviously well tolerated; the (long-term) maintenance treatment includes only anti-CD40 and mycophenolate mofetil (steroids are tapered off within seven post-operative weeks).

The genetically modified porcine hearts were however placed into the abdomen of the recipient, in a beating but non-working mode. Figure 9.2a shows the arrangement; the donor ascending aorta was connected to the recipient's abdominal aorta and the donor pulmonary artery to the inferior caval vein. Further orthotopic experiments are needed to verify these excellent results; orthotopic procedures will however necessitate the use of a heart-lung-machine, incurring some technical challenges and inducing a non-specific (complement driven) inflammatory reaction [65], which is not trivial to overcome under these circumstances. Effective inhibition of C3 binding with Compstatin is therefore planned from our side.

In the field of preclinical macro-encapsulated islet transplantation, efficacy and safety studies (carried out by the Dresden team) are underway.

Since glutaraldehyde fixed/decellularised porcine valves still present α-Gal and non-Gal antigens, tissues from (multiple) knock-out animals will prove beneficial.

9.4 Safety Preconditions, First Preclinical Results

Among the possible infectious agents perceived as xenotransplantation risks, porcine endogenous retroviruses (PERV-A, B, C) have received the most attention because they are integrated in the germ line and transmitted vertically to offspring, and cannot thus be eliminated by raising pigs under DPF conditions [66]. Initial studies showed the human-tropic potential of PERV in vitro [67] and revealed their predisposition for retroviral recombination [68]. Recombined PERV-A/C has higher infectious potential than PERV-C [69]. But most importantly, there was no evidence of cross-species transmission in the first clinical trials of islet xenotransplantation (see also above; [34]).

Regarding safety issues, the International Xenotransplantation Association (IXA) has defined and regularly updates the conditions for xenotransplantation [70, 71] (current update on the Second International Conference on Clinical Islet Xenotransplantation, August 2014, San Francisco, USA).

9.5 Conclusion

The great Norman Shumway, pioneer of human heart transplantation, used to say: "Xenotransplantation is the future and always will be!" Recent breakthroughs in preclinical xenotransplantation experiments suggest that, fortunately, this pessimistic view may no longer be true.

Funding The Transregio Collaborative Research Center 127 is funded by the German Research Foundation (DFG).

References

1. Lund LH, et al. The registry of the International Society for Heart and Lung Transplantation: thirty-first official adult heart transplant report–2014; focus theme: retransplantation. J Heart Lung Transplant. 2014;33(10):996–1008.
2. Rahmel AO. Eurotransplant annual report 2013. www.eurotransplant.org, 2014.
3. Tonelli M, et al. Systematic review: kidney transplantation compared with dialysis in clinically relevant outcomes. Am J Transplant. 2011;11(10):2093–109.
4. Reichenspurner H, et al. Optimalization of immunosuppression after xenogeneic heart transplantation in primates. J Heart Transplant. 1989;8(3):200–7. discussion 207–8.
5. Sautermeister J. Themen der Katholischen Akademie in Bayern. Zur Debatte. 2014;44: 19–44.
6. Leveque X, et al. Intracerebral xenotransplantation: recent findings and perspectives for local immunosuppression. Curr Opin Organ Transplant. 2011;16(2):190–4.
7. Choi HJ, et al. Efficacy of pig-to-rhesus lamellar corneal xenotransplantation. Invest Ophthalmol Vis Sci. 2011;52(9):6643–50.
8. Kim MK, et al. The International Xenotransplantation Association consensus statement on conditions for undertaking clinical trials of xenocorneal transplantation. Xenotransplantation. 2014;21(5):420–30.
9. International Diabetes Federation. Diabetes Atlas. 6th ed. Brussels: International Diabetes Federation; 2014.
10. Bellin MD, et al. Potent induction immunotherapy promotes long-term insulin independence after islet transplantation in type 1 diabetes. Am J Transplant. 2012;12(6):1576–83.
11. Tiwari JL, et al. Islet cell transplantation in type 1 diabetes: an analysis of efficacy outcomes and considerations for trial designs. Am J Transplant. 2012;12(7):1898–907.
12. Collaborative islet transplant registry. Annual report 2013. www.citeregistry.org
13. Ludwig B, et al. Islet transplantation at the Dresden diabetes center: five years' experience. Horm Metab Res. 2015;47(1):4–8.
14. Cardona K, et al. Long-term survival of neonatal porcine islets in nonhuman primates by targeting costimulation pathways. Nat Med. 2006;12(3):304–6.
15. Dufrane D, Goebbels RM, Gianello P. Alginate macroencapsulation of pig islets allows correction of streptozotocin-induced diabetes in primates up to 6 months without immunosuppression. Transplantation. 2010;90(10):1054–62.
16. van der Windt DJ, et al. Long-term controlled normoglycemia in diabetic non-human primates after transplantation with hCD46 transgenic porcine islets. Am J Transplant. 2009;9(12): 2716–26.
17. Collas V, Philipsen T, Rodrigues J, Vrints C, Paelinck BP, Bosmans J. Transcatheter aortic valve implantation: review and current state of the art. EMJ Int Cardiol. 2014;1:52–61.
18. Kouchoukos NT, Blackstone E, Hanley FL, Kirklin JK. Kirklin/Barratt-Boyes cardiac surgery. 4th ed. London: Churchill Livingstone; 2012.
19. Rieder E, et al. Tissue engineering of heart valves: decellularized porcine and human valve scaffolds differ importantly in residual potential to attract monocytic cells. Circulation. 2005;111(21):2792–7.
20. Mueller NJ, et al. Microbial safety in xenotransplantation. Curr Opin Organ Transplant. 2011; 16(2):201–6.
21. Diswall M, et al. Structural characterization of alpha1,3-galactosyltransferase knockout pig heart and kidney glycolipids and their reactivity with human and baboon antibodies. Xenotransplantation. 2010;17(1):48–60.
22. Byrne GW, et al. Identification of new carbohydrate and membrane protein antigens in cardiac xenotransplantation. Transplantation. 2011;91(3):287–92.
23. Ierino FL, et al. Disseminated intravascular coagulation in association with the delayed rejection of pig-to-baboon renal xenografts. Transplantation. 1998;66(11):1439–50.

24. Buhler L, et al. Coagulation and thrombotic disorders associated with pig organ and hematopoietic cell transplantation in nonhuman primates. Transplantation. 2000;70(9):1323–31.
25. Houser SL, et al. Thrombotic microangiopathy and graft arteriopathy in pig hearts following transplantation into baboons. Xenotransplantation. 2004;11(5):416–25.
26. Cozzi E, et al. Alterations in the coagulation profile in renal pig-to-monkey xenotransplantation. Am J Transplant. 2004;4(3):335–45.
27. Kuwaki K, et al. Heart transplantation in baboons using alpha1,3-galactosyltransferase gene-knockout pigs as donors: initial experience. Nat Med. 2005;11(1):29–31.
28. Cowan PJ, Roussel JC, d'Apice AJ. The vascular and coagulation issues in xenotransplantation. Curr Opin Organ Transplant. 2009;14(2):161–7.
29. Lin CC, et al. Recipient tissue factor expression is associated with consumptive coagulopathy in pig-to-primate kidney xenotransplantation. Am J Transplant. 2010;10(7):1556–68.
30. Roussel JC, et al. Pig thrombomodulin binds human thrombin but is a poor cofactor for activation of human protein C and TAFI. Am J Transplant. 2008;8(6):1101–12.
31. van der Windt DJ, et al. Rapid loss of intraportally transplanted islets: an overview of pathophysiology and preventive strategies. Xenotransplantation. 2007;14(4):288–97.
32. Hawthorne WJ, et al. Control of IBMIR in neonatal porcine islet xenotransplantation in baboons. Am J Transplant. 2014;14(6):1300–9.
33. Calafiore R, et al. Grafts of microencapsulated pancreatic islet cells for the therapy of diabetes mellitus in non-immunosuppressed animals. Biotechnol Appl Biochem. 2004;39:159–64.
34. Wynyard S, et al. Microbiological safety of the first clinical pig islet xenotransplantation trial in New Zealand. Xenotransplantation. 2014;21(4):309–23.
35. Rokstad AM, et al. Advances in biocompatibility and physico-chemical characterization of microspheres for cell encapsulation. Adv Drug Deliv Rev. 2014;67–68:111–30.
36. Dolgin E. Encapsulate this. Nat Med. 2014;20(1):9–11.
37. Ludwig B, et al. Transplantation of human islets without immunosuppression. Proc Natl Acad Sci U S A. 2013;110(47):19054–8.
38. Ludwig B, et al. Improvement of islet function in a bioartificial pancreas by enhanced oxygen supply and growth hormone releasing hormone agonist. Proc Natl Acad Sci U S A. 2012;109(13):5022–7.
39. Galili U. Discovery of the natural anti-Gal antibody and its past and future relevance to medicine. Xenotransplantation. 2013;20(3):138–47.
40. Phelps CJ, et al. Production of alpha 1,3-galactosyltransferase-deficient pigs. Science. 2003;299(5605):411–4.
41. Yamada K, et al. Marked prolongation of porcine renal xenograft survival in baboons through the use of alpha1,3-galactosyltransferase gene-knockout donors and the cotransplantation of vascularized thymic tissue. Nat Med. 2005;11(1):32–4.
42. Cozzi E, White DJ. The generation of transgenic pigs as potential organ donors for humans. Nat Med. 1995;1(9):964–6.
43. Cowan PJ, et al. Renal xenografts from triple-transgenic pigs are not hyperacutely rejected but cause coagulopathy in non-immunosuppressed baboons. Transplantation. 2000;69(12):2504–15.
44. Diamond LE, et al. A human CD46 transgenic pig model system for the study of discordant xenotransplantation. Transplantation. 2001;71(1):132–42.
45. Miyagawa S, et al. Complement regulation in the GalT KO era. Xenotransplantation. 2010;17(1):11–25.
46. Ekser B, et al. Clinical xenotransplantation: the next medical revolution? Lancet. 2012;379(9816):672–83.
47. Pierson 3rd RN. Antibody-mediated xenograft injury: mechanisms and protective strategies. Transpl Immunol. 2009;21(2):65–9.
48. Petersen B, et al. Transgenic expression of human heme oxygenase-1 in pigs confers resistance against xenograft rejection during ex vivo perfusion of porcine kidneys. Xenotransplantation. 2011;18(6):355–68.

49. Weiss EH, et al. HLA-E/human beta2-microglobulin transgenic pigs: protection against xenogeneic human anti-pig natural killer cell cytotoxicity. Transplantation. 2009;87(1):35–43.
50. Petersen B, et al. Pigs transgenic for human thrombomodulin have elevated production of activated protein C. Xenotransplantation. 2009;16(6):486–95.
51. Mohiuddin MM, et al. One-year heterotopic cardiac xenograft survival in a pig to baboon model. Am J Transplant. 2014;14(2):488–9.
52. Wuensch A, et al. Regulatory sequences of the porcine THBD gene facilitate endothelial-specific expression of bioactive human thrombomodulin in single- and multitransgenic pigs. Transplantation. 2014;97(2):138–47.
53. Mohiuddin MM, et al. Genetically engineered pigs and target-specific immunomodulation provide significant graft survival and hope for clinical cardiac xenotransplantation. J Thorac Cardiovasc Surg. 2014;148(3):1106–13. discussion 1113–4.
54. Heinzelmann F, et al. Immunosuppressive total lymphoid irradiation-based reconditioning regimens enable engraftment after graft rejection or graft failure in patients treated with allogeneic hematopoietic stem cell transplantation. Int J Radiat Oncol Biol Phys. 2008;70(2):523–8.
55. Hering BJ, et al. Prolonged diabetes reversal after intraportal xenotransplantation of wild-type porcine islets in immunosuppressed nonhuman primates. Nat Med. 2006;12(3):301–3.
56. Klymiuk N, et al. Xenografted islet cell clusters from INSLEA29Y transgenic pigs rescue diabetes and prevent immune rejection in humanized mice. Diabetes. 2012;61(6):1527–32.
57. Palumbo A, Anderson K. Multiple myeloma. N Engl J Med. 2011;364(11):1046–60.
58. Bauer A, et al. First experience with heterotopic thoracic pig-to-baboon cardiac xenotransplantation. Xenotransplantation. 2010;17(3):243–9.
59. McGregor C, et al. Cardiac function after preclinical orthotopic cardiac xenotransplanation. Am J Transplant. 2009;9(Suppl):380.
60. Badin R, et al. Long-term clinical recovery in Parkinsonian monkey recipients of CTLA4-Ig transgenic porcine neural precursors. Transplantation. 2010;90 Suppl 2:47.
61. Baldan N, et al. Ureteral stenosis in HDAF pig-to-primate renal xenotransplantation: a phenomenon related to immunological events? Am J Transplant. 2004;4:475–81.
62. Ramirez P, et al. Life-supporting human complement regulator decay accelerating factor transgenic pig liver xenograft maintains the metabolic function and coagulation in the nonhuman primate for up to 8 days. Transplantation. 2000;70(7):989–98.
63. Cantu E, et al. Prolonged function of macrophage, von Willebrand factor-deficient porcine pulmonary xenografts. Am J Transplant. 2007;7(1):66–75.
64. Mohiuddin M. Key note lecture: recent breakthroughs in pig-to-primate cardiac xenotransplantation. Sharing advances on large animal models-SALAAM, COST Action BM1308, Munich, December 15–17, 2014.
65. Nilsson B, et al. The role of complement in biomaterial-induced inflammation. Mol Immunol. 2007;44(1-3):82–94.
66. Scobie L, Takeuchi Y. Porcine endogenous retrovirus and other viruses in xenotransplantation. Curr Opin Organ Transplant. 2009;14(2):175–9.
67. Le Tissier P, et al. Two sets of human-tropic pig retrovirus. Nature. 1997;389(6652):681–2.
68. Klymiuk N, et al. Characterization of porcine endogenous retrovirus gamma pro-pol nucleotide sequences. J Virol. 2002;76(22):11738–43.
69. Bartosch B, et al. Evidence and consequence of porcine endogenous retrovirus recombination. J Virol. 2004;78:13880–90.
70. First and second WHO global consultations on regulatory requirements for xenotransplantation clinical trials. Changsha, China, November 19–21, 2008; Geneva, Switzerland, October 17–19, 2011. http://www.who.int/transplantation/xeno/en/
71. Denner J, Schuurman HJ, Patience C. The International Xenotransplantation Association consensus statement on conditions for undertaking clinical trials of porcine islet products in type 1 diabetes – chapter 5: strategies to prevent transmission of porcine endogenous retroviruses. Xenotransplantation. 2009;16(4):239–48.

Chapter 10
Macroencapsulated Pig Islets Correct Induced Diabetes in Primates up to 6 Months

Pierre Gianello

Abstract A bioartificial pancreas, in which islets of Langerhans are encapsulated within a semipermeable membrane, may be an alternative therapeutic device for diabetic patients. It may constitute another safe and simple method of transplanting islets without the need for immunosuppressive therapy. Since the semipermeable membrane protects the islets from the host immune system, the islets are likely to survive and release insulin for a long period of time, thereby controlling glucose metabolism in the absence of immunosuppressive medication. Recent data using macroencapsulation of pig islets in primate seems encouraging. In fact, a "mono/bilayer" configuration of macroencapsulated pig islets implanted subcutaneously has been found to significantly improve diabetes control in primates for 6 months without any immunosuppression.

Keywords Pig islets • Alginate • Encapsulation • Primates • Subcutaneous

Abbreviations

HACM	Human acellular collagen matrix
IEQ	Islet equivalent
IS	Immunosuppression
RGD	Arginylglycylaspartic acid
SLG	Sterile lyophilized high-guluronate
SLM	Sterile lyophilized high-mannuronate
STZ	Streptozotocin
T1DM	Type 1 diabetes mellitus
VLDG	Very low density guluronate
VLDM	Very low density mannuronate

P. Gianello, M.D., Ph.D. (✉)
Université catholique de Louvain, SSS/IREC/CHEX,
Avenue Hippocrate 55, Bte B1.55.04, 1200 Brussels, Belgium
e-mail: pierre.gianello@uclouvain.be

J.D. Lambris et al. (eds.), *Immune Responses to Biosurfaces*,
Advances in Experimental Medicine and Biology 865,
DOI 10.1007/978-3-319-18603-0_10

Allogeneic transplantation is today the only successful therapy for several life-threatening diseases. However, organ donation only partially meets the demand and many patients still die while waiting for transplantation. Cellular transplantation represents a very successful tool to treat type 1 diabetes mellitus (T1DM) by transplantation of human islets [1, 2]. Unfortunately, islet allotransplantation suffers from comparable limitations which is even aggravated by the fact that more than one donor is regularly needed to treat one T1DM recipient. Although human stem cells may solve these problems in the future, there are still several major hurdles that preclude their use for clinical applications. Therefore, like in the 60s when, in the absence of dialysis, clinicians referred to xenogeneic (non-human primates) organs to treat human beings (a renal xenograft survived up to 9 months), the scientific community has today reconsidered the possibility of using *porcine cells* to cure specific diseases by xenogeneic cellular transplantation. In fact, (1) pig cells have a stable function and differentiation pattern and are not tumorigenic; (2) pig cells have been shown to meet the physiological needs in large animal models (primates); (3) the source of pig cells can be scaled-up to meet all demands on a highly standardized manner, in the respect of animal welfare rules; (4) Designated Pathogen-Free (DPF) pig lines can be produced and could result in a higher safety profile than allotransplantation itself; (5) the risk of zoonosis, which was raised years ago as the major hurdle, has been recently circumvented and is actually viewed as a controlled risk and (6) the pig insulin has been used during decades for treating T1DM patients since its differs from human insulin (52aa) only at one amino-acid. The use of xenogeneic cells, however, raises a major difficulty which is the need for a heavy systemic immunosuppression (IS). In order to avoid this heavy IS, mechanical immunoprotection has been investigated to be used in preclinical models. An attractive alternative to immunosuppressive drugs is cell immunoisolation by encapsulation in a semipermeable matrix to protect transplanted tissues against immune cells from the recipient as well as against antibodies (autoimmunity of T1DM, ABO/human leukocyte antigen incompatibility, preformed antibodies against α-Gal and other antigens in xenotransplantation).

Macroencapsulation and microencapsulation systems have been proposed for cell immunoisolation [3–9]. However, the lack of biocompatibility [3, 4, 10–14], the nonselective permeability (cytokines, antibodies), the implant degradation, and the limitation of nutrient diffusion are also reported as major causes of encapsulated islet dysfunction [15]. Although several materials have been assessed (agarose, chitosan, copolymers of acetonitrile, AN69, poly(2-hydroxyethyl methacrylate), polyurethane, monomethoxy poly(ethylene glycol), Biodritin) [16–22], alginate is currently one of the major material used in the field of islet transplantation to provide immunoisolation of encapsulated cells [23–27]. This material, extracted from brown alga, is a polysaccharide composed of subunits of mannuronic (M) and guluronic (G) acids. The M/G ratio directly affects physical and biocompatible properties of implants. High-G alginates are more stable and therefore more resistant to mechanical stresses than high-M alginates after implantation [3]. In contrast, a smaller pore size, found in high-M alginates [28], can promote selective permeability for small molecules, avoiding immunoglobulins and immune cells. Alginates of high viscosity and high content in mannuronic (SLM) or guluronic acids (SLG) are the most commonly reported in the literature. New alginates with coupled peptide sequence (arginylglycylaspartic acid [RGD]) were also assessed to improve encapsulated cell adherence in the matrix [29]. Alginates with a very low density (very low density Mannuronate (VLDM) and very low density Guluronate (VLDG)) were similarly tested to reduce implant size by loading a higher number of islets per volume of polymer. The content in M and G acids as well as alginate viscosity and the use of peptidic sequences [30] may influence biocompatibility [4, 31].

As a first step, there is a need to select an encapsulation material that possesses ideal biocompatible properties for islets encapsulation such as (1) stability during the graft process, (2) immunologic protection (impermeability to molecules >150 kDa such as IgG) coupled with permeability to molecules of low molecular weight such as insulin, glucose, nutrients, and metabolites, and (3) promotion of angiogenesis to allow a sufficient oxygen pressure (pO_2) thereby ensuring encapsulated tissue survival and function. To avoid nonspecific immune response against alginates, each material is characterized by a low level of endotoxin content (<100 EU/g).

10.1 Choice of Encapsulating Alginate

First, it was investigated, in vivo, the biocompatible properties of different chemical alginates and their potential use for islet encapsulation and subcutaneous transplantation in both rat and primate models [23, 32].

Alginates composed of either high mannuronic (SLM) or high guluronic (SLG) content were tested (Tables 10.1 and 10.2). Three subtypes in each group were used: (1) high viscosity (SLM vs. SLG), (2) Very Low Density (VLDM vs. VLDG), and (3) peptide (arginine, glycine, aspartic acid for RGD)-coupled alginate (SLM-RGD vs. SLG-RGD) (Novamatrix, Drammen, Norway). Alginate implants of disc-like shape of about 1–1.5 cm^2 and a thickness of 3–6 mm were subcutaneously implanted in the paravertebral space of Wistar rats. Seven experimental groups of seven rats

Table 10.1 Characteristics of various alginates in vivo

Alginate	% W/V	% M	% G	Coupled peptide	Viscosity (mPa.s)	Endotoxin content (EU/g)
SLM	3	>50	–	No	>100	<100
SLG	3	–	>60	No	>100	<100
SLM RGD	3	>50	–	Yes	>100	<100
SLG RGD	3	–	>60	Yes	>100	<100
VLDM	7	>50	–	No	<20	<100
VLDG	7	–	>60	No	<20	<100
Ctrl+	3	>50	–	No	>100	<100

Table 10.2 Alginate selection: biocompatibility

		SLM	SLG	SLM RGD	SLG RGD	VLDM	VLDG	Ctrl+
Permeability to molecules of 150 kDa	Prior implantation	No	**Yes**	**Yes**	**Yes**	**Yes**	**Yes**	**Yes**
	At each explantation time	No	**Yes**	/	**Yes**	/	/	/
Degradation		No	**No**	**Yes**	**Yes**	**Yes**	**Yes**	**Yes**
Fibrosis		No	**Yes**	**Yes**	**Yes**	**Yes**	**Yes**	**Yes**
Lymphocytes recruitment		+	+	+++	++	++	+++	++++
Macrophages recruitment		+	+	+++	++	+++	+++	++
Angiogenesis		+++	++	+	++	+	+	++
Ideal pO_2	$pO_2 > 12$ mmHg	Yes	**Yes**	**No**	**Yes**	**No**	**No**	**No**
	$pO_2 \sim 40$ mmHg	Yes	**No**	**No**	**No**	**No**	**No**	**No**

($n=49$) were created: one group per alginate type (SLM, SLG, SLM-RGD, SLG-RGD, VLDM, VLDG) and one positive control group. Each animal from the seven experimental groups received two implants, which were placed in small subcutaneous pockets located on each side of the dorsal column. In each group, three rats were sacrificed after 2 weeks and two additional rats were euthanized at 4 weeks after implantation. After 12 weeks, the last four implants were explanted from the remaining rats in each group (14 rats/28 implants).

Alginate implants were weighed before and after implantation to assess the weight recovery and then the percentage of graft recovery. Surrounding tissues and implants (structured and destructured) were taken for investigations. Sections were thereafter routinely colored with silver methenamine (PASM) and Masson's trichrome to assess, respectively, the degree of fibrosis and angiogenesis. Lymphocyte (CD3) and macrophage (CD68) infiltrations were assessed by immunohistochemistry [23]. The numbers of macrophages, lymphocytes, and vessels were quantified histomorphologically. For characterization of the permeability of different alginates, before and after implantation, implants of each alginate were incubated with FITC-coupled lectins of different molecular weights: 36, 75 or 150 kDa [33]. In vivo biocompatibility was characterized by evaluation of graft stability, neoangiogenesis in periphery of implants, recruitment of lymphocytes and macrophages, and assessment of graft permeability to small molecules and to the immune system of the receiver.

Electronic paramagnetic resonance (EPR) oximetry was used to assess evolution in pO_2 inside grafts in vivo up to 4 weeks and to evaluate in vitro a possible gradient of pO_2 inside the SLM3 % grafts. The measurement is based on the oxygen-dependent broadening of the EPR spectrum of a paramagnetic oxygen sensor [34]. The pO_2 inside the alginate implants, placed subcutaneously in rats, was studied up to 4 weeks after transplantation. Paramagnetic carbon was used as the oxygen-sensitive probe. Adding carbon exclusively to alginate implants ensures the graft specificity of the signal measured. EPR spectra were recorded with a modulation amplitude less than one third of the peak-to-peak line width.

Implants were weighed before and after each explantation time to calculate the percentage of weight recovered after implantation (Tables 10.1 and 10.2). Control material was totally degraded 4 weeks after implantation. The percentage of weight recovery >100 % indicated serious fibrosis surrounding Ctrl + (after 2 weeks), SLM-RGD, and VLDG and SLG-RGD (after 12 weeks). Serious implant degradation was also observed for SLM-RGD at 12 weeks (−58 % of graft weight), VLDM from 2 weeks after implantation (−70 %), and VLDG (−52 %). Suitable implant stability, up to 12 weeks after implantation, was observed only for SLM (−27 %) and SLG (−16 %). The weight of the SLG implant, however, decreased significantly from 2 to 4 weeks, whereas the weight recovery of the SLM implant was stable during the complete graft course without a serious fibrosis process.

Angiogenesis is required to allow oxygenation of transplanted tissues. Therefore, angiogenesis was quantified by histomorphologic analysis of tissues surrounding alginate implants (number of vessels/0.16 mm^2) at each explantation time. Angiogenesis surrounding the alginate material was significantly higher in SLM than in other alginates at 2 and 4 weeks after implantation. Although SLG and Ctrl + demonstrated a transient angiogenesis at 2 weeks, it was not maintained at 4 and 12 weeks after implantation. Because the major cause of encapsulated cell death is probably hypoxia, pO_2 was assessed in vivo inside alginate implants at 1, 2, 3, and 4 weeks after implantation. Only SLM, SLG, and SLG-RGD alginates showed a $pO_2 > 10$ mmHg during the 4 weeks of follow-up and only SLM clearly demonstrated a constant and much higher oxygenation (~40 mmHg) during the entire 4-week follow-up.

Low lymphocyte infiltration (<35 lymphocytes/0.16 mm^2) was observed for all experimental alginates at each explantation time. However, a higher degree of lymphocyte infiltration was found at 2 weeks after implantation for SLM-RGD and Ctrl + (22.45 ± 5.85 and 34.55 ± 5.30, respectively, vs. a mean of 5.18 ± 0.61 cells/0.16 mm^2 for other alginates). At 4 and 12 weeks after implantation, VLDM and VLDG, respectively, demonstrated the highest lymphocyte recruitment (16.70 ± 1.46 and 10.10 ± 2.20, respectively, vs. a mean of 6.46 ± 0.68 cells/0.16 mm^2 for other alginates). In contrast, a lower recruitment of $CD3^+$ cells was observed for SLM and SLG at each explantation time (a mean of 4.98 ± 1.09 and 2.42 ± 0.48 cells/0.16 mm^2, respectively). Looking also at macrophage recruitment during the graft process, 2 weeks after implantation, SLM-RGD, VDLM, VDLG, and Ctrl + were characterized by significantly higher macrophage infiltration than that in SLM, SLG, and SLG-RGD. After 4 weeks, $CD68^+$ cell infiltration persisted at a higher level for VLDM and VLDG than other alginates. At 12 weeks after implantation, VLDG and even SLG-RGD demonstrated a significantly higher infiltration of macrophages than that in SLM. Throughout the whole graft process, SLM showed a constantly low level of macrophage infiltration similar to that in SLG and even SLM-RGD at 4 and 12 weeks.

The permeability of the 6 alginates and control material to lectins of 36, 75, and 150 kDa was tested in vitro before implantation. The 6 alginates and the control material were permeable to small-molecular-weight molecules (36 and 75 kDa[2]). In contrast, lectins of 150 kDa could not penetrate SLM alginate, whereas similar lectins penetrated all other tested materials. All alginate devices implanted in rats were explanted after 2, 4, and 12 weeks for permeability testing. Since the permeability assay for lectins requires well-structured alginates, permeability characterization was not performed on SLM-RGD, VLDM, and Ctrl+materials because they lost their structure after 2 weeks.

After explantation, each tested alginate maintained its permeability to molecules of low molecular weight at each explantation time. Only SLM and SLG maintained their permeability to molecules of 75 kDa during the entire graft process. SLM preserved the level of selective permeability to 150 kDa up to 12 weeks after implantation, whereas a significantly higher degree of permeability to such molecular weight molecules was evidenced for SLG and SLG-RGD.

All over, these data suggested to use SLM alginate to micro or macroencpasulate pig islets and evaluate the survival of these islets in vivo in a preclinical model i.e., pig to primate.

10.2 In Vivo Proof of Concept in Pig to Primate Model

As a second step, encapsulated pig islets in high-M alginate were implanted under the kidney capsula and the encapsulated material improved the graft survival (vs non encapsulated islets) after transplantation into several non-diabetic primates. A mean level of 0.14 ± 0.08 ng/ml of porcine C-peptide was detected until day 30 post-transplantation, in the sera of 7 primates. Level of C-peptide was significantly higher than the level obtained in animals receiving non-encapsulated pig islets (0.03 ± 0.02 ng/ml). Although no porcine C-peptide was detected in primate sera over 90, 135, 180 days post-transplantation, no graft fibrosis, no capsule overgrowth and insulin positive cells were observed. Dithizone positive cells were found inside grafts after 135 and 180 days of transplantation.

Capsules were removed 135 ($n=2$) and 180 ($n=3$) days after transplantation and were incubated in the presence of different concentrations of glucose to assess the function of pig islets from explanted capsules. An increase in insulin release, after exposure to glucose 15 mM supplemented with Forskolin, was observed for pig encapsulated islets removed at day 135: 6.6 ± 2.3 % vs. 2.9 ± 0.9 % of insulin content for glucose 15 mM+Fsk 1 μM vs. glucose 5 mM ($p=0.028$, $n=2$). The mean SI was calculated at 2.2 (range 2.0 – 2.7) (Fig. 10.1).

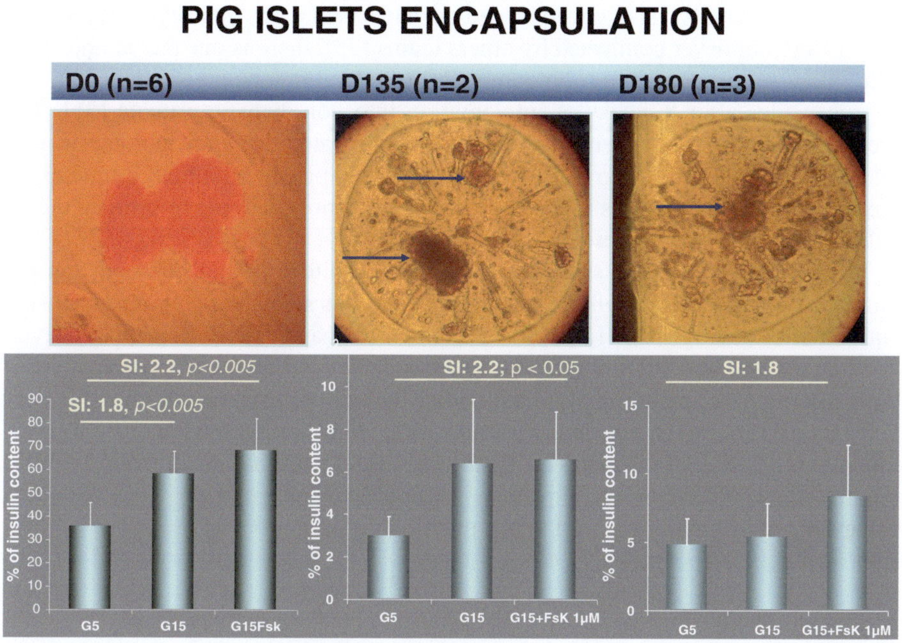

Fig. 10.1 An increase in insulin release, after exposure to glucose 15 mM supplemented with Forskolin, was observed for pig encapsulated islets removed at day 135 and 180 days of transplantation ($p<0.005$) as compared to those extracted from capsules prior to transplantation ($p<0.005$)

However, a significant decrease in insulin content was observed in capsules explanted from primates after 135 (2.2 ± 1.9 ng/islet) and 180 (1.1 ± 1.0 ng/islet) days of transplantation ($p<0.005$) as compared to those extracted from capsules prior to transplantation (32.2 ± 24.3 ng/islet) for capsules containing a mean of 2–3 pig islet cells ($p<0.005$).

In all primates, the presence of anti-pig antibodies (IgM and IgG) was detected prior to transplantation thereby confirming the presence of preformed anti-pig antibodies. No increase in IgM or IgG anti-pig antibodies was found in the sera of primates transplanted with empty capsules. In contrast, when primates were given non-encapsulated pig islets ($n=2$) the level of anti-pig IgM and IgG antibodies was strongly increased, therefore suggesting the sensitization by pig proteins or glycoproteins.

The first aim of this second step was to demonstrate the biocompatibility of encapsulated pig islets for long-term (6 months) in primates and overall, these data suggest that encapsulated pig islets must be embedded in very pure alginate, cultivated for 18 or 24 h in serum-free medium containing a concentration of 1.8 mM of $CaCl_2$. In addition, the ratio of well formed capsules must be over 90 % to obtain a long term in vivo biocompatibility in the pig to primate model.

Although the survival of encapsulated pig islets in diabetic monkeys was reported 9 years ago but never confirmed by others teams [35], there is one recent and casuistic manuscript describing biocompatibility of alginate/polyornithine/alginate microcapsules after 8 weeks of implantation into non-diabetic primate [36]. The present experimental work in vivo clearly demonstrated that implantation of optimised capsules might improve pig islet survival into primates without immunosuppression for up to 6 months in the most stringent xenogeneic pig to primate model without any immunosuppression.

Some of the pig islets survived long-term despite a strong humoral anti-pig immune response. In fact, all the primates used in this study had preformed anti-pig antibodies of both IgM and IgG types. Despite the encapsulation, all primates developed an elicited anti-pig immune response as evidenced by the significant shift of both anti-pig IgM and mainly IgG antibodies by Flow Cytometry. Despite this antibody production, no rejection or fibrosis was evidenced thereby demonstrating the immune protection of the pig islets by the capsules [37]. The immunization against pig proteins could be the consequence of a small percentage of pig islets not being encapsulated or simply prove that pig proteins might get out of the capsules [38], such as porcine C peptide [39].

10.3 Macroencapsulation of Pig Islets Can Control a Diabetes In Vivo up to 6 Months

To confirm these data in non-diabetic primates and evaluate how much these encapsulated pig islets could control a diabetes in the same preclinical model, it was crucial to use diabetic monkeys and to modify the graft now being designed as a mono/bi-layer graft to improve the oxygenation of beta cells (Fig. 10.2) and therefore avoid any lack of Oxygen diffusion.

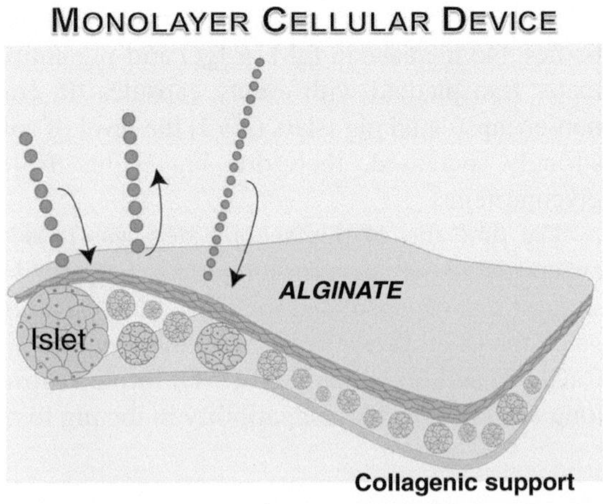

Fig. 10.2 The collagenic support (HACM) is covered by mono/bilayer of pig islets and embedded both size with SLM alginate 3 % to be implanted subcutaneously in in vivo models

After Streptozotocin treatment and prior to transplantation, six animals displayed clinical features of diabetes including polyuria, polydipsia, weight loss (−29 ± 13 % of initial weight prior to diabetes induction), persistent fasting hyperglycemia (271 ± 92 mg/dl), glycosuria (>1,000 mg/dl), and elevated glycosylated hemoglobin (>13 %). The absence of endogenous production of insulin was confirmed by an abnormal intravenous glucose tolerance test (IVGTT). When the animals were sacrificed, 94 % of beta cell mass in the native primate pancreas had been destroyed by streptozotocin (STZ).

Recipients of empty capsules (sham animals) showed no correction of diabetes. After transplantation of nonencapsulated pig islets under the kidney capsule (KC) of two primates, a peak in the porcine C-peptide level was observed 1 h after transplantation (range 2.438–6.525 ng/ml). The C-peptide level, however, was below the detection threshold (<0.1 ng/ml) 7 days after transplantation.

In addition, three to five monolayer cellular devices (MCDs formed of a collagenic support and embedded into alginate) were implanted in each primate's abdominal subcutaneous tissue containing a mean of 50,000 adult pig islet equivalents (IEQs) seeded on a 1-cm^2 human acellular collagen matrix and embedded in alginate 3 % w/v. A total amount of 30,000 IEQ/kg per primate was delivered. After MCD implantation, the diabetes was completely corrected for 20, 20, 23, 24, and 28 weeks (Fig. 10.3). Average FBG was 94 ± 11 mg/dl; basal levels of porcine C-peptide were detected (0.362 ± 0.392 ng/ml in fasted state); glycosuria, polyuria, and polydipsia disappeared; and body weight increased (+8.2 % of initial body

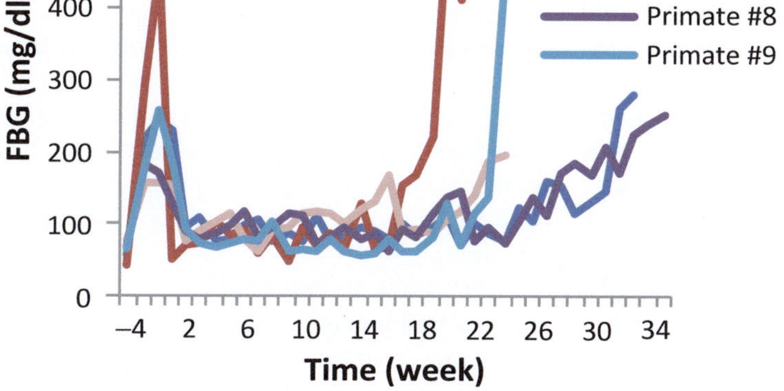

Fig. 10.3 FBG course in primates after diabetes induction by STZ and implantation of subcutaneous patches. Correction of FBG was obtained up to 6 months in some primates

weight). The control of diabetes was highlighted by correction of HbA1C, which normalized (<7 %) in primates 5, 8, and 9 up to 16 weeks after implantation. Although all transplanted primates had a decrease in HbA1C, primates 6 and 7 did not show a normal HbA1C <7 %.

Function of islets encapsulated in MCDs was assessed by IVGTT 12, 14, and 16 weeks after grafting in three animals. Whereas diabetic monkeys were unable to manage a glucose challenge after STZ treatment (insulin sera levels <1.5 μU/ml during IVGTT course), MCD implantation allowed normalization of the glucose course during IVGTT with six times more insulin release on average. In addition, the peak level of porcine C-peptide was measured in the primate sera. When total graft dysfunction was observed at 24 weeks post-transplantation (HbA1C >13 %), an assay again demonstrated a pathological arginine level. All implants were removed when diabetes completely reappeared, as indicated by elevated FBG, body weight lost (−24 %), and HbA1C >13 %.

Two animals underwent a second implant with fresh MCDs after failure of the first graft. After total dysfunction of the primary implants, the diabetic state was confirmed by an elevation of HbA1C at >13 and 12.9 % for primates 5 and 8, respectively. Secondary MCDs were placed in the same subcutaneous pouch as the first implants, and diabetes was then completely controlled again for an additional 16 and 20 weeks, as shown by normal FBG (91 ± 21 mg/dl and 68 ± 11 mg/dl), decreased HbA1C (9.6 and 7.4 %), and basal level of porcine C-peptide production (mean of 0.22 and 0.16 ng/ml) for primates 5 and 8, respectively. After this period of graft function, all signs of diabetes re-appeared with an elevation of HbA1C at 13 % but without any graft destruction.

Histologic examination revealed no alginate degradation and lower CD3 (64.4 ± 45.9 vs. 215.9 ± 15.5 cells/mm^2, P < 0.005) and CD68 (126.3 ± 23.1 vs. 496.2 ± 61.8 cells/mm^2, P < 0.005) infiltration for explanted MCD versus free pig islets (Ctrl + at day 7 after transplantation). No C3d/C9 deposition and some insulin-positive cells seeded between the human acellular collagen and alginate matrix were found in MCDs after total graft dysfunction.

Similarly to non-diabetic primates, in all animals receiving MCDs the presence of anti-pig antibodies (IgM and IgG) was detected before transplantation, and an increased level of anti-pig IgG antibodies after one, two and 6 months. These anti-pig antibodies were mainly directed against the Gal epitope and were highly cytotoxic.

Although the second transplant succeeded for primates 5 and 8, anti-pig antibodies again increased at 6 weeks after retransplantation with MCD. These newly induced antibodies were specific for the Gal epitope and highly cytotoxic. As observed for primary grafting, these secondary induced antibodies decreased during a long time course of transplantation.

The aim of this third study was to prove the concept of a subcutaneous macrodevice by demonstrating that encapsulated porcine islets can control diabetes up to 6 months after implantation into the most stringent xenogeneic model and without immunosuppression.

The long-term survival of the macroencapsulated graft in this work can be attributed to two major factors: (1) the metabolic activity of the MCD device in the subcutaneous tissue and (2) the selective permeability of the alginate against anti-pig antibodies.

The MCD was designed with a monolayer deposition of islets to provide biological support for the pig islets; immunoprotection was provided by alginate. The human acellular collagen matrix (HACM) used for islet support is a human decellularized collagen tissue. The freeze-dried structure of the HACM promotes islet adhesion and can improve the number of islets seeded per graft. A mean of 50,000 IEQ can be placed per 1 cm^2 of HACM if the purity of the islet preparation corresponds to a volume of 200 μl of the cellular pellet. Many improvements of our porcine islet isolation method have resulted in >85 % purity of the endocrine tissue, which avoids exocrine contamination [40] and positively affects duration of encapsulated graft function [41].

The most relevant factors for the implantation site for encapsulated islets are (1) physical and chemical stability of the graft after transplantation and (2) metabolic compatibility between the site and transplanted islets to control diabetes. The biocompatibility of the alginate capsules placed in subcutaneous tissue was confirmed in primates (up to 120 days post-implantation, data not shown) prior to testing MCD implantation [42]. The metabolic properties were determined by the response of encapsulated pig islets in MCDs to in vivo glucose and arginine stimulation. Although the subcutaneous tissue can be considered to have a lower physiological effect on insulin compared with portal drainage after transplantation into the liver, similar glucose courses were obtained in non-diabetic and transplanted states for primates. In addition, it was demonstrated that subcutaneous tissue allows a sufficient oxygen tension for survival of encapsulated islets MCD [43].

Although adult beta cells express a low level of Gal epitope (5.1 % of adult pig beta cells) [44–47] we confirm that Gal expression can persist after the isolation procedure (on endothelial cells) [23, 48], and therefore remains a target for humoral xenorejection against free pig islet xenotransplantation in humans and nonhuman primates. In contrast to immunosuppressed primate recipients in which no antibody response was elicited [49, 50], a high level of cytotoxic anti-Gal antibody was found in the sera of primates given transplants of encapsulated pig islets without immunosuppression. Therefore, the material for encapsulation must possess selective permeability for nutrients while preventing passage of immune cells and anti-Gal antibodies associated with pig islet xenotransplantation. The alginate 3 % w/v (used for MCD) demonstrated the selective permeability necessary to avoid the passage of IgG (150 kDa) prior and after transplantation.

10.4 Conclusion

Macroencapsulated adult pig islets transplanted into the subcutaneous tissue of diabetic cynomolgus monkeys (1) sustain long-term function without immunosuppression when placed on a collagen support with a monolayer deposition, (2) can treat diabetes with HbA1C correction <7 %, (3) can metabolically control the glucose course with an acute stimulation, and (4) are easy to transplant and retransplant into the subcutaneous space, which is a clinically applicable site involving a low-invasion procedure. Following the guidelines recently reported by Cooper and Casu [51], these data show that it is possible to meet International Xenotransplanatation Association (IXA) guidelines for a clinical pilot study. Following the properties of alginate 3 % w/v, the MCD failure at 6 months could be attributed to the lifespan of adult pig islets.

Now, SPF pigs, low in PERV needs to be selected to serve as a source of pig islets into human pilot studies.

Acknowledgement This work was supported by a European grant titled Xenoislet (FP7-HEALTH-F4-2013-601827).

References

1. Shapiro AM, Ricordi C, Hering BJ, Auchincloss H, Lindblad R, Robertson RP, et al. International trial of the Edmonton protocol for islet transplantation. N Engl J Med. 2006;355(13):1318–30. Pubmed PMID: 17005949.
2. Alejandro R, Barton FB, Hering BJ, Wease S. 2008 update from the Collaborative Islet Transplant Registry. Transplantation. 2008;86(12):1783–8. Pubmed PMID: 19104422.
3. de Groot M, Schuurs TA, van Schilfgaarde R. Causes of limited survival of microencapsulated pancreatic islet grafts. J Surg Res. 2004;121(1):141–50. Pubmed PMID: 15313388.
4. De Vos P, Hamel AF, Tatarkiewicz K. Considerations for successful transplantation of encapsulated pancreatic islets. Diabetologia. 2002;45(2):159–73. Pubmed PMID: 11935147.
5. Knazek RA, Gullino PM, Kohler PO, Dedrick RL. Cell culture on artificial capillaries: an approach to tissue growth in vitro. Science. 1972;178(4056):65–6. Pubmed PMID: 4560879.
6. Chick WL, Like AA, Lauris V. Beta cell culture on synthetic capillaries: an artificial endocrine pancreas. Science. 1975;187(4179):847–9. Pubmed PMID: 1114330.
7. Scharp DW. Isolation and transplantation of islet tissue. World J Surg. 1984;8(2):143–51. Pubmed PMID: 6428054.
8. Efrat S. Cell replacement therapy for type 1 diabetes. Trends Mol Med. 2002;8(7):334–9. Pubmed PMID: 12114113.
9. Lim F, Sun AM. Microencapsulated islets as bioartificial endocrine pancreas. Science. 1980;210(4472):908–10. Pubmed PMID: 6776628.
10. De Vos P, Van Straaten JF, Nieuwenhuizen AG, de Groot M, Ploeg RJ, De Haan BJ, et al. Why do microencapsulated islet grafts fail in the absence of fibrotic overgrowth? Diabetes. 1999;48(7):1381–8. Pubmed PMID: 10389842.
11. De Vos P, De Haan BJ, Wolters GH, Strubbe JH, van Schilfgaarde R. Improved biocompatibility but limited graft survival after purification of alginate for microencapsulation of pancreatic islets. Diabetologia. 1997;40(3):262–70. Pubmed PMID: 9084963.
12. King A, Lau J, Nordin A, Sandler S, Andersson A. The effect of capsule composition in the reversal of hyperglycemia in diabetic mice transplanted with microencapsulated allogeneic islets. Diabetes Technol Ther. 2003;5(4):653–63. Pubmed PMID: 14511420.
13. Fritschy WM, De Vos P, Groen H, Klatter FA, Pasma A, Wolters GH, et al. The capsular overgrowth on microencapsulated pancreatic islet grafts in streptozotocin and autoimmune diabetic rats. Transpl Int. 1994;7(4):264–71. Pubmed PMID: 7916926.
14. De Vos P, De Haan BJ, van Schilfgaarde R. Upscaling the production of microencapsulated pancreatic islets. Biomaterials. 1997;18(16):1085–90. Pubmed PMID: 9247345.

15. De Vos P, Smedema I, van Goor H, Moes H, van Zanten J, Netters S, et al. Association between macrophage activation and function of micro-encapsulated rat islets. Diabetologia. 2003;46(5):666–73. Pubmed PMID: 12750768.
16. Iwata H, Kobayashi K, Takagi T, Oka T, Yang H, Amemiya H, et al. Feasibility of agarose microbeads with xenogeneic islets as a bioartificial pancreas. J Biomed Mater Res. 1994;28(9):1003–11. Pubmed PMID: 7814428.
17. Zielinski BA, Aebischer P. Chitosan as a matrix for mammalian cell encapsulation. Biomaterials. 1994;15(13):1049–56. Pubmed PMID: 7888575.
18. Kessler L, Legeay G, Jesser C, Damge C, Pinget M. Influence of corona surface treatment on the properties of an artificial membrane used for Langerhans islets encapsulation: permeability and biocompatibility studies. Biomaterials. 1995;16(3):185–91. Pubmed PMID: 7748994.
19. Klomp GF, Hashiguchi H, Ursell PC, Takeda Y, Taguchi T, Dobelle WH. Macroporous hydrogel membranes for a hybrid artificial pancreas. II. Biocompatibility. J Biomed Mater Res. 1983;17(5):865–71. Pubmed PMID: 6413510.
20. Zondervan GJ, Hoppen HJ, Pennings AJ, Fritschy W, Wolters G, van Schilfgaarde R. Design of a polyurethane membrane for the encapsulation of islets of Langerhans. Biomaterials. 1992;13(3):136–44. Pubmed PMID: 1567937.
21. Sawhney AS, Hubbell JA. Poly(ethylene oxide)-graft-poly(L-lysine) copolymers to enhance the biocompatibility of poly(L-lysine)-alginate microcapsule membranes. Biomaterials. 1992;13(12):863–70. Pubmed PMID: 1457680.
22. Campos-Lisboa AC, Mares-Guia TR, Grazioli G, Goldberg AC, Sogayar MC. Biodritin microencapsulated human islets of Langerhans and their potential for type 1 diabetes mellitus therapy. Transplant Proc. 2008;40(2):433–5. Pubmed PMID: 18374092.
23. Dufrane D, Goebbels RM, Saliez A, Guiot Y, Gianello P. Six-month survival of microencapsulated pig islets and alginate biocompatibility in primates: proof of concept. Transplantation. 2006;81(9):1345–53. Pubmed PMID: 16699465.
24. De Vos P, De Haan BJ, de Haan A, van Zanten J, Faas MM. Factors influencing functional survival of microencapsulated islet grafts. Cell Transplant. 2004;13(5):515–24. Pubmed PMID: 15565864.
25. Schneider S, Feilen PJ, Brunnenmeier F, Minnemann T, Zimmermann H, Zimmermann U, et al. Long-term graft function of adult rat and human islets encapsulated in novel alginate-based microcapsules after transplantation in immunocompetent diabetic mice. Diabetes. 2005;54(3):687–93. Pubmed PMID: 15734844.
26. Elliott RB, Escobar L, Tan PL, Muzina M, Zwain S, Buchanan C. Live encapsulated porcine islets from a type 1 diabetic patient 9.5 yr after xenotransplantation. Xenotransplantation. 2007;14(2):157–61. Pubmed PMID: 17381690.
27. De Vos P, Faas MM, Strand B, Calafiore R. Alginate-based microcapsules for immunoisolation of pancreatic islets. Biomaterials. 2006;27(32):5603–17. Pubmed PMID: 16879864.
28. Ertesvag H, Valla S. Biosynthesis and applications of alginates. Polym Degrad Stab. 1998; 59(1):85–91(7).
29. Burdick JA, Anseth KS. Photoencapsulation of osteoblasts in injectable RGD-modified PEG hydrogels for bone tissue engineering. Biomaterials. 2002;23(22):4315–23. Pubmed PMID: 12219821.
30. Evangelista MB, Hsiong SX, Fernandes R, Sampaio P, Kong HJ, Barrias CC, et al. Upregulation of bone cell differentiation through immobilization within a synthetic extracellular matrix. Biomaterials. 2007;28(25):3644–55. Pubmed PMID: 17532040.
31. Orive G, Tam SK, Pedraz JL, Halle JP. Biocompatibility of alginate-poly-L-lysine microcapsules for cell therapy. Biomaterials. 2006;27(20):3691–700. Pubmed PMID: 16574222.
32. Abstracts of the Joint Meeting of the International Xenotransplantation Association (IXA), the International Pancreas and Islet Transplant Association (IPITA), and the Cell Transplant Society (CTS), Minneapolis, Minnesota, USA, September 15-20, 2007. Xenotransplantation. 2007;14(5):375–549. Pubmed PMID: 17803509.
33. Barnett BP, Kraitchman DL, Lauzon C, Magee CA, Walczak P, Gilson WD, et al. Radiopaque alginate microcapsules for X-ray visualization and immunoprotection of cellular therapeutics. Mol Pharm. 2006;3(5):531–8. Pubmed PMID: 17009852.

34. Gallez B, Baudelet C, Jordan BF. Assessment of tumor oxygenation by electron paramagnetic resonance: principles and applications. NMR Biomed. 2004;17(5):240–62. Pubmed PMID: 15366026.
35. Sun Y, Ma X, Zhou D, Vacek I, Sun AM. Normalization of diabetes in spontaneously diabetic cynomologus monkeys by xenografts of microencapsulated porcine islets without immunosuppression. J Clin Invest. 1996;98(6):1417–22. Pubmed PMID: 8823307.
36. Elliott RB, Escobar L, Calafiore R, Basta G, Garkavenko O, Vasconcellos A, et al. Transplantation of micro- and macroencapsulated piglet islets into mice and monkeys. Transplant Proc. 2005;37(1):466–9. Pubmed PMID: 15808678.
37. Duvivier-Kali VF, Omer A, Lopez-Avalos MD, O'Neil JJ, Weir GC. Survival of microencapsulated adult pig islets in mice in spite of an antibody response. Am J Transplant. 2004;4(12):1991–2000. Pubmed PMID: 15575901.
38. De Vos P, De Haan BJ, Pater J, van Schilfgaarde R. Association between capsule diameter, adequacy of encapsulation, and survival of microencapsulated rat islet allografts. Transplantation. 1996;62(7):893–9. Pubmed PMID: 8878380.
39. Lanza RP, Beyer AM, Chick WL. Xenogenic humoral responses to islets transplanted in biohybrid diffusion chambers. Transplantation. 1994;57(9):1371–5. Pubmed PMID: 8184477.
40. Dufrane D, D'hoore W, Goebbels RM, Saliez A, Guiot Y, Gianello P. Parameters favouring successful adult pig islet isolations for xenotransplantation in pig-to-primate models. Xenotransplantation. 2006;13(3):204–14. Pubmed PMID: 16756563.
41. Sun YL, Ma X, Zhou D, Vacek I, Sun AM. Porcine pancreatic islets: isolation, microencapsulation, and xenotransplantation. Artif Organs. 1993;17(8):727–33. Pubmed PMID: 8215955.
42. Dufrane D, van Steenberghe M, Goebbels RM, Saliez A, Guiot Y, Gianello P. The influence of implantation site on the biocompatibility and survival of alginate encapsulated pig islets in rats. Biomaterials. 2006;27(17):3201–8. Pubmed PMID: 16497373.
43. Vériter S, Mergen J, Goebbels RM, Aouassar N, Grégoire C, Jordan B, et al. In Vivo selection of biocompatible alginates for islet encapsulation and subcutaneous transplantation. Tissue Eng A. 2010;16(5):1503–13.
44. Rayat GR, Rajotte RV, Hering BJ, Binette TM, Korbutt GS. In vitro and in vivo expression of Galalpha-(1,3)Gal on porcine islet cells is age dependent. J Endocrinol. 2003;177(1):127–35. Pubmed PMID: 12697044.
45. Bennet W, Sundberg B, Lundgren T, Tibell A, Groth CG, Richards A, et al. Damage to porcine islets of Langerhans after exposure to human blood in vitro, or after intraportal transplantation to cynomologus monkeys: protective effects of sCR1 and heparin. Transplantation. 2000; 69(5):711–9. Pubmed PMID: 10755515.
46. Dor Y, Brown J, Martinez OI, Melton DA. Adult pancreatic beta-cells are formed by self-duplication rather than stem-cell differentiation. Nature. 2004;429(6987):41–6. Pubmed PMID: 15129273.
47. Dor FJ, Cheng J, Alt A, Cooper DK, Schuurman HJ. Gal alpha 1,3Gal expression on porcine pancreatic islets, testis, spleen, and thymus. Xenotransplantation. 2004;11(1):101–6. Pubmed PMID: 14962299.
48. Dufrane D, Goebbels RM, Guiot Y, Gianello P. Is the expression of Gal-alpha1,3Gal on porcine pancreatic islets modified by isolation procedure? Transplant Proc. 2005;37(1):455–7. Pubmed PMID: 15808674.
49. Hering BJ, Wijkstrom M, Graham ML, Hardstedt M, Aasheim TC, Jie T, et al. Prolonged diabetes reversal after intraportal xenotransplantation of wild-type porcine islets in immunosuppressed nonhuman primates. Nat Med. 2006;12(3):301–3. Pubmed PMID: 16491083.
50. Cardona K, Korbutt GS, Milas Z, Lyon J, Cano J, Jiang W, et al. Long-term survival of neonatal porcine islets in nonhuman primates by targeting costimulation pathways. Nat Med. 2006;12(3):304–6. Pubmed PMID: 16501570.
51. Cooper DK, Casu A. The International Xenotransplantation Association consensus statement on conditions for undertaking clinical trials of porcine islet products in type 1 diabetes–chapter 4: Pre-clinical efficacy and complication data required to justify a clinical trial. Xenotransplantation. 2009;16(4):229–38. Pubmed PMID: 19799763.

Chapter 11
Regulation of Instant Blood Mediated Inflammatory Reaction (IBMIR) in Pancreatic Islet Xeno-Transplantation: Points for Therapeutic Interventions

Ioannis Kourtzelis, Peetra U. Magnusson, Klara Kotlabova, John D. Lambris, and Triantafyllos Chavakis

Abstract Xeno-transplantation of pancreatic islets represents a promising therapeutic alternative for the treatment of type 1 diabetes mellitus. However, potent innate immune responses induced shortly after the transplantation of donor islets to the recipient, comprising the Instant Blood Mediated Immune Reaction (IBMIR), exert detrimental actions on islet graft function. The coagulation and complement cascades together with the leukocyte and platelet populations are the major players in IBMIR. This innate immune attack affects dramatically islet integrity and leads to significant loss of function of the xenograft. In the present review, we focus on the mechanisms contributing to IBMIR components and address therapeutic intervention approaches to limit IBMIR by administering inhibitors in circulation, by coating the islet surface with inhibitors or by generating transgenic donor animals; these approaches could result in improved xenograft survival.

Keywords Instant blood mediated immune reaction (IBMIR) • Coagulation • Complement • Islet xenotransplantation • Compstatin

---

I. Kourtzelis (✉) • K. Kotlabova • T. Chavakis
Department of Clinical Pathobiochemistry, Medical Faculty, Technische Universität Dresden, Fetscherstrasse 74, Dresden 01307, Germany
e-mail: Iannis.Kourtzelis@uniklinikum-dresden.de

P.U. Magnusson
Department of Immunology, Genetics and Pathology, Uppsala University, Uppsala SE-751 85, Sweden

J.D. Lambris
Perelman School of Medicine, University of Pennsylvania, Philadelphia, PA, USA

J.D. Lambris et al. (eds.), *Immune Responses to Biosurfaces*, Advances in Experimental Medicine and Biology 865,
DOI 10.1007/978-3-319-18603-0_11

11.1 Introduction

Cell or organ transplantation (Tx) is a promising therapeutic approach for the treatment of patients suffering from end-stage organ deficiency [1, 2]. Type 1 diabetes mellitus (T1DM), which is a disease resulting from an autoimmune reaction, is linked to high morbidity and mortality rates, especially because of its micro- and macrovascular complications. Patients suffering from T1DM can only be treated with exogenous administration of insulin. Allotransplantation of isolated islets of Langerhans or whole pancreas transplantation has been established as a therapeutic option, however, the shortage of appropriate donor islets is a major limiting factor [3, 4].

Transplantation between different species, termed as xenotransplantation (xeno-Tx), may provide a promise to bypass the issue of shortage of human donor organs [5]. Until now, a variety of xenogeneic applications have been described including the transplantation of heart, kidneys, lungs or liver [6]. In the same context, xenogeneic islet transplantation to T1D patients may represent an alternative therapeutic approach to treat T1DM [7, 8], especially due to the unlimited availability of xenografts.

Pigs are currently considered the preferred xenograft donor species for several reasons. This species share physiological similarities with humans, while their low reproduction time together with the high number of progeny are further obvious advantages [9]. Moreover, the major advances in genetics in recent years have yielded the generation of transgenic pigs feasible; these tools are engaged to generate xenogeneic grafts with optimal function and protection from the host immune attack [10–12].

Despite the organ similarities between human and pigs, inter-species incompatibilities give rise to immune and thrombotic reactions that result in the xenograft rejection [13, 14]. Besides rejection reactions that are based on adaptive immunity [15, 16], in the context of islet xeno-Tx, a major potential compromise in graft function may derive from a group of innate immune responses that are termed Instant Blood Mediated Inflammatory Reaction (IBMIR). IBMIR is triggered by the xenogeneic contact between blood and islets and includes a complex interplay between activation of coagulation and the complement system, as well as leukocyte and platelet activation and accumulation (Fig. 11.1), thereby dramatically influencing the function and the survival of the xenograft, thus affecting adversely the outcome of islet xeno-Tx [17, 18]. The present review will focus on the mechanisms and interactions that regulate the pathophysiology of IBMIR, with a special emphasis on innate immunity and will address treatment strategies and points of therapeutic intervention that could ameliorate the adverse responses following islet xeno-Tx.

11.2 The Complement System

Complement system, a major component of immunity, consists of a complex network of soluble and membrane-bound proteins that cooperate in the recognition and elimination of microbial pathogens as well as foreign materials [19]. In recent years, the classical view of the complement system has been extended to include a variety

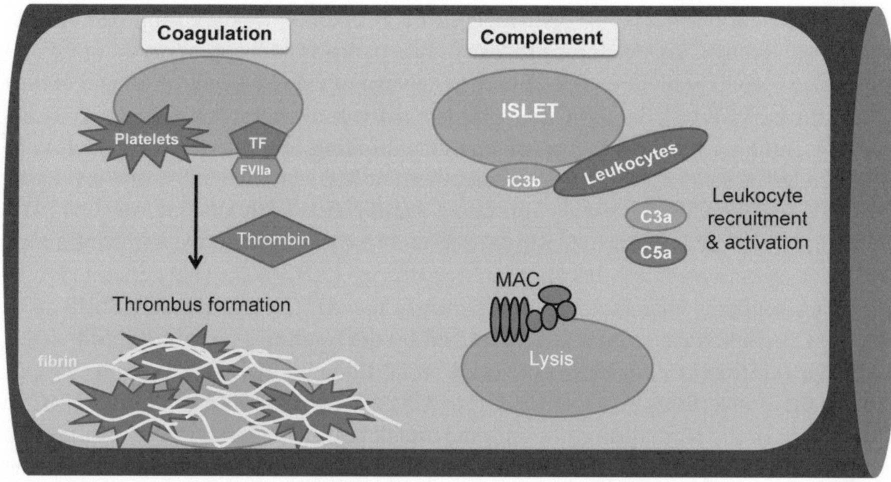

Fig. 11.1 Simplified overview of the key steps occurring during IBMIR in islet xeno-Tx. The xenogeneic contact between blood and islets triggers the activation of the extrinsic pathway of coagulation that is mediated through tissue factor (TF). As a result, the downstream effector thrombin is generated thus leading to fibrin deposition and the entrapment of islets into thrombi. Attachment of platelets to islets further supports the procoagulant effects. Activated fragments of complement (iC3b) are deposited on the islet surface, C3a and C5a anaphylatoxins activate and attract leukocytes and formation of MAC mediates the lysis of islets (*FVIIa*: activated coagulation factor VII, *MAC*: membrane attack complex)

of immune and non-immunological responses, including processes linked with tissue degeneration and regeneration, such as age-related macular degeneration (AMD) [20], liver regeneration [21, 22], and wound healing [23]. Further physiologic and pathophysiologic processes regulated by the complement system include angiogenesis [24], the mobilization of stem cells [25–27], lipid metabolism and inflammation in metabolic organs leading to insulin resistance and diabetes [28–30].

Impaired or excessive complement activation has been associated with the adverse effects observed after biomaterial applications in clinical practice [31, 32], in neurological disorders [33] and several autoimmune diseases [34, 35].

Activation of the complement system occurs via three different loops (termed the classical, lectin and alternative pathways) depending on the nature of the initial trigger. Irrespective of the activation loop, all pathways converge to the cleavage of C3 by C3 convertases. C3 convertases cleave the central component C3 to the anaphylatoxin C3a and C3b [19], the primary function of which, as well as of its split product iC3b is opsonophagocytosis. Moreover, iC3b can bind to the complement receptors CR3 and CR4 and lead to immune cell adhesion and cell activation [36, 37]. In addition, since activation of complement is associated with proteolytic cleavage of its components, proteases represent a further "non-traditional" pathway of complement activation [38, 39].

The classical pathway (CP) is triggered by antigen-antibody complexes, which are recognized by C1q. A major process in this pathway is the generation of CP C3 convertase C4b2b, which results from the cleavage of C4 to C4a and C4b and in turn the splitting of C2 to C2a and C2b [40]. The activation of the lectin pathway (LP), which is initiated by the binding of mannose-binding lectin (MBL) or ficolins to pathogen surfaces and subsequent involvement of MBL-associated serine proteases MASP-1 and MASP-2, shares significant similarities with that of the CP [41]. Spontaneous C3 hydrolysis to C3(H_2O) accounts for the constitutive and continuous low level of activation of the alternative pathway (AP) [42]. The generated C3b assembles together with factor B and factor D the AP C3 convertase C3bBb [43]. The AP C3 convertase complex is stabilized by the binding of properdin [44–46].

The generation of C3b by C3 cleavage in all three pathways is a major component of C5 convertases that cleave C5 to C5a anaphylatoxin and C5b [47]. C5b participates in the formation of membrane attack complex (MAC) by recruiting the complement components C6, C7, C8 and C9 with a main function to mediate the lysis of pathogens or targeted cells [48].

On the other hand, the C3a and C5a anaphylatoxins are very potent chemoattractants, via interaction with their G-protein-coupled C3a- and C5a-receptors, respectively, and thereby contribute to inflammatory cell recruitment to the site of injury or infection. In addition, C3a and C5a can activate immune cells and upregulate expression and release of inflammatory cytokines and mediators [19, 49].

Several soluble and membrane-bound complement regulators ensure that excessive complement activation is prevented [50]. The membrane-bound regulators of complement activity include the decay accelerating factor (DAF or CD55), the membrane cofactor protein (MCP or CD46), the complement receptor type 1 (CR1 or CD35), and the CD59 (or protectin) [50]. CD55 inhibits C3 and C5 convertases [51], whereas CD46 acts as a cofactor with factor I to promote the cleavage of C3b to iC3b [52]. The complement regulatory role of CD59 is mediated by the blockade of the polymerization of C9, thus interfering with the MAC formation [53]. Membrane complement regulators have been chosen as therapeutic strategies to block the function of complement in the context of IBMIR. For that reason, transgenic pigs that overexpress human CD55 (hCD55), human CD46 (hCD46) or human CD59 (hCD59) have been generated. Islet xenografts from these animals were protected from complement-dependent lysis and displayed better engraftment, as will be outlined in detail under the paragraph "Therapeutic targeting of IBMIR" [54–59]. Soluble regulators of complement activity include factor H that affects negatively the AP, the C1 inhibitor (C1INH) that inhibits serine proteases involved in the activation of CP [60] and the C4 binding protein that targets effectively both CP and LP [50, 61].

11.3 The Coagulation Cascade

The coagulation cascade participates in both hemostasis and thrombosis [62]. Tissue factor of the so-called extrinsic cascade is the central player for coagulation [63] and participates in thrombotic pathologies, including cardiovascular disease

[64, 65], and biomaterial-associated processes [32]. An inflammatory stimulus or endothelial cell activation results in generation of the extrinsic Xase complex consisting of TF and activated factor VII (FVIIa) [66]. The Xase-complex in turn promotes the activation of factor X (FX), which together with activated FVa and Ca^{2+} forms the prothrombinase complex that mediates the conversion of prothrombin to thrombin [67]. Thrombin can activate platelets and cleave fibrinogen to fibrin, thereby resulting in the formation of insoluble fibrin clot [68].

Coagulation and thrombosis participate in acute reactions to islet allo- [69] or xeno-Tx [70]. Notably, the exposure of islets of human or porcine origin to human blood results in the rapid activation of coagulation, as evidenced by up-regulation of TF levels [71] and by generation of high amounts of thrombin [72]. Moreover, islet Tx has been associated with thrombotic manifestations, such as fibrin deposition, and localization of the transplanted islets within thrombi [70]. Endogenous antithrombotic agents are therefore of major importance as potential beneficial modulators of IBMIR. The fine tuning of the coagulation cascade [73] is mediated by antithrombin III (ATIII), which inactivates thrombin, FXa and FIXa [74], the activated protein C (APC), which together with Protein S blocks FVa and FVIIIa [75], the tissue factor pathway inhibitor (TFPI) as well as thrombomodulin (TM). TFPI binds to and inhibits either FXa or the TF/FVIIa complex [76]. The anticoagulant activity of TM is mediated by its binding to thrombin. The TM-thrombin complex further promotes the generation of APC [77]. However, thrombin bound to TM can cleave and activate thrombin-activatable fibrinolysis inhibitor (TAFI) [78] that exerts procoagulant properties by blocking fibrinolysis. In the context of islet xeno-Tx, genetically modified pigs that overexpress hemostasis-regulatory molecules have been generated. To this end, expression of hTFPI [79] protected xenografts and promoted the achievement of normoglycemia after xeno-Tx. Porcine TM has shown to be a poor co-factor for human thrombin and its protective function is therefore lost, which leads to increased coagulation [80]. For that reason, transgenic overexpression of hTM in pigs could avert the thrombotic manifestations observed after islet xeno-Tx [81].

11.3.1 Interactions Between Coagulation and Complement

Several connections between complement and the coagulation systems including their mutual regulation have been suggested [38, 39]. On the one hand, coagulation proteases can cleave complement components, thus providing an additional extrinsic way of complement activation. The coagulation factors FIXa, FXa, FXIa as well as thrombin cleave C3 and C5 and as a result C3a and C5a are generated [38]. TM is capable of negatively regulating the activation of complement system [82]. On the other hand, MASP-2 promotes the activation of coagulation by cleaving prothrombin to thrombin [83], while the complement regulator C1INH can inhibit coagulation factors XIa and XIIa [84]. Of interest, C5a either generated as a result of biomaterial-induced complement activation [32] or in antiphospholipid syndrome [85], induces the up-regulation of TF expression. C5a may also promote coagulation indirectly by up-regulating plasminogen activator inhibitor-1, thus inhibiting fibrinolysis [86]. Therefore, complement and coagulation should be considered as two closely interacting and mutually regulated systems.

11.4 Leukocyte-Endothelial Interactions

Upon tissue inflammation, infection or injury, the interaction of leukocytes with the activated endothelium ensures a proper host response and provides the platform for the recruitment of immune cells to the site of injury or inflammation [87]. The leukocyte adhesion cascade includes multiple steps, such as rolling, adhesion, crawling and the subsequent leukocyte transmigration [88]. Initially, the rolling of leukocytes is mediated by interaction between the endothelial E- and P-selectins and their ligands CD44 and P-selectin glycoprotein ligand-1 (PSGL-1) [89]. The leukocyte adhesion and crawling to the endothelium takes places via the interaction between adhesion molecules present on the endothelial surface and leukocyte integrins. To this end, the β2 integrins Mac-1 (αMβ2) and LFA-1 (αLβ2) bind to intercellular adhesion molecule-1 and 2 (ICAM-1, 2) [87, 90]. Mac-1 has also specificity for binding to the receptor for advanced glycation end products (RAGE) [91]. Moreover, the β1 integrin VLA-4 (α4β1) binds to the adhesion molecule VCAM-1 [92]. Following this step, the adherent leukocytes transmigrate through the endothelium and accumulate within the inflamed tissues [93, 94].

Leukocyte adhesion and infiltration to the transplanted tissues has been associated with xenograft dysfunction and subsequent rejection [15]. Importantly, many of the interactions between leukocyte integrins and adhesion molecules remain operative in pig to human xeno-Tx settings [16]. Several studies have addressed the capacity of human leukocytes to roll and adhere to porcine endothelial cells (pECs), thus demonstrating the functionality of the selectin- and integrin-dependent interactions between the two species [13, 95, 96]. More specifically, the adhesion of human lymphocytes to pECs was shown to be dependent on LFA-1 and VLA-4 [97] and the adhesion of human monocytes to pECs was prevented with combined inhibition of E-selectin, LFA-1 and VLA-4 [98]. In addition, the inhibition of VLA-4, LFA-1 and Mac-1 resulted in decreased adhesion of human NK cells to pECs [99]. Besides the adhesion step, the human leukocyte transmigration across the porcine endothelium has also been studied [100]. Inhibition of β2 integrins, CD99 [96] or VCAM-1 [100] led to the reduction of the xenogeneic leukocyte transmigration. Of note, the activation of complement system has been associated with the upregulation of selectins [101] and adhesion molecules and the blockade of this system was associated with a dramatic decrease of leukocyte adhesion to pECs in a xenogeneic whole blood model [102].

In the context of islet xeno-Tx, leukocyte-endothelial interactions are however less relevant. In the native pancreas, the islets are highly vascularized and upon enzymatic isolation, islets are disconnected from the donor vasculature. It should be noted that in cultured islets the endothelial cells regress or lose their vascular markers [103]. Furthermore, the detection of endothelial cells is decreased after in vitro culture of porcine islets [104]. The low levels of remaining islet EC may cause a rather low direct involvement of interactions between recipient leukocytes and EC in porcine islets transplanted to the portal vein [105]. Thus, in contrast to other xeno-Tx settings, leukocyte-endothelial interactions may be less operative in the context of islet xeno-Tx.

The smooth function of the leukocyte adhesion cascade is controlled by endogenous negative regulators [106, 107]. These molecules include Del-1, pentraxin-3, growth differentiation factor-15, galectin-1 and annexin 1 that block at several points the cascade [87]. The integrin inhibitor Del-1 (or epidermal growth factor (EGF)-like repeats and discoidin-I-like domains 3; EDIL3) is an endothelial-derived glycoprotein [90]. Del-1 blocks the interaction between the leukocyte integrin LFA-1 and ICAM-1. As a consequence, the absence of Del-1 can result in increased leukocyte recruitment [108, 109]. In addition, Del-1 inhibits the binding of the complement fragment iC3b to Mac-1 integrin [110], thus further enhancing its anti-inflammatory properties. The inhibitory role of Del-1 on leukocyte recruitment has been suggested by the enhanced severity of chronic inflammatory diseases in Del-1-deficient mice [111, 112]. The relevance of endogenous inhibitors of the leukocyte adhesion cascade in the context of xeno-Tx merits further examination.

11.5 Modulation of IBMIR

IBMIR takes place shortly after transplantation of isolated islets into the portal vein of diabetic recipients [18], or after xeno-transplantation of islets (e.g. from pig) to a different species (e.g. non-human primates). The coagulation cascade, the complement system and innate immune cells together with platelets turn out to be main drivers of the IBMIR (Fig. 11.1) [18].

The contact of host blood with the transplanted islets elicits rapidly a series of thrombo-inflammatory reactions, including upregulation of TF expression [71] and thrombin generation [72]. Moreover, the induction of TAFI further propagates pro-coagulant effects [113]. Intravascular clotting is induced [56] and thrombi, that entrap the islets, are formed [70]. In parallel, activation of complement CP and AP occurs, anaphylatoxins are generated, resulting in inflammatory cell recruitment to the graft. Moreover, active complement fragments are deposited on grafts, thus promoting the complement-dependent lysis of islets [114]. In addition, platelets and leukocytes infiltrate the site of transplantation and bind to the surface of the islets [72, 115]. As a consequence, the integrity of islet grafts is disrupted leading to an early massive loss of transplanted islets [116, 117]. The acute destruction of a significant proportion of transplanted islets by IBMIR is the major reason that the number of islets required for effective Tx is very high [118]. Interestingly, the extent of islet damage increases with decreased compatibility between the donor and recipient species. Thus, in the context of xeno-Tx, IBMIR becomes more relevant, as the recipient cannot control the IBMIR induced by xeno-Tx due to the incompatibility observed between the regulators and the effector molecules that are present on the xenograft and on the cells of the recipient, respectively [8]. Moreover, regulatory proteins are rather absent from porcine islets preparations [119]. It is therefore imperative to develop efficient therapeutic options targeting the parameters orchestrating IBMIR [18].

11.5.1 Therapeutic Targeting of IBMIR

In an effort to protect islet xenografts from the harmful effects of IBMIR, several approaches have been followed, such as strategies to inhibit coagulation, complement, leukocyte recruitment or combinations thereof. Such strategies involve the application of soluble inhibitors, the immobilization of inhibitory molecules on the surface of the graft or the generation of donor animals that lack antigenic molecules or overexpress regulatory elements.

In a xenogeneic in vitro whole blood system that simulates IBMIR, administration of a recombinant form of APC, either alone or in combination with the platelet activation inhibitor tirofiban, protected islet viability via reduced coagulation and IBMIR [120]. In islet-Tx in vivo, APC decreased the degree of IBMIR, as assessed by reduced inflammation and coagulation markers and thereby promoted graft viability and function [121].

The glycosaminoglycan LMW-DS (low-molecular weight dextran sulfate) has been reported to inhibit effectively both complement and coagulation cascades [122]. Thereby, LMW-DS attenuates significantly the activation of complement [123] and coagulation cascades, thus affecting the degree and outcome of IBMIR in both in vitro and in vivo models [115, 124, 125]. To further ensure the potent inhibition of IBMIR following islet xeno-Tx, LMW-DS was used in combination with inhibitors of complement, such as compstatin [126, 127].

Generation of thrombin has not only a major impact on thrombotic effects, but can also regulate complement activation [128]. The thrombin inhibitor melagatran blocked the activation of plasmatic coagulation and complement and decreased the activation of leukocytes after the exposure of islets to whole blood, thus suggesting a beneficial role for thrombin inhibition in IBMIR [72].

CD39 (ectonucleoside triphosphate diphosphohydrolase 1; ENTPD1) has also served as a target to minimize IBMIR effects. This molecule plays an important role in the regulation of thromboinflammation by degrading ATP and ADP, thus exerting anti-inflammatory and anti-coagulant properties [129]. Incubation of islets overexpressing CD39 with human blood induced a prolongation in clotting time, thereby suggesting a protective role for CD39 in islet xeno-Tx [130].

The specific inhibition of complement system has been extensively tested in the context of IBMIR. The AP seems to be the predominant complement pathway in the course of IBMIR. More specifically, treatment of isolated islets with factor H, or an antibody against factor B resulted in decreased complement activation upon their exposure to human serum, while C1INH did not block the generation of complement effectors in the same context [131]. The contribution of AP in IBMIR was further confirmed in a xenogeneic model of islet-Tx, whereby administration of factor H resulted in the blockade of complement and protected the islets from damage [131]. A peptide blocking complement effector C5a, alone [132] or in combination with the synthetic protease inhibitor gabexate mesilate [133] was shown to eliminate the detrimental effects of IBMIR, as coagulation activity was decreased and the islet function was improved. Of note, the suppressive effects of C5a-blocking peptide on both complement and coagulation pathways further support the interplay between these two cascades. Compstatin, a potent peptidic inhibitor that blocks complement system at the point of C3 [134], blocked the binding of active fragments of complement to islets exposed to human plasma, diminished complement activation in fluid phase [123, 126] and protected islets from lysis [114].

In further studies targeting IBMIR, Bennet et al. incubated isolated islets with whole blood in the presence of a soluble form of CR1 (sCR1). They demonstrated that treatment with sCR1 blocked the IBMIR-associated complement activation and protected the islets from damage. Simultaneous inhibition with sCR1 and heparin eliminated IBMIR adverse effects as depicted by the decreased activation of coagulation, complement and leukocytes. Interestingly, the protective role of sCR1 was confirmed in vivo, since administration of the inhibitor supported islet integrity, as assessed by reduced insulin release shortly after Tx [70].

It is worth mentioning that isolated islets can serve as a source of procoagulant factors. TF, the main initiator of coagulation in vivo, was found to be present in isolated islets [69, 71] and its knock-down [135, 136] or its inhibition with specific antibodies [137] has been proven beneficial for the blocking of IBMIR. Interestingly, nicotinamide, a vitamin B derivative, was used to decrease the expression levels of TF and coagulation, thereby ameliorating IBMIR [138] and leading to improved islet function after islet-Tx [139].

Islet xenografts can be assumed as foreign biosurfaces, which exposed to recipient blood trigger vigorous innate immune responses. Therefore, an emerging treatment strategy to eliminate the adverse effects of IBMIR is the coating of inhibitory molecules on the surface of isolated islets, thereby suppressing coagulation and complement systems locally at the site of transplantation.

In this context, heparin has been extensively studied as an inhibitor of IBMIR-associated detrimental effects. Several techniques of heparin immobilization have been introduced [140, 141]. Coating of islets with heparin abrogated the thrombotic manifestations during IBMIR [141] and was associated with increased graft survival [142]. Heparin coating of islets in combination with angiogenic growth factor increased the interaction with co-cultured EC and could be beneficial for islet vascularization [143]. Moreover, sCR1-coated islets displayed less release of insulin upon their exposure to serum, as a result of decreased complement-dependent lysis [144, 145], which led to overall better survival and function post transplantation [146]. The simultaneous immobilization of sCR1 with heparin inhibited IBMIR and further increased the frequency of normoglycemia observed after Tx [147]. The plasminogen activator urokinase has also been immobilized on the islet surface [148, 149], either alone or in combination with soluble thrombomodulin [150] or heparin [151]. Furthermore, administration of liposome carriers with TM contributed to the achievement of normoglycemia after islet-Tx via decreasing levels of fibrin and immune cell accumulation in the transplanted tissues [152].

During recent years, the design and generation of genetically engineered pigs, either lacking or overexpressing a combination of molecules that can regulate complement and coagulation cascades, has opened new ways for the treatment of IBMIR.

Animals deficient in a1,3-galactosyltransferase, an enzyme promoting the synthesis of the Gal antigen, was a first approach to obtain genetically modified islets [153]. Existing xeno-reactive antibodies of the human recipients can bind to Gal that is present in the donor graft [1] and not in humans [154] and in turn induce rapid immune responses that are responsible for the graft dysfunction and loss [155]. In alternative approaches, the expression of human factors in pig islets, e.g. via adenoviral overexpression of the hCD55 or hCD59 made them less susceptible to complement-dependent lysis [156, 157]. These findings were further confirmed when hCD55-overexpressing islets isolated from transgenic pigs were also protected from lysis [158]. Interestingly, transplantation of islets overexpressing the CRP hCD46 resulted in the achievement of long-term normoglycemia in a xenogeneic model of Tx [57].

To further enhance the protection of xenografts from the innate immune mechanisms elicited during IBMIR, research efforts to generate multi-transgenic animals, which target multiple regulation points of complement and coagulation systems, have been undertaken. Pigs overexpressing a combination of human CRPs, such as hCD46, hCD55 or hCD59 in the presence of Gal deficiency have been generated [55, 56, 58, 59]. In addition, the human anticoagulant proteins TFPI and CD39 were simultaneously introduced to the porcine genome [79]. The use of these animals in models of xenogeneic islet-Tx resulted in protection of islet engraftment thus increasing the possibility to treat T1D-associated hyperglycemia.

11.6 Conclusions

Significant efforts are being undertaken to treat type 1 diabetes by applying islet xeno-Tx. Before that is translated into clinical studies, the adverse effects of IBMIR, which is the main culprit for the early damage and loss of islet xenografts, should be effectively bypassed. To unravel the mechanisms that orchestrate IBMIR, ex vivo whole blood models that simulate IBMIR as well as in vivo Tx models are utilized. The combined therapeutic approaches targeting complement, coagulation or leukocyte activation may ameliorate the IBMIR-related complications and bring the islet xeno-Tx closer to clinical practice.

Acknowledgment This work was supported in part by the European Community's Seventh Framework Programme under grant agreement n°602699 (DIREKT) and by the Deutsche Forschungsgemeinschaft (SFB-TRR 127 Project A3).

References

1. Auchincloss H, Sachs DH. Xenogeneic transplantation. Annu Rev Immunol. 1998;16:433–70.
2. Mulligan MS, Shearon TH, Weill D, Pagani FD, Moore J, Murray S. Heart and lung transplantation in the United States, 1997–2006. Am J Transplant. 2008;8:977–87.
3. Bruni A, Gala-Lopez B, Pepper AR, Abualhassan NS, Shapiro AJ. Islet cell transplantation for the treatment of type 1 diabetes: recent advances and future challenges. Diabetes Metab Syndr Obes. 2014;7:211–23.
4. Luan NM, Iwata H. Long-term allogeneic islet graft survival in prevascularized subcutaneous sites without immunosuppressive treatment. Am J Transplant. 2014;14:1533–42.
5. O'Connell PJ, Cowan PJ, Hawthorne WJ, Yi S, Lew AM. Transplantation of xenogeneic islets: are we there yet? Curr Diab Rep. 2013;13:687–94.
6. Cooper DKC, Satyananda V, Ekser B, van der Windt DJ, Hara H, Ezzelarab MB, Schuurman H-J. Progress in pig-to-non-human primate transplantation models (1998–2013): a comprehensive review of the literature. Xenotransplantation. 2014;21:397–419.
7. Reichart B, Niemann H, Chavakis T, et al. Xenotransplantation of porcine islet cells as a potential option for the treatment of type 1 diabetes in the future. Horm Metab Res. 2015;47:31–5.
8. van der Windt DJ, Bottino R, Kumar G, et al. Clinical islet xenotransplantation: how close are we? Diabetes. 2012;61:3046–55.
9. Zeyland J, Lipiński D, Słomski R. The current state of xenotransplantation. J Appl Genet. 2014. doi:10.1007/s13353-014-0261-6.
10. Cooper DKC, Gollackner B, Knosalla C, Teranishi K. Xenotransplantation – how far have we come? Transpl Immunol. 2002;9:251–6.
11. Klymiuk N, Aigner B, Brem G, Wolf E. Genetic modification of pigs as organ donors for xenotransplantation. Mol Reprod Dev. 2010;77:209–21.
12. Nagaraju S, Bottino R, Wijkstrom M, Hara H, Trucco M, Cooper DKC. Islet xenotransplantation from genetically engineered pigs. Curr Opin Organ Transplant. 2013;18:695–702.
13. Holgersson J, Ehrnfelt C, Hauzenberger E, Serrander L. Leukocyte endothelial cell interactions in pig to human organ xenograft rejection. Vet Immunol Immunopathol. 2002;87: 407–15.
14. Wang H, Yang Y-G. Innate cellular immunity and xenotransplantation. Curr Opin Organ Transplant. 2012;17:162–7.
15. Ekser B, Cooper DKC. Overcoming the barriers to xenotransplantation: prospects for the future. Expert Rev Clin Immunol. 2010;6:219–30.
16. Yang Y-G, Sykes M. Xenotransplantation: current status and a perspective on the future. Nat Rev Immunol. 2007;7:519–31.
17. Bennet W, Groth CG, Larsson R, Nilsson B, Korsgren O. Isolated human islets trigger an instant blood mediated inflammatory reaction: implications for intraportal islet transplantation as a treatment for patients with type 1 diabetes. Ups J Med Sci. 2000;105:125–33.
18. Nilsson B, Ekdahl KN, Korsgren O. Control of instant blood-mediated inflammatory reaction to improve islets of Langerhans engraftment. Curr Opin Organ Transplant. 2011;16:620–6.
19. Ricklin D, Hajishengallis G, Yang K, Lambris JD. Complement: a key system for immune surveillance and homeostasis. Nat Immunol. 2010;11:785–97.
20. Anderson DH, Radeke MJ, Gallo NB, et al. The pivotal role of the complement system in aging and age-related macular degeneration: hypothesis re-visited. Prog Retin Eye Res. 2010;29:95–112.
21. DeAngelis RA, Markiewski MM, Kourtzelis I, Rafail S, Syriga M, Sandor A, Maurya MR, Gupta S, Subramaniam S, Lambris JD. A complement-IL-4 regulatory circuit controls liver regeneration. J Immunol. 2012;188:641–8.

22. Markiewski MM, DeAngelis RA, Strey CW, Foukas PG, Gerard C, Gerard N, Wetsel RA, Lambris JD. The regulation of liver cell survival by complement. J Immunol. 2009;182:5412–8.
23. Rafail S, Kourtzelis I, Foukas PG, Markiewski MM, DeAngelis RA, Guariento M, Ricklin D, Grice EA, Lambris JD. Complement deficiency promotes cutaneous wound healing in mice. J Immunol. 2015;194:1285–91.
24. Langer HF, Chung K-J, Orlova VV, et al. Complement-mediated inhibition of neovascularization reveals a point of convergence between innate immunity and angiogenesis. Blood. 2010;116:4395–403.
25. Jalili A, Shirvaikar N, Marquez-Curtis L, Qiu Y, Korol C, Lee H, Turner AR, Ratajczak MZ, Janowska-Wieczorek A. Fifth complement cascade protein (C5) cleavage fragments disrupt the SDF-1/CXCR4 axis: further evidence that innate immunity orchestrates the mobilization of hematopoietic stem/progenitor cells. Exp Hematol. 2010;38:321–32.
26. Ratajczak MZ, Reca R, Wysoczynski M, Yan J, Ratajczak J. Modulation of the SDF-1-CXCR4 axis by the third complement component (C3) – implications for trafficking of CXCR4+ stem cells. Exp Hematol. 2006;34:986–95.
27. Reca R, Mastellos D, Majka M, et al. Functional receptor for C3a anaphylatoxin is expressed by normal hematopoietic stem/progenitor cells, and C3a enhances their homing-related responses to SDF-1. Blood. 2003;101:3784–93.
28. MacLaren R, Cui W, Cianflone K. Adipokines and the immune system: an adipocentric view. Adv Exp Med Biol. 2008;632:1–21.
29. Phieler J, Chung K-J, Chatzigeorgiou A, et al. The complement anaphylatoxin C5a receptor contributes to obese adipose tissue inflammation and insulin resistance. J Immunol. 2013;191:4367–74.
30. Phieler J, Garcia-Martin R, Lambris JD, Chavakis T. The role of the complement system in metabolic organs and metabolic diseases. Semin Immunol. 2013;25:47–53.
31. Kourtzelis I, Rafail S, DeAngelis RA, Foukas PG, Ricklin D, Lambris JD. Inhibition of biomaterial-induced complement activation attenuates the inflammatory host response to implantation. FASEB J. 2013;27:2768–76.
32. Kourtzelis I, Markiewski MM, Doumas M, et al. Complement anaphylatoxin C5a contributes to hemodialysis-associated thrombosis. Blood. 2010;116:631–9.
33. Orsini F, De Blasio D, Zangari R, Zanier ER, De Simoni M-G. Versatility of the complement system in neuroinflammation, neurodegeneration and brain homeostasis. Front Cell Neurosci. 2014;8:380.
34. Carroll MV, Sim RB. Complement in health and disease. Adv Drug Deliv Rev. 2011;63:965–75.
35. Ricklin D, Lambris JD. Complement in immune and inflammatory disorders: pathophysiological mechanisms. J Immunol. 2013;190:3831–8.
36. Ehlers MR. CR3: a general purpose adhesion-recognition receptor essential for innate immunity. Microbes Infect. 2000;2:289–94.
37. Lachmann PJ. The amplification loop of the complement pathways. Adv Immunol. 2009;104:115–49.
38. Amara U, Flierl MA, Rittirsch D, et al. Molecular intercommunication between the complement and coagulation systems. J Immunol. 2010;185:5628–36.
39. Oikonomopoulou K, Ricklin D, Ward PA, Lambris JD. Interactions between coagulation and complement – their role in inflammation. Semin Immunopathol. 2012;34:151–65.
40. Gaboriaud C, Ling WL, Thielens NM, Bally I, Rossi V. Deciphering the fine details of c1 assembly and activation mechanisms: "mission impossible"? Front Immunol. 2014;5:565.
41. Gál P, Barna L, Kocsis A, Závodszky P. Serine proteases of the classical and lectin pathways: similarities and differences. Immunobiology. 2007;212:267–77.
42. Thurman JM, Holers VM. The central role of the alternative complement pathway in human disease. J Immunol. 2006;176:1305–10.

43. Ricklin D. Manipulating the mediator: modulation of the alternative complement pathway C3 convertase in health, disease and therapy. Immunobiology. 2012;217:1057–66.
44. Alcorlo M, Tortajada A, Rodríguez de Córdoba S, Llorca O. Structural basis for the stabilization of the complement alternative pathway C3 convertase by properdin. Proc Natl Acad Sci U S A. 2013;110:13504–9.
45. Hourcade DE. The role of properdin in the assembly of the alternative pathway C3 convertases of complement. J Biol Chem. 2006;281:2128–32.
46. Lesher AM, Nilsson B, Song W-C. Properdin in complement activation and tissue injury. Mol Immunol. 2013;56:191–8.
47. Rawal N, Pangburn MK. Structure/function of C5 convertases of complement. Int Immunopharmacol. 2001;1:415–22.
48. Bubeck D. The making of a macromolecular machine: assembly of the membrane attack complex. Biochemistry. 2014;53:1908–15.
49. Klos A, Tenner AJ, Johswich K-O, Ager RR, Reis ES, Köhl J. The role of the anaphylatoxins in health and disease. Mol Immunol. 2009;46:2753–66.
50. Zipfel PF, Skerka C. Complement regulators and inhibitory proteins. Nat Rev Immunol. 2009;9:729–40.
51. Kim DD, Song W-C. Membrane complement regulatory proteins. Clin Immunol. 2006;118:127–36.
52. Barilla-LaBarca ML, Liszewski MK, Lambris JD, Hourcade D, Atkinson JP. Role of membrane cofactor protein (CD46) in regulation of C4b and C3b deposited on cells. J Immunol. 2002;168:6298–304.
53. Kimberley FC, Sivasankar B, Paul Morgan B. Alternative roles for CD59. Mol Immunol. 2007;44:73–81.
54. Bennet W, Sundberg B, Elgue G, Larsson R, Korsgren O, Nilsson B. A new in vitro model for the study of pig-to-human vascular hyperacute rejection. Xenotransplantation. 2001;8:176–84.
55. Bottino R, Wijkstrom M, van der Windt DJ, et al. Pig-to-monkey islet xenotransplantation using multi-transgenic pigs. Am J Transplant. 2014;14:2275–87.
56. Hawthorne WJ, Salvaris EJ, Phillips P, et al. Control of IBMIR in neonatal porcine islet xenotransplantation in baboons. Am J Transplant. 2014;14:1300–9.
57. van der Windt DJ, Bottino R, Casu A, et al. Long-term controlled normoglycemia in diabetic non-human primates after transplantation with hCD46 transgenic porcine islets. Am J Transplant. 2009;9:2716–26.
58. Zhou C-Y, McInnes E, Copeman L, Langford G, Parsons N, Lancaster R, Richards A, Carrington C, Thompson S. Transgenic pigs expressing human CD59, in combination with human membrane cofactor protein and human decay-accelerating factor. Xenotransplantation. 2005;12:142–8.
59. Zhou C-Y, McInnes E, Parsons N, et al. Production and characterization of a pig line transgenic for human membrane cofactor protein. Xenotransplantation. 2002;9:183–90.
60. Perry VH, O'Connor V. C1q: the perfect complement for a synaptic feast? Nat Rev Neurosci. 2008;9:807–11.
61. Rawal N, Rajagopalan R, Salvi VP. Stringent regulation of complement lectin pathway C3/C5 convertase by C4b-binding protein (C4BP). Mol Immunol. 2009;46:2902–10.
62. Williams JC, Mackman N. Tissue factor in health and disease. Front Biosci (Elite Ed). 2012;4:358–72.
63. Owens AP, Mackman N. Tissue factor and thrombosis: the clot starts here. Thromb Haemost. 2010;104:432–9.
64. Saha D, S S, Sergeeva EG, Ionova ZI, Gorbach AV. Tissue factor and atherothrombosis. Curr Pharm Des. 2015;21:1152–7.
65. Samad F, Ruf W. Inflammation, obesity, and thrombosis. Blood. 2013;122:3415–22.
66. Edgington TS, Mackman N, Brand K, Ruf W. The structural biology of expression and function of tissue factor. Thromb Haemost. 1991;66:67–79.

67. Walker RK, Krishnaswamy S. The activation of prothrombin by the prothrombinase complex. The contribution of the substrate-membrane interaction to catalysis. J Biol Chem. 1994;269:27441–50.
68. Mosesson MW. Fibrinogen and fibrin structure and functions. J Thromb Haemost. 2005;3:1894–904.
69. Moberg L, Johansson H, Lukinius A, et al. Production of tissue factor by pancreatic islet cells as a trigger of detrimental thrombotic reactions in clinical islet transplantation. Lancet. 2002;360:2039–45.
70. Bennet W, Sundberg B, Lundgren T, et al. Damage to porcine islets of Langerhans after exposure to human blood in vitro, or after intraportal transplantation to cynomologus monkeys: protective effects of sCR1 and heparin. Transplantation. 2000;69:711–9.
71. Johansson H, Lukinius A, Moberg L, et al. Tissue factor produced by the endocrine cells of the islets of Langerhans is associated with a negative outcome of clinical islet transplantation. Diabetes. 2005;54:1755–62.
72. Ozmen L, Ekdahl KN, Elgue G, Larsson R, Korsgren O, Nilsson B. Inhibition of thrombin abrogates the instant blood-mediated inflammatory reaction triggered by isolated human islets: possible application of the thrombin inhibitor melagatran in clinical islet transplantation. Diabetes. 2002;51:1779–84.
73. Davie EW, Fujikawa K, Kisiel W. The coagulation cascade: initiation, maintenance, and regulation. Biochemistry. 1991;30:10363–70.
74. Hirsh J. Current anticoagulant therapy – unmet clinical needs. Thromb Res. 2003;109 Suppl 1:S1–8.
75. Bouwens EAM, Stavenuiter F, Mosnier LO. Mechanisms of anticoagulant and cytoprotective actions of the protein C pathway. J Thromb Haemost. 2013;11 Suppl 1:242–53.
76. Wood JP, Ellery PER, Maroney SA, Mast AE. Biology of tissue factor pathway inhibitor. Blood. 2014;123:2934–43.
77. Sadler JE. Thrombomodulin structure and function. Thromb Haemost. 1997;78:392–5.
78. Mosnier LO, Meijers JC, Bouma BN. Regulation of fibrinolysis in plasma by TAFI and protein C is dependent on the concentration of thrombomodulin. Thromb Haemost. 2001;85:5–11.
79. Ayares D, Phelps C, Vaught T, et al. Multi-transgenic pigs for xenoislet transplantation. Xenotransplantation. 2013;20:46.
80. Roussel JC, Moran CJ, Salvaris EJ, Nandurkar HH, d'Apice AJF, Cowan PJ. Pig thrombomodulin binds human thrombin but is a poor cofactor for activation of human protein C and TAFI. Am J Transplant. 2008;8:1101–12.
81. Wuensch A, Baehr A, Bongoni AK, et al. Regulatory sequences of the porcine THBD gene facilitate endothelial-specific expression of bioactive human thrombomodulin in single- and multitransgenic pigs. Transplantation. 2014;97:138–47.
82. Wang H, Vinnikov I, Shahzad K, et al. The lectin-like domain of thrombomodulin ameliorates diabetic glomerulopathy via complement inhibition. Thromb Haemost. 2012;108:1141–53.
83. Krarup A, Wallis R, Presanis JS, Gál P, Sim RB. Simultaneous activation of complement and coagulation by MBL-associated serine protease 2. PLoS One. 2007;2, e623.
84. Caliezi C, Wuillemin WA, Zeerleder S, Redondo M, Eisele B, Hack CE. C1-Esterase inhibitor: an anti-inflammatory agent and its potential use in the treatment of diseases other than hereditary angioedema. Pharmacol Rev. 2000;52:91–112.
85. Ritis K, Doumas M, Mastellos D, Micheli A, Giaglis S, Magotti P, Rafail S, Kartalis G, Sideras P, Lambris JD. A novel C5a receptor-tissue factor cross-talk in neutrophils links innate immunity to coagulation pathways. J Immunol. 2006;177:4794–802.
86. Kastl SP, Speidl WS, Kaun C, et al. The complement component C5a induces the expression of plasminogen activator inhibitor-1 in human macrophages via NF-kappaB activation. J Thromb Haemost. 2006;4:1790–7.

87. Mitroulis I, Alexaki VI, Kourtzelis I, Ziogas A, Hajishengallis G, Chavakis T. Leukocyte integrins: role in leukocyte recruitment and as therapeutic targets in inflammatory disease. Pharmacol Ther. 2014;147:123–35. doi:10.1016/j.pharmthera.2014.11.008.
88. Ley K, Laudanna C, Cybulsky MI, Nourshargh S. Getting to the site of inflammation: the leukocyte adhesion cascade updated. Nat Rev Immunol. 2007;7:678–89.
89. Kuwano Y, Spelten O, Zhang H, Ley K, Zarbock A. Rolling on E- or P-selectin induces the extended but not high-affinity conformation of LFA-1 in neutrophils. Blood. 2010;116:617–24.
90. Chavakis E, Choi EY, Chavakis T. Novel aspects in the regulation of the leukocyte adhesion cascade. Thromb Haemost. 2009;102:191–7.
91. Chavakis T, Bierhaus A, Al-Fakhri N, et al. The pattern recognition receptor (RAGE) is a counterreceptor for leukocyte integrins: a novel pathway for inflammatory cell recruitment. J Exp Med. 2003;198:1507–15.
92. Gahmberg CG, Fagerholm SC, Nurmi SM, Chavakis T, Marchesan S, Grönholm M. Regulation of integrin activity and signalling. Biochim Biophys Acta. 2009;1790:431–44.
93. Choi EY, Santoso S, Chavakis T. Mechanisms of neutrophil transendothelial migration. Front Biosci (Landmark Ed). 2009;14:1596–605.
94. Woodfin A, Voisin M-B, Nourshargh S. Recent developments and complexities in neutrophil transmigration. Curr Opin Hematol. 2010;17:9–17.
95. Robinson LA, Tu L, Steeber DA, Preis O, Platt JL, Tedder TF. The role of adhesion molecules in human leukocyte attachment to porcine vascular endothelium: implications for xenotransplantation. J Immunol. 1998;161:6931–8.
96. Schneider MKJ, Ghielmetti M, Rhyner DM, Antsiferova MA, Seebach JD. Human leukocyte transmigration across Galalpha(1,3)Gal-negative porcine endothelium is regulated by human CD18 and CD99. Transplantation. 2009;87:491–9.
97. Birmele B, Thibault G, Nivet H, Gruel Y, Bardos P, Lebranchu Y. Human lymphocyte adhesion to xenogeneic porcine endothelial cells: modulation by human TNF-alpha and involvement of VLA-4 and LFA-1. Transpl Immunol. 1996;4:265–70.
98. Zhang XF, Feng MF. Adherence of human monocytes and NK cells to human TNF-alpha-stimulated porcine endothelial cells. Immunol Cell Biol. 2000;78:633–40.
99. Kwiatkowski P, Artrip JH, John R, Edwards NM, Wang SF, Michler RE, Itescu S. Induction of swine major histocompatibility complex class I molecules on porcine endothelium by tumor necrosis factor-alpha reduces lysis by human natural killer cells. Transplantation. 1999;67:211–8.
100. Hauzenberger E, Hauzenberger D, Hultenby K, Holgersson J. Porcine endothelium supports transendothelial migration of human leukocyte subpopulations: anti-porcine vascular cell adhesion molecule antibodies as species-specific blockers of transendothelial monocyte and natural killer cell migration. Transplantation. 2000;69:1837–49.
101. Saethre M, Schneider MKJ, Lambris JD, Magotti P, Haraldsen G, Seebach JD, Mollnes TE. Cytokine secretion depends on Galalpha(1,3)Gal expression in a pig-to-human whole blood model. J Immunol. 2008;180:6346–53.
102. Kourtzelis I, Ferreira A, Mitroulis I, Ricklin D, Bornstein SR, Waskow C, Lambris JD, Chavakis T. Complement inhibition in a xenogeneic model of interactions between human whole blood and porcine endothelium. Horm Metab Res. 2015;47:36–42.
103. Nyqvist D, Köhler M, Wahlstedt H, Berggren P-O. Donor islet endothelial cells participate in formation of functional vessels within pancreatic islet grafts. Diabetes. 2005;54:2287–93.
104. Heald KA, Carless N, Jay TR, Boucher N, Downing R. Expression of the GALalpha(1-3) GAL epitope on pig islets. J Mol Med. 1999;77:169–71.
105. Henriksnäs J, Lau J, Zang G, Berggren P-O, Köhler M, Carlsson P-O. Markedly decreased blood perfusion of pancreatic islets transplanted intraportally into the liver: disruption of islet integrity necessary for islet revascularization. Diabetes. 2012;61:665–73.

106. Chavakis T. Leucocyte recruitment in inflammation and novel endogenous negative regulators thereof. Eur J Clin Invest. 2012;42:686–91.
107. Hajishengallis G, Chavakis T. Endogenous modulators of inflammatory cell recruitment. Trends Immunol. 2013;34:1–6.
108. Choi EY, Chavakis E, Czabanka MA, et al. Del-1, an endogenous leukocyte-endothelial adhesion inhibitor, limits inflammatory cell recruitment. Science. 2008;322:1101–4.
109. Kang Y-Y, Kim D-Y, Lee S-H, Choi EY. Deficiency of developmental endothelial locus-1 (Del-1) aggravates bleomycin-induced pulmonary fibrosis in mice. Biochem Biophys Res Commun. 2014;445:369–74.
110. Mitroulis I, Kang Y-Y, Gahmberg CG, Siegert G, Hajishengallis G, Chavakis T, Choi E-Y. Developmental endothelial locus-1 attenuates complement-dependent phagocytosis through inhibition of Mac-1-integrin. Thromb Haemost. 2014;111:1004–6.
111. Choi EY, Lim J-H, Neuwirth A, et al. Developmental endothelial locus-1 is a homeostatic factor in the central nervous system limiting neuroinflammation and demyelination. Mol Psychiatry. 2014. doi:10.1038/mp.2014.146.
112. Eskan MA, Jotwani R, Abe T, et al. The leukocyte integrin antagonist Del-1 inhibits IL-17-mediated inflammatory bone loss. Nat Immunol. 2012;13:465–73.
113. Wang S, Zhao Z, Cong Z, Suo G. Thrombin-activatable fibrinolysis inhibitor is activated in an instant blood-mediated inflammatory reaction after intraportal islet transplant. Exp Clin Transplant. 2014;12:62–6.
114. Tjernberg J, Ekdahl KN, Lambris JD, Korsgren O, Nilsson B. Acute antibody-mediated complement activation mediates lysis of pancreatic islets cells and may cause tissue loss in clinical islet transplantation. Transplantation. 2008;85:1193–9.
115. Goto M, Johansson H, Maeda A, Elgue G, Korsgren O, Nilsson B. Low molecular weight dextran sulfate prevents the instant blood-mediated inflammatory reaction induced by adult porcine islets. Transplantation. 2004;77:741–7.
116. Biarnés M, Montolio M, Nacher V, Raurell M, Soler J, Montanya E. Beta-cell death and mass in syngeneically transplanted islets exposed to short- and long-term hyperglycemia. Diabetes. 2002;51:66–72.
117. van der Windt DJ, Bottino R, Casu A, Campanile N, Cooper DKC. Rapid loss of intraportally transplanted islets: an overview of pathophysiology and preventive strategies. Xenotransplantation. 2007;14:288–97.
118. Harlan DM, Kenyon NS, Korsgren O, Roep BO, Immunology of Diabetes Society. Current advances and travails in islet transplantation. Diabetes. 2009;58:2175–84.
119. Tai JH, Sun H, Liu W, Melling CWJ, Hasilo C, White DJG. Isolating human islets of Langerhans causes loss of decay accelerating factor (CD55) on beta-cells. Cell Transplant. 2008;17:1349–59.
120. Akima S, Hawthorne WJ, Favaloro E, Patel A, Blyth K, Mudaliar Y, Chapman JR, O'Connell PJ. Tirofiban and activated protein C synergistically inhibit the Instant Blood Mediated Inflammatory Reaction (IBMIR) from allogeneic islet cells exposure to human blood. Am J Transplant. 2009;9:1533–40.
121. Contreras JL, Eckstein C, Smyth CA, et al. Activated protein C preserves functional islet mass after intraportal transplantation: a novel link between endothelial cell activation, thrombosis, inflammation, and islet cell death. Diabetes. 2004;53:2804–14.
122. Wuillemin WA, te Velthuis H, Lubbers YT, de Ruig CP, Eldering E, Hack CE. Potentiation of C1 inhibitor by glycosaminoglycans: dextran sulfate species are effective inhibitors of in vitro complement activation in plasma. J Immunol. 1997;159:1953–60.
123. van der Windt DJ, Marigliano M, He J, et al. Early islet damage after direct exposure of pig islets to blood: has humoral immunity been underestimated? Cell Transplant. 2012;21:1791–802.
124. Goto M, Johansson H, Maeda A, Elgue G, Korsgren O, Nilsson B. Low-molecular weight dextran sulfate abrogates the instant blood-mediated inflammatory reaction induced by adult porcine islets both in vitro and in vivo. Transplant Proc. 2004;36:1186–7.

125. Johansson H, Goto M, Dufrane D, Siegbahn A, Elgue G, Gianello P, Korsgren O, Nilsson B. Low molecular weight dextran sulfate: a strong candidate drug to block IBMIR in clinical islet transplantation. Am J Transplant. 2006;6:305–12.
126. Goto M, Tjernberg J, Dufrane D, et al. Dissecting the instant blood-mediated inflammatory reaction in islet xenotransplantation. Xenotransplantation. 2008;15:225–34.
127. Rood PPM, Bottino R, Balamurugan AN, Smetanka C, Ayares D, Groth C-G, Murase N, Cooper DKC, Trucco M. Reduction of early graft loss after intraportal porcine islet transplantation in monkeys. Transplantation. 2007;83:202–10.
128. Huber-Lang M, Sarma JV, Zetoune FS, et al. Generation of C5a in the absence of C3: a new complement activation pathway. Nat Med. 2006;12:682–7.
129. Kanthi YM, Sutton NR, Pinsky DJ. CD39: interface between vascular thrombosis and inflammation. Curr Atheroscler Rep. 2014;16:425.
130. Dwyer KM, Mysore TB, Crikis S, Robson SC, Nandurkar H, Cowan PJ, D'Apice AJF. The transgenic expression of human CD39 on murine islets inhibits clotting of human blood. Transplantation. 2006;82:428–32.
131. Kang HJ, Lee H, Ha J-M, et al. The role of the alternative complement pathway in early graft loss after intraportal porcine islet xenotransplantation. Transplantation. 2014;97:999–1008.
132. Tokodai K, Goto M, Inagaki A, Nakanishi W, Okada N, Okada H, Satomi S. C5a-inhibitory peptide combined with gabexate mesilate prevents the instant blood-mediated inflammatory reaction in a rat model of islet transplantation. Transplant Proc. 2010;42:2102–3.
133. Tokodai K, Goto M, Inagaki A, et al. Attenuation of cross-talk between the complement and coagulation cascades by C5a blockade improves early outcomes after intraportal islet transplantation. Transplantation. 2010;90:1358–65.
134. Qu H, Ricklin D, Bai H, et al. New analogs of the clinical complement inhibitor compstatin with subnanomolar affinity and enhanced pharmacokinetic properties. Immunobiology. 2013;218:496–505.
135. Ji M, Yi S, Smith-Hurst H, Phillips P, Wu J, Hawthorne W, O'Connell P. The importance of tissue factor expression by porcine NICC in triggering IBMIR in the xenograft setting. Transplantation. 2011;91:841–6.
136. Ma X, Ye B, Gao F, Liang Q, Dong Q, Liu Y, Rong P, Wang W, Yi S. Tissue factor knockdown in porcine islets: an effective approach to suppressing the instant blood-mediated inflammatory reaction. Cell Transplant. 2012;21:61–71.
137. Berman DM, Cabrera O, Kenyon NM, et al. Interference with tissue factor prolongs intrahepatic islet allograft survival in a nonhuman primate marginal mass model. Transplantation. 2007;84:308–15.
138. Moberg L, Olsson A, Berne C, et al. Nicotinamide inhibits tissue factor expression in isolated human pancreatic islets: implications for clinical islet transplantation. Transplantation. 2003;76:1285–8.
139. Jung D-Y, Park JB, Joo S-Y, Joh J-W, Kwon C-H, Kwon G-Y, Kim S-J. Effect of nicotinamide on early graft failure following intraportal islet transplantation. Exp Mol Med. 2009;41:782–92.
140. Cabric S, Eich T, Sanchez J, Nilsson B, Korsgren O, Larsson R. A new method for incorporating functional heparin onto the surface of islets of Langerhans. Tissue Eng Part C Methods. 2008;14:141–7.
141. Cabric S, Sanchez J, Lundgren T, et al. Islet surface heparinization prevents the instant blood-mediated inflammatory reaction in islet transplantation. Diabetes. 2007;56:2008–15.
142. Vaithilingam V, Kollarikova G, Qi M, Larsson R, Lacik I, Formo K, Marchese E, Oberholzer J, Guillemin GJ, Tuch BE. Beneficial effects of coating alginate microcapsules with macromolecular heparin conjugates-in vitro and in vivo study. Tissue Eng Part A. 2014;20:324–34.
143. Cabric S, Sanchez J, Johansson U, Larsson R, Nilsson B, Korsgren O, Magnusson PU. Anchoring of vascular endothelial growth factor to surface-immobilized heparin on

pancreatic islets: implications for stimulating islet angiogenesis. Tissue Eng Part A. 2010; 16:961–70.
144. Luan NM, Teramura Y, Iwata H. Immobilization of soluble complement receptor 1 on islets. Biomaterials. 2011;32:4539–45.
145. Luan NM, Teramura Y, Iwata H. Immobilization of the soluble domain of human complement receptor 1 on agarose-encapsulated islets for the prevention of complement activation. Biomaterials. 2010;31:8847–53.
146. Luan NM, Iwata H. Xenotransplantation of islets enclosed in agarose microcapsule carrying soluble complement receptor 1. Biomaterials. 2012;33:8075–81.
147. Luan NM, Iwata H. Inhibition of instant blood-mediated inflammatory responses by co-immobilization of sCR1 and heparin on islets. Biomaterials. 2013;34:5019–24.
148. Teramura Y, Iwata H. Improvement of graft survival by surface modification with poly(ethylene glycol)-lipid and urokinase in intraportal islet transplantation. Transplantation. 2011;91:271–8.
149. Totani T, Teramura Y, Iwata H. Immobilization of urokinase on the islet surface by amphiphilic poly(vinyl alcohol) that carries alkyl side chains. Biomaterials. 2008;29:2878–83.
150. Chen H, Teramura Y, Iwata H. Co-immobilization of urokinase and thrombomodulin on islet surfaces by poly(ethylene glycol)-conjugated phospholipid. J Control Release. 2011;150:229–34.
151. Teramura Y, Iwata H. Islets surface modification prevents blood-mediated inflammatory responses. Bioconjug Chem. 2008;19:1389–95.
152. Cui W, Wilson JT, Wen J, Angsana J, Qu Z, Haller CA, Chaikof EL. Thrombomodulin improves early outcomes after intraportal islet transplantation. Am J Transplant. 2009;9:1308–16.
153. Thompson P, Badell IR, Lowe M, et al. Islet xenotransplantation using gal-deficient neonatal donors improves engraftment and function. Am J Transplant. 2011;11:2593–602.
154. Galili U, Shohet SB, Kobrin E, Stults CL, Macher BA. Man, apes, and Old World monkeys differ from other mammals in the expression of alpha-galactosyl epitopes on nucleated cells. J Biol Chem. 1988;263:17755–62.
155. Puga Yung G, Schneider MKJ, Seebach JD. Immune responses to alpha1,3 galactosyltransferase knockout pigs. Curr Opin Organ Transplant. 2009;14:154–60.
156. Omori T, Nishida T, Komoda H, Fumimoto Y, Ito T, Sawa Y, Gao C, Nakatsu S, Shirakura R, Miyagawa S. A study of the xenoantigenicity of neonatal porcine islet-like cell clusters (NPCC) and the efficiency of adenovirus-mediated DAF (CD55) expression. Xenotransplantation. 2006;13:455–64.
157. Schmidt P, Goto M, Le Mauff B, Anegon I, Korsgren O. Adenovirus-mediated expression of human CD55 or CD59 protects adult porcine islets from complement-mediated cell lysis by human serum. Transplantation. 2003;75:697–702.
158. Bennet W, Björkland A, Sundberg B, Brandhorst D, Brendel MD, Richards A, White DJ, Nilsson B, Groth CG, Korsgren O. Expression of complement regulatory proteins on islets of Langerhans: a comparison between human islets and islets isolated from normal and hDAF transgenic pigs. Transplantation. 2001;72:312–9.

Chapter 12
Cell Surface Engineering for Regulation of Immune Reactions in Cell Therapy

Yuji Teramura, Sana Asif, Kristina N. Ekdahl, and Bo Nilsson

Abstract Transplantation of the pancreatic islets of Langerhans (islets) is a promising cell therapy for treating insulin-dependent type 1 diabetes mellitus. Islet transplantation is a minimally-invasive technique involving relatively simple surgery. However, after intraportal transplantation, the transplanted islets are attacked by the recipient's immune system, because they activate a number of systems, including coagulation, complement response, inflammation, immune rejection, and recurrence of autoimmune disease. We have developed a surface modification and microencapsulation technique that protects cells and islets with biomaterials and bioactive substances, which may be useful in clinical settings. This approach employs amphiphilic polymers, which can interact with lipid bilayer membranes, without increasing cell volume. Molecules attached to these polymers can protect transplanted cells and islets from attack by the host immune system. We expect that this surface modification technique will improve graft survival in clinical islet transplantation.

Keywords Surface modification • Poly(ethylene glycol) (PEG) • PEG-lipid • Islet transplantation • Instant blood-mediated inflammatory reaction (IBMIR) • Diabetes • Biomaterials • Bioartificial pancreas • Microencapsulation

Y. Teramura (✉)
Department of Bioengineering, The University of Tokyo,
7-3-1 Hongo, Bunkyo-ku, Tokyo 113-8656, Japan
e-mail: teramura@bioeng.t.u-tokyo.ac.jp

S. Asif
Department of Immunology, Genetics and Pathology (IGP), Uppsala University,
Dag Hammarskjölds väg 20, Uppsala SE-751 85, Sweden

K.N. Ekdahl • B. Nilsson
Immunology, Genetics and Pathology, University of Uppsala, Uppsala, Sweden

J.D. Lambris et al. (eds.), *Immune Responses to Biosurfaces*,
Advances in Experimental Medicine and Biology 865,
DOI 10.1007/978-3-319-18603-0_12

12.1 Introduction

Insulin-dependent (type 1) diabetes mellitus is a chronic disease that develops during childhood. This disease involves an autoimmune disorder that destroys insulin-secreting cells, called beta cells, of the Islets of Langerhans (islets) in the pancreas, which results in insulin deficiency [1, 2]. Thus, these patients cannot control glucose metabolism. The number of patients with type 1 diabetes is highly variable from area to area throughout the world. In Europe and the US, the prevalence of type 1 diabetes is much higher than in Asian countries, such as Japan and China, and Scandinavian countries have a much higher prevalence than other countries. Current therapies for type 1 diabetes are not ideal; therefore, there is a need to establish an effective therapy for type 1 diabetes.

To date, three approaches are available for treating type 1 diabetes. The first is insulin therapy. For this treatment, patients must inject themselves with insulin subcutaneously before every meal and before sleeping to control their glucose metabolism. These insulin injections must be continued throughout the patient's life, because the disease develops during childhood. This treatment is very difficult for patients. Moreover, it is not easy to maintain accurate glucose control throughout life. Thus, as these patients grow up, they run an increased risk of hypoglycemia and various complications, including retinopathy, nephropathy, and neuropathy.

The second approach is a whole pancreas transplantation. The pancreas can be isolated from donors after brain death or once the heart stops beating; then, it is transplanted into the recipient. Techniques are constantly improving for procuring and preserving excised pancreas, and for performing the surgical transplant procedure; thus, the success rates for pancreas transplantations increase each year. However, despite the increasing success rates, the shortage of donors is a serious problem. Consequently, many patients do not get the opportunity to receive a pancreas transplantation. Also, except when the donor and recipient are identical twins, most recipients must take immune-suppressive drugs after a pancreas transplantation. Currently, the side effects of immune-suppressive drugs are not fully understood for long term treatments; thus, safety issues must be carefully discussed between the physician and the patient.

The third approach is the transplantation of pancreatic islet cells, which are isolated from the pancreas of a donor. Islets are aggregates of 1,000–2,000 endocrine cells, that form a cluster of 100–300 μm in diameter, and they coexist with the exocrine tissues of the pancreas. Islets comprise four kinds of endocrine cells; beta cells, alpha cells, delta cells, and pancreatic polypeptide (PP) cells, which secrete insulin, glucagon, somatostatin, and pancreatic polypeptide, respectively. After the islets are isolated from the donor pancreas with digestion enzymes, the resulting islet cell suspension is infused into the recipient's body. This procedure is quite simple. A number of studies have described islet isolation and islet transplantation in small animals. Moreover, various clinical reports have been published since the beginning of 1970s [3], when we discovered that diabetes could be cured by transplanting only pancreatic islets into recipient patients. Although islet transplantation has many features similar to organ transplantation, the surgical procedure is simpler and it is minimally-invasive for the patient. Thus, islet transplantation is a promising treatment for curing type 1 diabetes, and it represents an alternative therapy to pancreas transplantation [4, 5].

However, like any transplantation therapy, transplanted islets face host immune rejection responses; thus, the patient must take immune-suppressive drugs to protect grafts. Additionally, graft survival is lower with transplanted islets compared to a transplanted pancreas. Currently, in the clinical setting, the islet suspension is infused into the liver of the recipient through the portal vein with a catheter. This technique has been accepted as a safe, effective therapy for patients with type 1 diabetes. Islets are carried by the blood stream to the liver, where they become entrapped inside hepatic blood vessels. After infusion into the liver, the islet surface is exposed to recipient blood, and this activates blood coagulation and a complement response, which subsequently induces inflammation [6–8]. This series of reactions is termed an instant blood-mediated inflammatory reaction (IBMIR). The IBMIR acts to destroy islets immediately after transplantation into the liver. This issue remains unresolved in clinical islet transplantation.

Some studies have examined the prevention of early unfavorable reactions to protect islets from the IBMIR. The results have shown that IBMIR could be regulated with systemic administration of anticoagulants, anti-thrombin inhibitor, melagatran [9], low-molecular weight dextransulfate [10], and some complement inhibitors [6, 11–13]. Promising results have been achieved with these approaches in experimental islet transplantation [14]. However, these techniques are difficult to apply in the clinical setting, due to a high risk of bleeding. Alternatively, our group has examined immobilization of bioactive substances onto the islet surface, which provides local regulation of unfavorable reactions [15–21]. This technique avoids the risk of bleeding after intraportal islet transplantation. In the following sections, we discuss how this approach can be implemented for regulating IBMIRs.

In islet transplantation, various approaches have been examined that used synthetic and natural polymers to improve graft survival [22]. To date, hydrogel and semi-permeable membrane-encapsulating islet devices have been developed to produce a bioartificial pancreas. The strategy is to construct a physical barrier to isolate islets from the host's immune system, and thus, protect islets from immune rejection reactions. The different bioartificial pancreas devices can be classified into microcapsular and macrocapsular types. These devices have shown promising results, particularly when they were transplanted into small animals with diabetes, such as mice and rats. However, it is quite difficult to find a transplantation site in the human body, due to the increase in total volume and/or size of the implant after the islets have been micro- or macroencapsulated [22]. The average diameter of islets is roughly 100–300 μm. The average diameter of encapsulated islets is approximately three times larger than that of freshly isolated islets. Thus, because volume is the radius to the third power, the microencapsulated islet volume is estimated to be 27 times larger than the freshly isolated islet volume. Additionally, empty capsules might remain in solution, although most would be removed in purification. In clinical settings, the fresh islet suspension volume is approximately 10 mL; thus, it would increase to about 270 mL after microencapsulation. A transplantation site for such a large volume is difficult to find in the human body.

To overcome this issue, we have proposed an alternative to encapsulation, which involves cell surface modification with various biomaterials. This approach enabled the fabrication of a new bioartificial pancreas without appreciably increasing particle size or volume [22, 23]. Our cell surface modification technique could protect islets from various immune responses during intraportal islet transplantation. Thus, patients could reduce or eliminate immune suppressive drugs, because the modified islets would suppress immune rejection. In the following sections, we describe our recent techniques for cell surface modification, and their applications to islet enclosure with synthetic polymers and a relatively thin membrane. We also discuss the advantages and disadvantages of these techniques.

12.2 Cell Surface Modifications with Synthetic Polymers

To date, several approaches have been proposed for cell surface modification with synthetic polymers. These methods involve attachment of various functional groups and bioactive substances onto the cell surface, which facilitates the formation of an immune-isolation membrane. Generally, cell surface modifications are classified into three types: (1) covalent conjugation of polymers to the amino groups of membrane proteins; (2) electrostatic interactions between cationic polymers and the negatively charged-cell surface; and (3) hydrophobic interactions between a hydrophobic domain on amphiphilic polymers and the lipid bilayer of the cell membrane.

Various synthetic polymers have been covalently conjugated to the amino groups of membrane proteins through reactions with *N*-hydroxyl-succinimide (NHS) and cyanuric chloride [24–30]. Polyethylene glycol (PEG) conjugated with NHS can be covalently attached to the cell membrane by mixing it into a cell suspension, without the need for organic solvents; this technique results in living cells coated with single PEG chains. However, this chemical reaction is difficult to control, because it occurs randomly with all cell membrane proteins. This unselective conjugation may disturb membrane protein function. Thus, cell modification with PEG-NHS should be carefully controlled. Additionally, the NHS group is readily hydrolyzed in buffer solution; consequently, the conjugation reaction must constantly compete with the inactivation reaction.

Surface modification by electrostatic interaction is based on multiple electrostatic interactions between cationic polymers and the negatively charged cell surface. Polyethyleneimine (PEI), poly-L-lysine (PLL), and polyallylamine (PAA) are often used for this method. The cell surface is modified by depositing alternating layers of anionic and cationic polymers. The thickness of the modified membrane can be controlled by repeating polymer depositions. After polymer deposition, the cell surface takes on the properties of the outermost polymer layer. Polycations, such as PEI, PAA, and PLL assemble into polyion complexes with polystyrene sulfate (PSS) [31–34]. However, the cell membrane is easily destroyed in interactions with most cationic polymers; consequently, most cationic polymers cause severe cytotoxicity and damage to islets and other cells. Although human islets were successfully modified with a layer-by-layer membrane composed of PAA/PSS/PAA [33], it was difficult to cover the entire islet surface.

Our group has focused on hydrophobic interactions with amphiphilic polymers for cell surface modifications. To date, we have employed amphiphilic polymers (Fig. 12.1) made of PEG-conjugated phospholipid derivatives (PEG-lipid) and poly(vinyl alcohol) carrying long alkyl side chains (PVA-alkyl) [22, 23]. We confirmed the properties of these amphiphilic polymers with surface plasmon resonance (SPR), which allowed us to monitor their interaction with a supported lipid membrane (Fig. 12.2a(i)). The SPR signal increased when a PEG-lipid solution was added to a supported lipid membrane on a SPR sensor surface. This signal reflected the incorporation of PEG-lipid into the lipid membrane. We measured three PEG-lipids with different alkyl chain lengths and found that the incorporation rates decreased with increasing alkyl chain lengths (Fig. 2a(ii)). When a PEG-lipid solution was mixed with cells, the hydrophobic alkyl chains on the PEG-lipid spontaneously anchored to the lipid bilayer of the cell membrane via hydrophobic interactions in aqueous solution. The spontaneous anchoring of PEG-lipid into cell membranes was also demonstrated in cultures of a human cell line derived from T-cell leukemia cells (CCRF-CEM cells). When a solution of fluorescein isothiocyanate (FITC)-conjugated

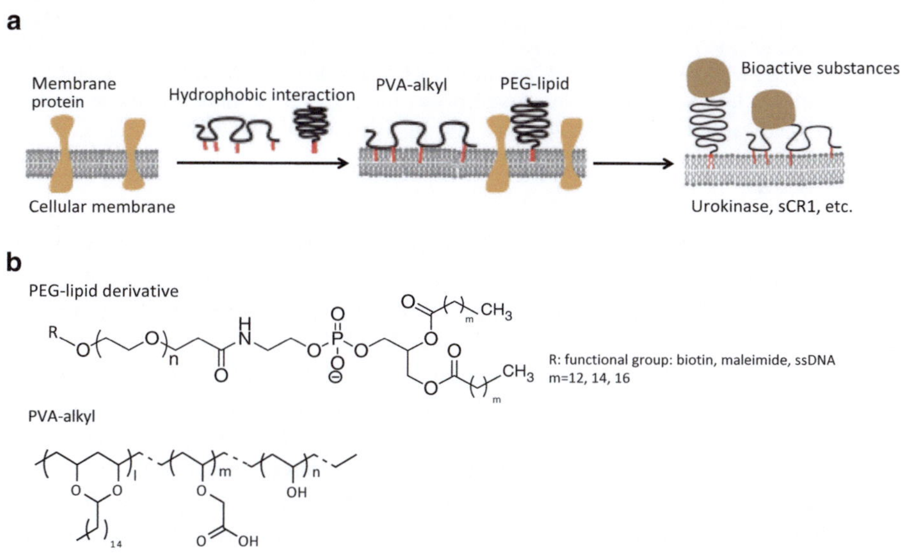

Fig. 12.1 Amphiphilic polymers employed for cell surface modifications. (**a**) Cell surfaces can be modified with amphiphilic polymers that interact with the membrane through hydrophobic bonds. Bioactive molecules can be attached to the polymers. (**b**) Chemical structures of amphiphilic polymers:polyethylene glycol-conjugated phospholipid (PEG-lipid) and poly(vinyl alcohol) carrying alkyl chains (PVA-alkyl). (Partially modified from [23])

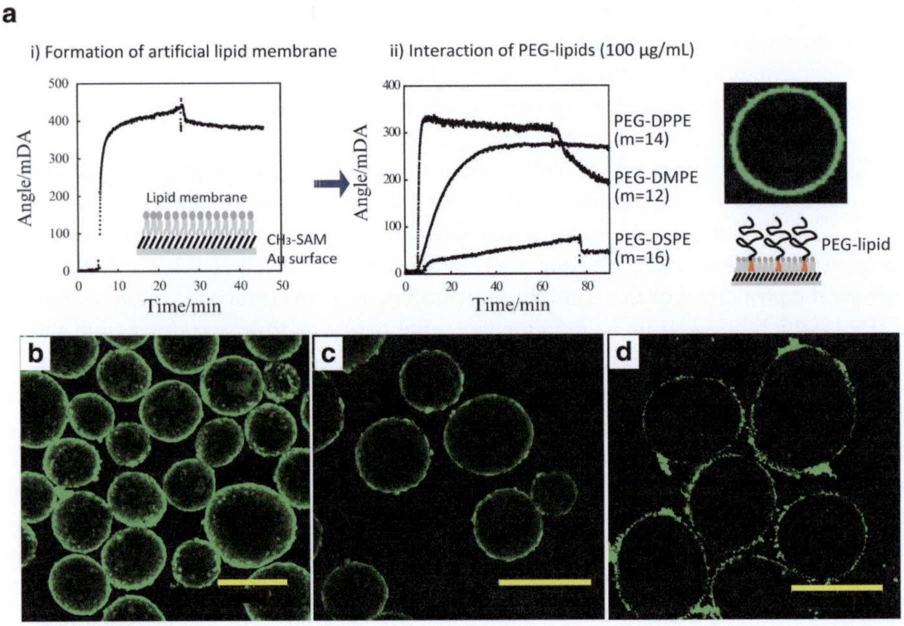

Fig. 12.2 Detection of pancreatic islet surfaces modified with an interaction between PEG-lipids and the lipid bilayer membrane. (**a**) *Left*, *middle*: Surface plasmon resonance (SPR) used to monitor the interaction between an artificial lipid membrane (*i*) and different PEG-lipids (*ii*). *Right*: Confocal laser scanning microscope image of a CCRF-CEM cell (human T-leukemia cell line) that was modified with FITC-PEG-lipid (*green*). (**b**) Pancreatic islets modified with FITC-PEG-lipid. (**c**) Islets coated with dA20-PEG-lipids and then reacted with FITC-dT20. (**d**) Islets microencapsulated within a layered PVA membrane which was formed with the thiol-maleimide reaction. PVA was labeled with FITC in advance. Bars: 200 µm. (Partially modified from [23])

PEG-lipid was added to CCRF-CEM cells, we observed a corona of FITC fluorescence at the peripheries of all cells with confocal laser scanning fluorescence microscopy (Fig. 2a). This result indicated that PEG-lipids were selectively located on the cell surface. Additionally, we found that PEG-lipid retention on the cell membrane depended on the alkyl chain length of the PEG-lipid. SPR measurements showed that the dissociation rate of PEG-lipids with long alkyl chains was much lower than that of PEG-lipids with short alkyl chains. The same phenomenon was observed in cell culture experiments. PEG-lipids with longer alkyl chains dissociated slowly from the cell surface into the medium compared to PEG-lipids with shorter alkyl chains [35]. This modification method can be applied to cell lines, primary cells, and islets, because clear fluorescence signals were detectable on all cell surfaces after treatment with FITC-PEG-lipids.

PEG-lipid derivatives can also be used to modify islet surfaces (Fig. 12.2b). For example, single stranded DNA molecules (ssDNA) can be incorporated into a PEG-lipid derivative (Fig. 12.2c), and further polymer modifications can be made with this approach (Fig. 12.2d). For example, this approach has been used to immobilize various bioactive substances on the cell surface, including urokinase [19, 21], argatroban-loaded liposomes [16], the soluble domain of complement receptor 1 (sCR1) [17, 18], and factor H-binding peptide. When these anticoagulants and complement regulators were attached to islet and cell surfaces, cells were protected from host immune responses, and graft survival was improved. In addition, this approach has been extended to fabricate an ultra-thin immune-isolation membrane composed of PVA and PEG and to microencapsulate islets with living cells. This new approach of creating an immune-isolation capsule holds potential for islet transplantation into the liver, because it does not appreciably increase graft volume.

12.3 Fibrinolytic Urokinase Immobilized on the Islet Surface

Urokinase is a serine protease that transforms inactive plasminogen into active plasmin, which then dissolves the fibrin gel in blood clots. When urokinase was immobilized onto islet surfaces, it was expected to dissolve blood clots surrounding the islets transplanted in the liver, thus inhibiting the reactions in the IBMIR cascade. The fibrinolytic enzyme, urokinase was immobilized on islets with the use of ssDNA hybridization between ssDNA-PEG-lipid and the complementary sequence, ssDNA′, attached to urokinase (Fig. 12.3) [19, 21]. To test this hybridization, a solution of dT20-PEG-lipid was incorporated onto a SPR sensor surface, and then either dA20-urokinase (ssDNA′) or dT20-urokinase (control) was added (Fig. 12.3c). A signal increase reflected the dA20-urokinase binding to the dT20-PEG-lipid modified surface; in contrast, no signal increase was observed when dT20-urokinase was added. These results indicated that urokinase could be specifically conjugated to the surface by the hybridization of dA20 to the dT20 molecules. Next, we used the same chemistry to modify the islet surface. Urokinase was immobilized onto islet surfaces by first coating the surface with dT20-PEG-lipid and then adding dA20-urokinase. The urokinase-immobilized islets were analyzed with an anti-urokinase antibody stain (Fig. 12.4a). Fluorescence could be observed on the urokinase-immobilized islets when dT20-PEG-lipids were used to immobilize urokinase, but little fluorescence was observed on unmodified islets or on islets treated with dA20-urokinase, but without dT20-PEG-lipid treatment. These results indicated that urokinase could be specifically immobilized on islet surfaces through DNA hybridization. PEG-lipids with different alkyl chain lengths were used to examine the stability of urokinase on the islet surface. As described above, dT20-PEG-lipid retention on the cell membrane depended on the alkyl chain length of the PEG-lipid [35]. Fluorescence was observed throughout 2 days of culture when islets were treated with dT20-PEG-lipid (C18), but the fluorescence faded on islets treated with dT20-PEG-lipid (C16).

To determine whether chemical modification and immobilization had disturbed urokinase function, we employed a fibrin-plate based assay to examine the function

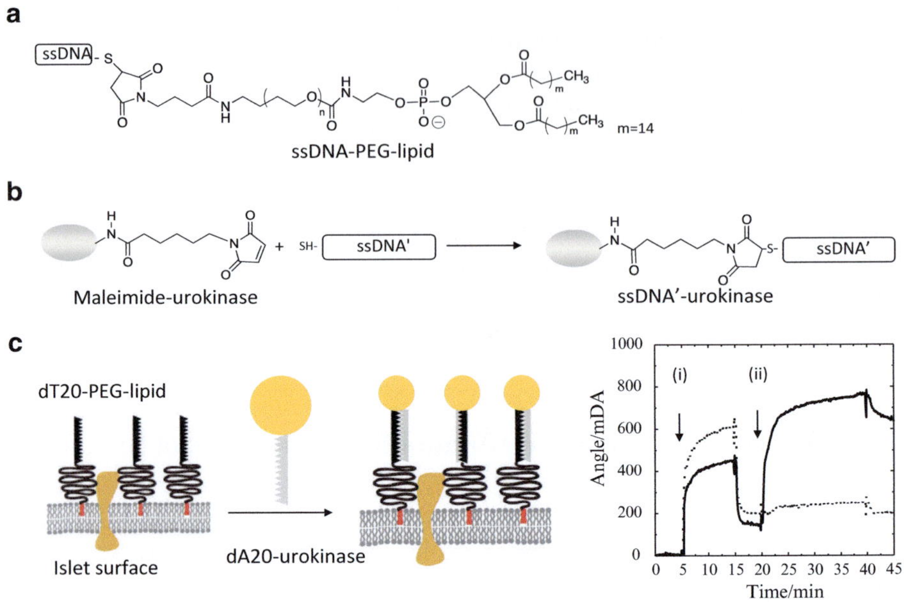

Fig. 12.3 Immobilization of urokinase onto a cell surface with ssDNA-PEG-lipids. (**a**) Chemical structure of ssDNA-PEG-lipid. (**b**) Chemical conjugation of urokinase (*grey oval*) to ssDNA′, where ssDNA′ is a complementary sequence to the ssDNA on the PEG-lipid. (**c**) Immobilization of urokinase onto a cell surface. *Left*: dT20-PEG-lipid (*black comb*) was incorporated into the plasma membrane through hydrophobic interactions; *arrow*: the dA20-urokinase (*grey comb with yellow ball*) was added; *middle*: urokinase was immobilized to the membrane through hybridization of dT20 and dA20 ssDNAs. *Right*: SPR was used to monitor immobilization. *Solid line*: dA20-urokinase immobilization to a dT20-immobilized sensor surface was detected by an increase in the signal. *Dotted line*: On the other hand, no reaction could be observed on a dA20-immobilized surface. These results indicate that the immobilization occurred through DNA hybridization. (*i*) BSA was used to block the non-specific binding of urokinase and (*ii*) dA20-urokinase. (Partially modified from [19])

of urokinase immobilized on islet surfaces (Fig. 12.4a). When placed in a fibrin gel supplemented with plasminogen, the urokinase-immobilized islets were surrounded by dissolved areas that were clearly larger than those surrounding unmodified islets. These results indicated that urokinase could be immobilized on islet surfaces and maintain its ability to activate plasminogen. We also used the fibrin-plate based assay to examine urokinase-immobilized islets cultured in the presence of serum. Urokinase activity decreased over 2 days in culture, which suggested that urokinase might be released from islet surfaces in the early stage of transplantation. Nevertheless, urokinase-immobilized islets maintained their morphology after 7 days in culture medium, and very few damaged cells were observed (Fig. 12.4a). Moreover, those islets retained the ability to regulate insulin release in response to glucose concentration changes.

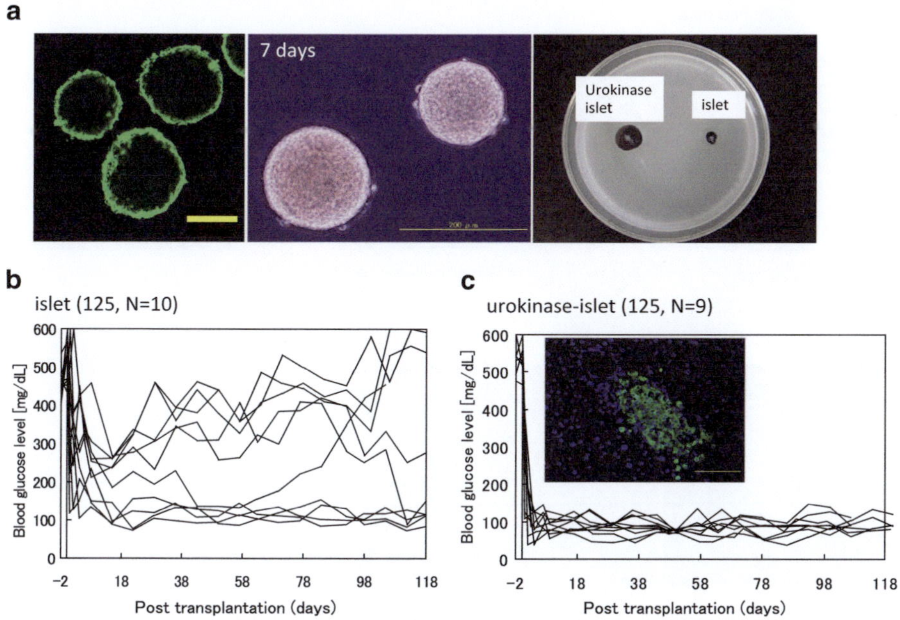

Fig. 12.4 Functional evaluation of urokinase-immobilized islets. (**a**) *Left*: Fluorescence images of urokinase-immobilized islets stained with an anti-urokinase antibody (*green*). *Middle*: Phase-contrast image of urokinase-immobilized islets after 7 days of incubation. *Right*: Fibrin-gel plate assay shows fibrinolytic activity (*dark areas*) of urokinase-immobilized islet and unmodified islet. (**b**, **c**) Changes in blood glucose levels in mice with type 1 diabetes after transplantation of (**b**) unmodified islets (125 islets) or (**c**) urokinase-immobilized islets (125 islets) into the liver. This model is a syngenic model (BALB/c islets into BALB/c diabetic mice). Type 1 diabetes was induced by an intraperitoneal injection of streptzotocin. The *inset* in (**c**) shows an image of insulin staining (*green*) in the liver of a recipient mouse 64 days after transplanting urokinase-immobilized islets. (Partially modified from [19, 21])

To evaluate the efficacy of urokinase on IBMIR *in vivo*, we performed animal experiments with a syngenic transplantation mouse model. Urokinase-immobilized islets (from BALB/c mice, 125 islets) or unmodified islets (125 islets) were infused into the livers of diabetic BALB/c mice through the portal vein. The unmodified islets did not alter diabetes in half the recipient mice (Fig. 12.5b). In contrast, the urokinase-immobilized islets led to normalized blood glucose levels in all recipient mice (Fig. 12.5c). Insulin concentration in blood increased sharply immediately after transplantation with non-modified islets, on the other hand, insulin remained low in recipients with urokinase-immobilized islets. These results indicated that a large number of islets without urokinase were immediately destroyed. When the livers of recipient mice were stained for insulin at 64 days, we observed insulin-positive grafts in liver sections (Fig. 12.5c, inset). These data suggested that immobilizing urokinase onto islets could be a promising method for evading IBMIR and improving graft survival in a clinical setting.

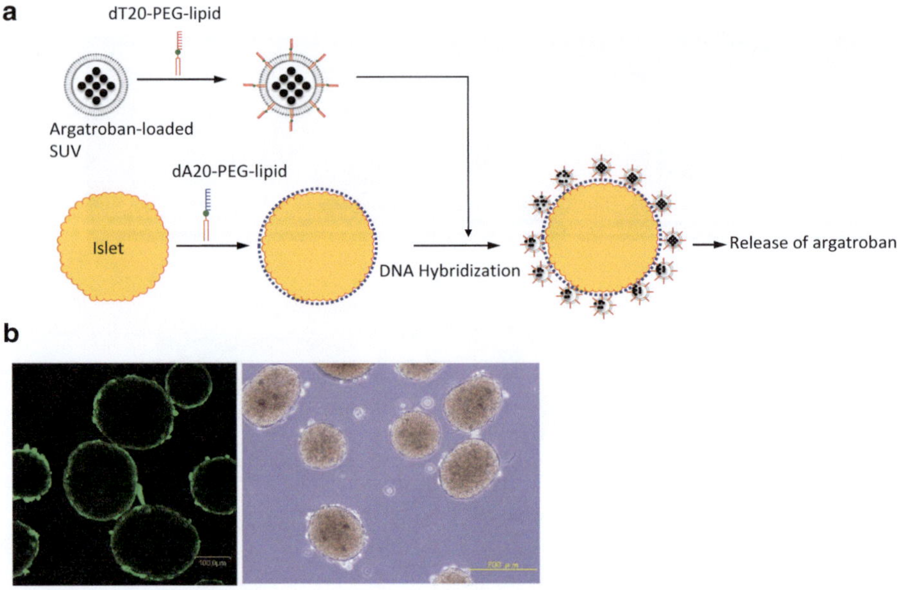

Fig. 12.5 Immobilization of argatroban-loaded liposomes onto the islet surface. (**a**) Schematic illustration of the immobilization of argatroban-loaded liposomes onto an islet surface. The surfaces of liposomes (*grey spheres*) loaded with argatroban (*black spots*) and islets (*yellow balls*) were modified with dA20-PEG-lipid and dT20-PEG-lipid, respectively. When combined, the liposomes were attached to islets via the hybridization of dA20 to dT20 ssDNAs. (**b**) *Left*: Confocal laser scanning microscope image of fluorescence-labeled liposomes immobilized on islets. *Right*: Phase-contrast image of liposome-immobilized islets. (Partially modified from [16])

12.4 Low-Molecular-Weight Drugs Inside Liposomes Immobilized on the Islet Surface

Low-molecular-weight drugs are suitable for effectively regulating blood coagulation enzymes, platelets, and complement, which are responsible for IBMIR. In this section, we will introduce a method for immobilizing a low-molecular-weight anticoagulant, argatroban, onto the islet surface. First, we attempted to attach argatroban to the islet surface [16]. However, it is not possible to apply the same methods developed for immobilizing high-molecular-weight proteins on the islet surface, because low-molecular-weight molecules typically have few or no functional groups available for immobilization. Even when drugs do have feasible functional groups, immobilization on the end of a PEG-lipid might disturb their function. Therefore, we used liposomes, which can encapsulate small drugs inside the membrane, and therefore, they serve as drug carriers. We immobilized drug-loaded liposomes onto the islet surface with the ssDNA-PEG-lipid method (Fig. 12.5a). First, liposomes that contained the thrombin inhibitor, argatroban, were modified with dA20-PEG-lipids, and islets were modified with dT20-PEG-lipids. Then, they were mixed to allow DNA hybridization between the dT20 and dA20 surface molecules. To monitor the immobilization of liposomes onto islet surfaces, we used fluorescence-labeled liposomes. Dynamic light scattering measurements showed that the liposome diameter was 105 ± 30 nm. Confocal laser scanning fluorescence microscopy showed that fluorescent liposomes were located at the periphery of each islet (Fig. 12.5b); thus, liposomes were immobilized on the islet surfaces. Moreover, islet morphology was retained after liposome immobilization (Fig. 12.5b). The release of argatroban from liposome-modified islets was evaluated by measuring anti-thrombin activity. We found that anti-thrombin activity increased with increasing culture periods; this indicated that argatroban was gradually released from the liposome-immobilized islets. In contrast, anti-thrombin activity was quite low in the media of control (unmodified) islets. In addition, glucose stimulation indexes were not significantly different between islets and liposome-modified islets, which suggested that liposome immobilization did not interfere with islet regulation of insulin release in response to glucose concentration changes.

As described above, IBMIR is involved in early islet graft loss during intraportal transplantation. Therefore, it is important to inhibit thrombotic reaction sat an early stage. Liposome immobilization is an attractive approach for releasing argatroban from the islet surface, because it provides effective local suppression of thrombotic reactions around liposome-immobilized islets. Moreover, the total dose of argatroban needed in liposomes is lower than that needed with systemic administration. This study demonstrated a first examination of the release of small drugs from liposomes attached to the islet surface. With this approach, liposomes can be loaded with a variety of anticoagulants and drugs that regulate the complement system to provide effective suppression of IBMIR in islet transplantation.

12.5 sCR1 Immobilized on Islets for Complement Regulation

Complement receptor type 1 (CR1) is a membrane glycoprotein, which is expressed on the surfaces of various blood cells, kidney podocytes, and dendritic cells [36]. CR1 is a potent inhibitor of both the classical and alternative pathways of complement activation, which are involved in IBMIR and in thrombotic reactions. CR1 suppresses complement activation by inducing dissociation of the C3 and C5 convertases and by acting as a cofactor for the proteolytic cleavage of C3b and C4b by Factor I.

We prepared the soluble domain of CR1 (sCR1) from gene-transfected Chinese hamster ovary cells [37]. To immobilize sCR1 onto islet surfaces [17], we modified islets with maleimide-conjugated PEG-lipid (Mal-PEG-lipid) and we introduced thiol groups onto sCR1 (sCR1-SH) with Traut's reagent. Then, we reacted sCR1-SH with Mal-PEG-lipid-modified islets (Fig. 12.6). To test this reaction, we attached Mal-PEG-lipid to the SPR sensor surface, and we monitored the reaction between sCR1-SH and the maleimide group. The SPR signal increased with time when a solution of sCR1-SH with 10 thiol groups per molecule was added to the

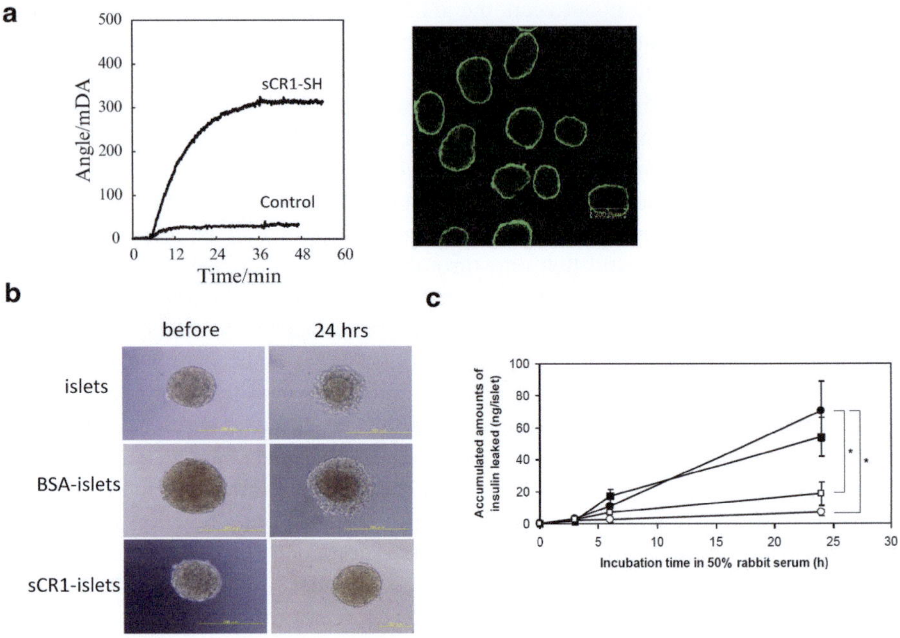

Fig. 12.6 Functional evaluation of sCR1-immobilized islets. (**a**) *Left*: SPR signals of sCR1-SH immobilization onto Mal-PEG-lipid modified surfaces. After an artificial lipid membrane was modified with Mal-PEG-lipid, sCR1-SH was added. For the control, cysteine was added first to the Mal-PEG-lipid modified surface, followed by sCR1-SH. *Right*: Images of sCR1-immobilized islets immunostained with anti-sCR1 antibodies (*green*). (**b**) Phase-contrast images of unmodified rat islets, BSA-immobilized islets, and sCR1-immobilized islets. Images show all three rat islets before and 24 h after incubation in rabbit serum. (**c**) Insulin leaked from damaged cells into the rabbit serum during incubation with unmodified rat islets (*black circles*), BSA-immobilized islets (*black squares*), or sCR1-immobilized islets (*white squares*). As a control, rat islets were incubated in 50 % heat-inactivated serum (*white circles*). The concentration of insulin released from rat islets was determined with the ELISA method. (Partially modified from [17])

Mal-PEG-lipid-immobilized sensor surface (Fig. 12.6a). On the other hand, only a slight increase in the SPR signal was detected when the modified surface was treated with cysteine in advance (Fig. 12.6a); cysteine occupied the available maleimide groups and blocked sCR1-SH immobilization. These results indicated that sCR1-SH could be immobilized with the thiol-maleimide reaction. We used the same chemistry for immobilization of sCR1 onto the islet surface. Islets were treated with Mal-PEG-lipid, and then, sCR1-SH was added. We detected sCR1-immobilized onto islets by treating with fluorescence-labeled anti-sCR1 antibody. sCR1 was detected at the periphery of the modified islets (Fig. 12.6a), but not on control islets. Moreover, sCR1 immobilization did not alter islet morphology or islet insulin secretion in response to changes in glucose concentration.

Next, we incubated sCR1-immobilized rat islets in rabbit serum to examine the protective effect of sCR1 against cellular destruction by xenoreactive antibodies and complement activation. As a control, we used bovine serum albumin (BSA)-immobilized islets (Fig. 12.6b). Because rat and rabbit are a discordant animal combination [38], rabbit serum contains xenoreactive antibodies against rat cells. We observed swollen and damaged cells in unmodified islets and BSA-immobilized islets that were incubated in rabbit serum, and the destruction increased with time. However, swollen and damaged cells were rarely observed when sCR1-immobilized islets were incubated in rabbit serum (Fig. 12.6b). In fact, immunohistochemistry showed that sCR1 was still detectable on sCR1-immobilized islets after 24 h incubation in rabbit serum; thus, these islets were protected from attacking antibodies and subsequent cellular destruction (Fig. 12.6b). When the islets are damaged, β-cells leak insulin into the medium (Fig. 12.6c). However, when we used heat-inactivated medium, which lacked complement activity, there was no insulin leakage from β-cells. This result supported the conclusion that islets had been destroyed by xenoreactive antibodies and subsequent complement activation. Incubation in normal rabbit serum caused significantly larger amounts of insulin leakage from unmodified and BSA-immobilized islets than from sCR1-immobilized islets at 24 h (Fig. 12.6c). These results suggested that sCR1 immobilized on islets effectively inhibited complement activation and protected islets from attack by xenoreactive antibodies and complement response. This method is expected to control IBMIR in islet transplantations in clinical settings and future xenotransplantations.

12.6 Other Complement System Regulators Immobilized on the Islet Surface

Some microorganisms, such as *Neisseria meningitidis* and *Yersinia enterocolitica*, express various molecules on their surfaces that can bind specifically to human complement regulators. The primary complement regulators in human blood are factor H (480 μg/mL) and C4 binding protein (C4BP; 300 μg/mL). Both these proteins inhibit complement activation. When microorganisms enter the blood, their surface molecules bind to these complement regulators in human blood until their whole surface is covered [39, 40]. The recruited complement regulators suppress innate immunity attack by the host. We reasoned that, if we could coat cells with molecules to mimic the microorganism surface, transplanted cells could escape the innate immunity of the host.

Here, we focused on the functional peptide, ASSSRCTYDHWCSH (5C6), which can specifically interact with factor H [41]. This factor H-binding peptide was conjugated onto the end of a PEG-lipid. Then, we modified cell surfaces with 5C6-conjugated PEG-lipid [42]. The 5C6 was expected to interact with and recruit factor H to the surface immediately after cells were placed in contact with human blood (Fig. 12.7a). We used porcine aortic endothelial cells (PAECs) to test this approach. When unmodified PAECs were mixed with human whole blood, platelet aggregation was induced, as shown by the increase in thrombin-antithrombin complexes (TAT), a marker of coagulation, and the increase in C3a, a marker of

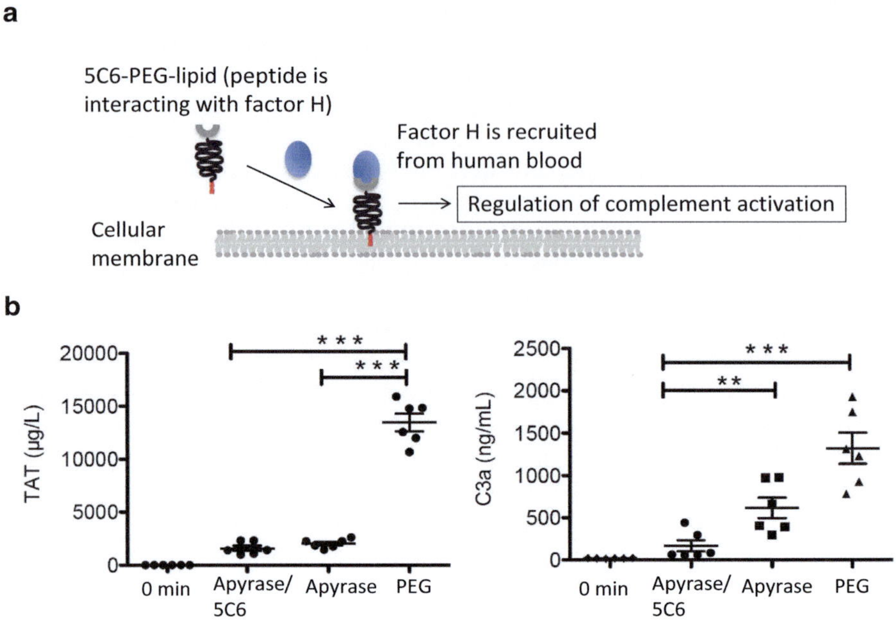

Fig. 12.7 Complement activation regulators in human blood recruited onto the cell surface. (**a**) The cell surface was modified with 5C6-conjugated PEG-lipids. The 5C6 (*grey crescents*) were selected to interact with complement regulators, like factor H (*blue oval*) and C4BP (not shown). Immobilized peptides recruit endogenous regulators onto the cell surface. The immobilized regulators inhibit complement activation. (**b**) Porcine aortic endothelial cells activated coagulation (TAT) and the complement pathway (C3a) in human whole blood. Cells were modified with PEG-lipid alone, immobilized apyrase, or co-immobilized 5C6 and apyrase. Cells were incubated with human whole blood for 30 min at 37 °C (n=6). **$p<0.01$, ***$p<0.001$, based on repeated measures of the one-way analysis of variance with Dunnett's post hoc-test. (Partially modified from [42])

complement activation. Eventually, the PAECs were damaged by the immune attack. Also, when we modified PAECs with PEG-lipid alone, to provide a PEG coating, we observed no inhibition of platelet aggregation, and the TAT and C3a signals were similar to those observed with unmodified PAECs (Fig. 12.7b). On the other hand, complement activation was suppressed when PAECs were modified with 5C6-conjugated PEG-lipid (Fig. 12.7b). Immunostaining with anti-factor H antibody showed that factor H was bound to the cell surface. These results indicated that factor H was specifically recruited to the cell surface from human blood, and that it could regulate complement activation. Coagulation activation (from TAT) and platelet aggregation could also be inhibited when PAECs were modified with apyrase (Fig. 12.7b), an enzyme that degrades ADP, which is involved in secondary platelet aggregation. Although several clinical anticoagulants are available, drugs that inhibit complement activation are limited. Therefore, it is advantageous to recruit and use endogenous complement regulators that exist in human blood. In particular, when cells are transplanted into humans, this is a useful approach for suppressing IBMIR.

12.7 Islet Microencapsulation with an Ultra-Thin Membrane

We have described regulators of innate immunity activation and various bioactive substances that were attached to the surfaces of cells and islets with PEG-lipid, which were active immediately after intraportal transplantation [21, 43]. However, these cell surface-immobilized materials tend to be taken up and/or detached from the surface over time, because the cell surface is in a dynamic state (e.g., endocytosis). Therefore, in the long run, single polymer chains and nanometer membranes might not be sufficient to protect cells from host immune attacks and immune rejection, which is primarily carried out by T cells. We reasoned that protection from host T cell attacks might be achieved by isolating islets within a polymer membrane that has high integrity and stability. The alternate deposition of cationic and anionic polymers has previously been used to microencapsulate cells and islets [20, 33, 44–46]. That method made it possible to produce membranes with thicknesses of several tens of nanometers [20]. In addition, there was no practical increase in membrane thickness after coating the cell surface with polymers. However, it is difficult to produce polymer membranes of micrometer thicknesses, even with several repetitions of polymer deposition, because the deposition eventually destroyed the cell structure. Although some improvements have been made to these approaches, there remains a risk of cell damage. It is likely that multiple direct interactions with the cell surface may influence the cell structure. To overcome these issues, we developed a method for microencapsulating islets that produces a stable membrane of micrometer thickness [47].

With our method, a polymer membrane is formed on the islet surface by immobilizing 8-arm PEG-SH and 4-arm PEG-Malpolymers in the presence of Mal-PEG-lipid micelles (Fig. 12.8a). Both 8-Arm PEG-SH and 4-arm PEG-maleimide are commercially available. It is important that gelation among these molecules should take place only at the periphery of the islet surface. Ideally, membrane thickness should be limited to the micrometer range. Otherwise, there will be an unnecessary volume increase after surface modification. As shown in Fig. 12.8a, first, the islets are mixed with Mal-PEG-lipid, which attaches to the surface. Then, the 8-arm PEG-SH is added in the presence of Mal-PEG-lipid micelles to achieve gelation. The gelation between 8-arm PEG-SH and Mal-PEG-lipid micelles takes place around the islet surface, which substantially increases the thickness of the islet periphery. After washing with buffer, the 4-arm PEG-Mal is added as a cross-linker between the SH groups. This cross-linking stabilizes the micelle coating. These procedures can be repeated to achieve the desired thickness, which can be controlled by monitoring the polymer membrane with confocal laser scanning microscopy (Fig. 12.8b). Thus, it was possible to encapsulate the islet within a thin membrane.

Although this membrane is stabilized by random thiol-maleimide cross-linking, many functional groups remain available for further modification with substances that carry SH or maleimide. When Alexa488-labeled apyrase-SH was used for this procedure, the polymer membrane was fluorescently labeled as apyrase was immobilized.

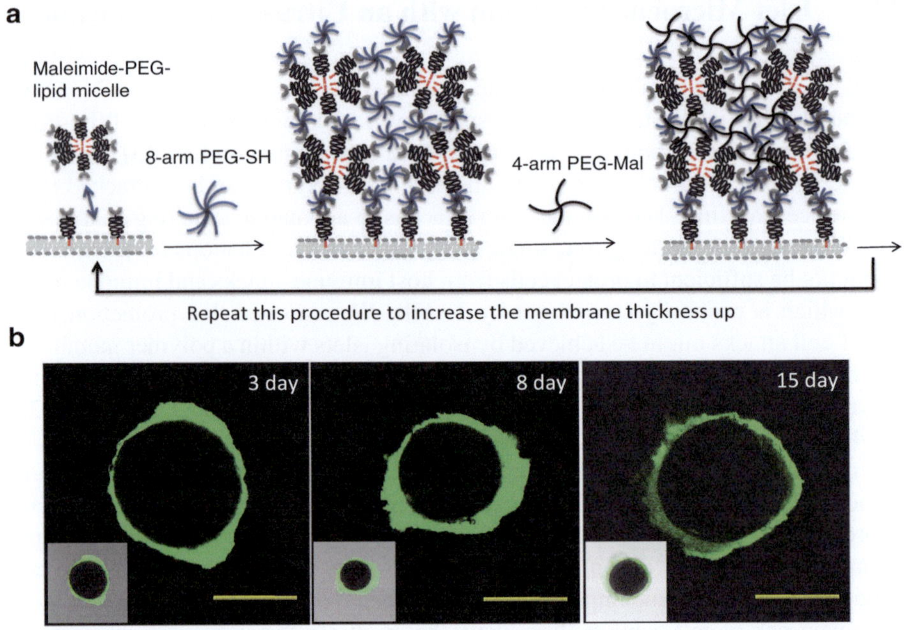

Fig. 12.8 Microencapsulation of pancreatic islets with a ultra-thin membrane. (**a**) The polymer membrane was formed by first allowing Mal-PEG-lipid to incorporate into the membrane; then, 8-arm PEG-SH and micelles of Mal-PEG-lipid are added to increase membrane thickness. Finally, 4-arm PEG-Mal is added to cross-link the polymer layer, which stabilizes the polymer membrane. This procedure is repeated to build up a capsule of micrometer thickness. (**b**) Confocal laser scanning microscope images show that islets remained stably encapsulated at 3, 8, and 15 days after microencapsulation. The membrane was labeled with Alexa488-apyrase-SH (*green*). (Partially modified from [47])

The role of apyrase was described in the previous section. The membrane thickness was estimated to be several micrometers in Fig. 12.8b. The islet surface can be completely covered with this thin membrane, which can protect the islet from immune attack.

Our method can be applied to various sizes of cells; e.g., erythrocytes and PAECs. Polymer membrane formation is mainly based on the hydrophobic interaction between PEG-lipids and the cell surface and the molecules that interact with the PEG-lipid; therefore, cell size does not affect membrane formation. The membrane thickness was achieved with the PEG-lipid micelles (size, 27 ± 0.2 nm), which were covalently cross-linked with branched polymers (4-arm PEG-Mal and 8-arm PEG-SH) attached to the cell surface. This membrane was substantially thicker than those formed with conventional methods that rely on layers of interactive polymers.

12.8 Islet Microencapsulation with Living Cells

In this section, we introduce the method of islet microencapsulation with living cells. This approach is based on the notion that transplanted islets could be made significantly more compatible with host blood and immune responses by covering the islet surface with recipient cells. In particular, IBMIR destruction and graft rejection could be ameliorated by coating the transplant with recipient vascular endothelial cells. This technique requires attaching recipient cells to donor cells. Here, we used a complementary pair of ssDNA-PEG-lipids (dT20-PEG-lipid and dA20-PEG-lipid) to modify the surfaces of cells to induce the cell-cell attachment [48, 49] (Fig. 12.9a). Accordingly, dT20 was incorporated onto the surfaces of single recipient cells with dT20-PEG-lipid, and dA20 was incorporated onto the surfaces of donor islets with dA20-PEG-lipid. When the dA20-PEG-lipid-modified islets were mixed with the dT20-PEG-lipid-treated cells, the hybridization of the dA20 and dT20 molecules immobilized the cells onto the islet surface (Fig. 12.9b). We used the human endoderm kidney cell line (HEK293) for immobilization onto isolated islets. When we cultured these HEK293 cell-immobilized islets, the HEK293 cells proliferated on the islet surface without detaching from the surface. After 3 days in culture, the whole islet surface was fully covered with a layer of HEK293 cells (Fig. 12.9b). Moreover, we observed no central necrosis inside the islet surrounded with cells. These islets were also analyzed by immunostaining. After 3 days in culture, we sliced islets into thin sections, and stained them with an anti-insulin antibody (Fig. 12.9b). We found that insulin was clearly observed in the islets. In addition, the insulin response was not altered by the cell encapsulation; changing the glucose concentration in the medium could stimulate insulin release. However, insulin secretion was reduced compared to that observed with control islets. Although HEK293 cells are a cell line, not primary cells, this study showed that it was possible to encapsulate islets inside living cells. The method proposed here may lead to a clinical procedure for encapsulating donor islets with cells derived from recipient patients with type 1 diabetes.

12.9 Conclusions and Outlook

We have developed techniques for protecting transplanted cells and islets with surface modifications and microencapsulation with PEG-lipid derivatives, bioactive substances, and living cells. Coating cells with bioactive substances, such as urokinase, apyrase, sCR1, and factor H, suppressed activation of blood coagulation cascades and the complement system. In the near future, these techniques will be implemented to inhibit IBMIR-mediated early islet loss during intraportal transplantation in the clinical setting. These surface modifications persist for several days after the islets are infused into the liver, which should be sufficient for protection from early graft loss.

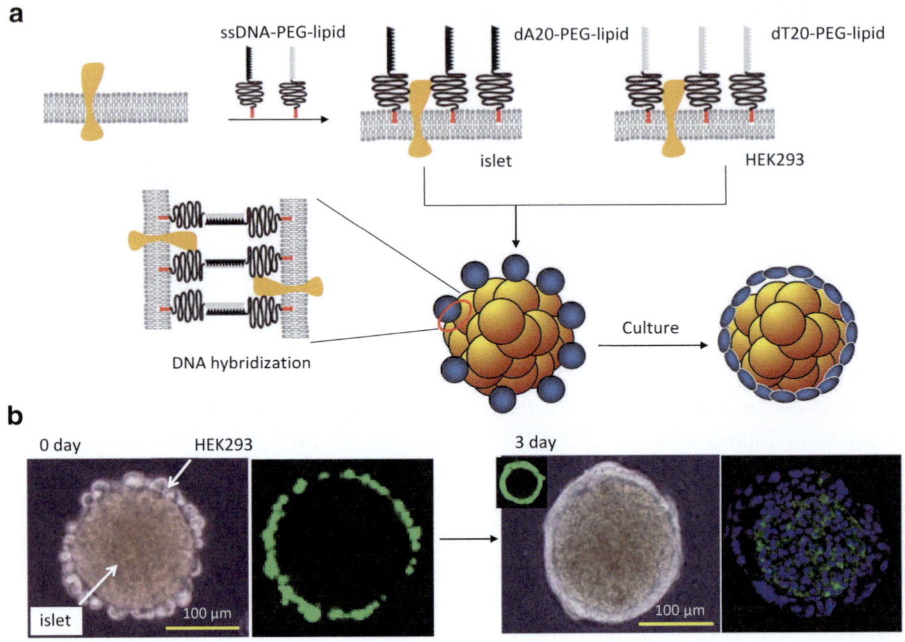

Fig. 12.9 Microencapsulation of a pancreatic islet with living cells via DNA hybridization. (**a**) Schematic illustration of the microencapsulation of an islet within living cells. Both the cell and islet surfaces are modified with polyDNA. dT20-PEG-lipid was immobilized on HEK293 cells (expressing GFP). dA20-PEG-lipid was immobilized on the surface of islets. When the modified cells and islets are mixed, DNA hybridization causes a shell of HEK293 cells (*blue*) to form on the islet surfaces (*orange*). After several days of culture, HEK293 proliferation encloses the islet within a cellular capsule. (**b**) *Left*: Phase-contrast and confocal microscope images show islets with HEK293 cells (*green*) attached at 0 and 3 days. *Right*, *Phase-contrast image*: after 3 days in culture, islets are completely encapsulated with GFP-expressing HEK293 (*green* in the *inset*). *Right*, *fluorescence image*: Image of a sliced section of encapsulated islet stained with an anti-insulin antibody. *Green*: insulin, *blue*: nuclei. (Partially modified from [48])

Although a major obstacle to cell transplantation therapy is the shortage of human donors, the shortage may not limit therapy for type 1 diabetes, because in the near future, we may be able to prepare insulin-secreting β cells or tissues from embryonic stem (ES) cells or induced pluripotent stem (iPS) cells [50]. However, the problem of immune reactions against grafts remains to be overcome. Cells differentiated from ES cells should be protected from the recipient's immune system, because they are allogeneic. Cells prepared from a patient's own iPS cells are expected to survive without immunosuppressive therapy. Nonetheless, type 1 diabetes is an autoimmune disease; thus, β cells derived from the patient's iPS cells may be destroyed by recurrent autoimmune reactions that are not fully controlled, even with immunosuppressive drugs. However, iPS cells might also be available from a cell bank, and these would be allogeneic. The idea of a bioartificial pancreas is important, because both stem cell-derived β cells and donated islets require protection from immune attack when they are transplanted into humans. While currently realizable, surface modification techniques must be improved before they can promote long-term graft survival.

Acknowledgements Y.T. would like to thank Prof Hiroo Iwata, Kyoto University, for useful discussions. This work was supported in part by a Grant-in-Aid for Young Scientists (A) (No. 26702017) and a Grant-in-Aid for Scientific Research on Innovative Areas (No. 26106709) from the Ministry of Education, Culture, Sports, Science, and Technology (MEXT) of Japan; a research grant from the Mukai Science and Technology Foundation (2014); and a research grant from the Japan Diabetes Foundation (2013).

References

1. Tisch R, McDevitt H. Insulin-dependent diabetes mellitus. Cell. 1996;85:291–7.
2. Mathis D, Vence L, Benoist C. Beta-cell death during progression to diabetes. Nature. 2001;414:792–8.
3. Ballinge WF, Lacy PE. Transplantation of intact pancreatic-islets in rats. Surgery. 1972;72:175–86.
4. Ryan EA, Lakey JR, Rajotte RV, Korbutt GS, Kin T, Imes S, et al. Clinical outcomes and insulin secretion after islet transplantation with the Edmonton protocol. Diabetes. 2001;50:710–9.
5. Shapiro AM, Lakey JR, Ryan EA, Korbutt GS, Toth E, Warnock GL, et al. Islet transplantation in seven patients with type 1 diabetes mellitus using a glucocorticoid-free immunosuppressive regimen. N Engl J Med. 2000;343:230–8.
6. Bennet W, Sundberg B, Groth CG, Brendel MD, Brandhorst D, Brandhorst H, et al. Incompatibility between human blood and isolated islets of Langerhans: a finding with implications for clinical intraportal islet transplantation? Diabetes. 1999;48:1907–14.
7. Bennet W, Sundberg B, Lundgren T, Tibell A, Groth CG, Richards A, et al. Damage to porcine islets of Langerhans after exposure to human blood in vitro, or after intraportal transplantation to cynomologus monkeys: protective effects of sCR1 and heparin. Transplantation. 2000;69:711–9.
8. Moberg L, Johansson H, Lukinius A, Berne C, Foss A, Kallen R, et al. Production of tissue factor by pancreatic islet cells as a trigger of detrimental thrombotic reactions in clinical islet transplantation. Lancet. 2002;360:2039–45.
9. Ozmen L, Ekdahl KN, Elgue G, Larsson R, Korsgren O, Nilsson B. Inhibition of thrombin abrogates the instant blood-mediated inflammatory reaction triggered by isolated human islets: possible application of the thrombin inhibitor melagatran in clinical islet transplantation. Diabetes. 2002;51:1779–84.
10. Goto M, Johansson H, Maeda A, Elgue G, Korsgren O, Nilsson B. Low molecular weight dextran sulfate prevents the instant blood-mediated inflammatory reaction induced by adult porcine islets. Transplantation. 2004;77:741–7.
11. Goto M, Tjernberg J, Dufrane D, Elgue G, Brandhorst D, Ekdahl KN, et al. Dissecting the instant blood-mediated inflammatory reaction in islet xenotransplantation. Xenotransplantation. 2008;15:225–34.
12. Kang HJ, Lee H, Ha JM, Lee JI, Shin JS, Kim KY, et al. The role of the alternative complement pathway in early graft loss after intraportal porcine islet xenotransplantation. Transplantation. 2014;97:999–1008.
13. Nilsson B, Ekdahl KN, Korsgren O. Control of instant blood-mediated inflammatory reaction to improve islets of Langerhans engraftment. Curr Opin Organ Transplant. 2011;16:620–6.
14. Naziruddin B, Iwahashi S, Kanak MA, Takita M, Itoh T, Levy MF. Evidence for instant blood-mediated inflammatory reaction in clinical autologous islet transplantation. Am J Transplant. 2014;14:428–37.

15. Chen H, Teramura Y, Iwata H. Co-immobilization of urokinase and thrombomodulin on islet surfaces by poly(ethylene glycol)-conjugated phospholipid. J Control Release. 2011;150: 229–34.
16. Chen H, Teramura Y, Iwata H. Immobilization of anticoagulant-loaded liposomes on cell surfaces by DNA hybridization. Biomaterials. 2011;32:7971–7.
17. Luan NM, Teramura Y, Iwata H. Immobilization of soluble complement receptor 1 on islets. Biomaterials. 2011;32:4539–45.
18. Luan NM, Teramura Y, Iwata H. Layer-by-layer co-immobilization of soluble complement receptor 1 and heparin on islets. Biomaterials. 2011;32:6487–92.
19. Takemoto N, Teramura Y, Iwata H. Islet surface modification with urokinase through DNA hybridization. Bioconjug Chem. 2011;22:673–8.
20. Teramura Y, Iwata H. Islets surface modification prevents blood-mediated inflammatory responses. Bioconjug Chem. 2008;19:1389–95.
21. Teramura Y, Iwata H. Improvement of graft survival by surface modification with poly(ethylene glycol)-lipid and urokinase in intraportal islet transplantation. Transplantation. 2011;91:271–8.
22. Teramura Y, Iwata H. Bioartificial pancreas microencapsulation and conformal coating of islet of Langerhans. Adv Drug Deliv Rev. 2010;62:827–40.
23. Teramura Y, Iwata H. Cell surface modification with polymers for biomedical studies. Soft Matter. 2010;6:1081–91.
24. Cabric S, Sanchez J, Lundgren T, Foss A, Felldin M, Kallen R, et al. Islet surface heparinization prevents the instant blood-mediated inflammatory reaction in islet transplantation. Diabetes. 2007;56:2008–15.
25. Contreras JL, Eckstein C, Smyth CA, Bilbao G, Vilatoba M, Ringland SE, et al. Activated protein C preserves functional islet mass after intraportal transplantation: a novel link between endothelial cell activation, thrombosis, inflammation, and islet cell death. Diabetes. 2004;53:2804–14.
26. Hashemi-Najafabadi S, Vasheghani-Farahani E, Shojaosadati SA, Rasaee MJ, Armstrong JK, Moin M, et al. A method to optimize PEG-coating of red blood cells. Bioconjug Chem. 2006;17:1288–93.
27. Lee DY, Lee S, Nam JH, Byun Y. Minimization of immunosuppressive therapy after islet transplantation: combined action of heme oxygenase-1 and PEGylation to islet. Am J Transplant. 2006;6:1820–8.
28. Nacharaju P, Boctor FN, Manjula BN, Acharya SA. Surface decoration of red blood cells with maleimidophenyl-polyethylene glycol facilitated by thiolation with iminothiolane: an approach to mask A, B, and D antigens to generate universal red blood cells. Transfusion. 2005;45:374–83.
29. Scott MD, Murad KL, Koumpouras F, Talbot M, Eaton JW. Chemical camouflage of antigenic determinants: stealth erythrocytes. Proc Natl Acad Sci U S A. 1997;94:7566–71.
30. Stabler CL, Sun XL, Cui W, Wilson JT, Haller CA, Chaikof EL. Surface re-engineering of pancreatic islets with recombinant azido-thrombomodulin. Bioconjug Chem. 2007;18: 1713–5.
31. Elbert DL, Herbert CB, Hubbell JA. Thin polymer layers formed by polyelectrolyte multilayer techniques on biological surfaces. Langmuir. 1999;15:5355–62.
32. Germain M, Balaguer P, Nicolas JC, Lopez F, Esteve JP, Sukhorukov GB, et al. Protection of mammalian cell used in biosensors by coating with a polyelectrolyte shell. Biosens Bioelectron. 2006;21:1566–73.
33. Krol S, del Guerra S, Grupillo M, Diaspro A, Gliozzi A, Marchetti P. Multilayer nanoencapsulation. New approach for immune protection of human pancreatic islets. Nano Lett. 2006;6:1933–9.
34. Veerabadran NG, Goli PL, Stewart-Clark SS, Lvov YM, Mills DK. Nanoencapsulation of stem cells within polyelectrolyte multilayer shells. Macromol Biosci. 2007;7:877–82.
35. Inui O, Teramura Y, Iwata H. Retention dynamics of amphiphilic polymers PEG-lipids and PVA-alkyl on the cell surface. ACS Appl Mater Interfaces. 2010;2:1514–20.

36. Krych-Goldberg M, Atkinson JP. Structure-function relationships of complement receptor type 1. Immunol Rev. 2001;180:112–22.
37. Weisman HF, Bartow T, Leppo MK, Marsh Jr HC, Carson GR, Concino MF, et al. Soluble human complement receptor type 1: in vivo inhibitor of complement suppressing post-ischemic myocardial inflammation and necrosis. Science. 1990;249:146–51.
38. Auchincloss Jr H, Sachs DH. Xenogeneic transplantation. Annu Rev Immunol. 1998; 16:433–70.
39. Hessing M, Vlooswijk RA, Hackeng TM, Kanters D, Bouma BN. The localization of heparin-binding fragments on human C4b-binding protein. J Immunol. 1990;144:204–8.
40. Schmidt CQ, Herbert AP, Hocking HG, Uhrin D, Barlow PN. Translational mini-review series on complement factor H: structural and functional correlations for factor H. Clin Exp Immunol. 2008;151:14–24.
41. Wu YQ, Qu H, Sfyroera G, Tzekou A, Kay BK, Nilsson B, et al. Protection of nonself surfaces from complement attack by factor H-binding peptides: implications for therapeutic medicine. J Immunol. 2011;186:4269–77.
42. Nilsson PH, Ekdahl KN, Magnusson PU, Qu HC, Iwata H, Ricklin D, et al. Autoregulation of thromboinflammation on biomaterial surfaces by a multicomponent therapeutic coating. Biomaterials. 2013;34:985–94.
43. Teramura Y, Iwata H. Surface modification of islets with PEG-lipid for improvement of graft survival in intraportal transplantation. Transplantation. 2009;88:624–30.
44. Teramura Y, Kaneda Y, Iwata H. Islet-encapsulation in ultra-thin layer-by-layer membranes of poly(vinyl alcohol) anchored to poly(ethylene glycol)-lipids in the cell membrane. Biomaterials. 2007;28:4818–25.
45. Kozlovskaya V, Zavgorodnya O, Chen Y, Ellis K, Tse HM, Cui W, et al. Ultrathin polymeric coatings based on hydrogen-bonded polyphenol for protection of pancreatic islet cells. Adv Funct Mater. 2012;22:3389–98.
46. Zhi ZL, Kerby A, King AJ, Jones PM, Pickup JC. Nano-scale encapsulation enhances allograft survival and function of islets transplanted in a mouse model of diabetes. Diabetologia. 2012;55:1081–90.
47. Teramura Y, Oommen OP, Olerud J, Hilborn J, Nilsson B. Microencapsulation of cells, including islets, within stable ultra-thin membranes of maleimide-conjugated PEG-lipid with multifunctional crosslinkers. Biomaterials. 2013;34:2683–93.
48. Teramura Y, Minh LN, Kawamoto T, Iwata H. Microencapsulation of islets with living cells using polyDNA-PEG-lipid conjugate. Bioconjug Chem. 2010;21:792–6.
49. Teramura Y, Chen H, Kawamoto T, Iwata H. Control of cell attachment through polyDNA hybridization. Biomaterials. 2010;31:2229–35.
50. Takahashi K, Tanabe K, Ohnuki M, Narita M, Ichisaka T, Tomoda K, et al. Induction of pluripotent stem cells from adult human fibroblasts by defined factors. Cell. 2007;131:861–72.

Chapter 13
Complement Interception Across Humoral Incompatibility in Solid Organ Transplantation: A Clinical Perspective

Ali-Reza Biglarnia, Kristina N. Ekdahl, and Bo Nilsson

Abstract The humoral barrier in transplant biology is the result of preformed donor-specific antibodies (DSAs), directed either against human leukocyte antigens (HLA) or non-HLA antigens such as blood group (ABO) molecules. The term "sensitization" applies to patients carrying these antibodies. Transplantation is widely accepted as a life-saving opportunity for patients with terminal end-organ disease. However, in sensitized patients, transplant outcome is hampered by antibody-mediated rejection (AMR) as a consequence of DSA exposure. Furthermore, sensitized patients have limited access to "matched" organs from the both living and deceased donor pool.

Considering the crucial role of the complement system in the pathophysiology of AMR and the availability of complement intervention therapeutics, there is a growing interest in complement-targeting strategies. This review highlights the emerging importance of monitoring and modulation of the complement system in the context of enabling transplantation across humoral incompatibility in sensitized recipients with preformed anti-HLA or natural anti-ABO antibodies. It also discusses the significance of the complement system in the induction of accommodation and further emphasizes current and future perspectives of novel complement therapeutics.

Keywords ABO- and HLA-incompatibility • Antibody-mediated rejection • Terminal complement inhibition • Transplantation • Complement therapeutics • Desensitization

---

A.-R. Biglarnia (✉)
Surgical Science, Department of Transplantation, Uppsala University Hospital, Uppsala University, Uppsala, Sweden
e-mail: alireza.biglarnia@surgsci.uu.se

K.N. Ekdahl • B. Nilsson
Immunology, Genetics and Pathology, University of Uppsala, Uppsala, Sweden

J.D. Lambris et al. (eds.), *Immune Responses to Biosurfaces*, Advances in Experimental Medicine and Biology 865, DOI 10.1007/978-3-319-18603-0_13

13.1 Introduction and Historical Overview

The history of transplantation has consistently been shaped by the need to overcome surgical and immunological barriers. Towards the beginning of the twentieth century, the era of solid organ transplantation began with the pioneer working of Emerich Ullman (1861–1937) who was the first to perform a successful kidney auto-transplantation in a dog. Encouraged by this success, he soon thereafter performed the first kidney transplantation (xenogeneic) from a pig donor to a female human patient who suffered from end-stage renal disease [1]. This surgical milestone was, however, overshadowed by a fatal biological reaction that decades later was identified as the immunological response to non-self antigen, an antibody-mediated rejection (AMR).

The modern era of solid organ transplantation started in Boston at Brigham Hospital on December 23, 1954 when Joseph Murray and colleagues performed a successful live donor kidney transplant with excellent long-term results. Unlike their predecessors [2], the Boston group achieved success because the transplant was performed in monocygotic identical twins, which enabled them to circumvent the immunological barrier, rather than challenge it [3]. However, given the shortage of genetically identical donors and the urgent need for the expansion of transplantation between genetically disparate individuals, the development of immune modulation strategies quickly became a necessity. Initial immunosuppressive protocols included sublethal total body irradiation for the induction of bone marrow aplasia and lymphocyte depletion [4]. This treatment modality, however, was not only ineffective in preventing destructive allogeneic immune reaction but also harbored a life-threatening risk of severe infectious complications.

The era of immunopharmacologic interventions began with the clinical use of 6-mercaptopurine and its pro-drug azathioprine in combination with glucocorticosteroids [5, 6] and it further evolved with the discovery of calcineurin inhibitors in the 1980s; these inhibitors are still the cornerstone of the current immunosuppressive treatment after organ transplantation [7–10].

Thanks to the evolution of immunopharmacology, transplantation has been transformed from a surgical adventure into a life-saving clinical practice that is broadly implemented [11]. However, this progress is also associated with challenges. In the United States and Europe, the number of patients on the waiting list with end-stage organ disease is increasing [12, 13]. Also, the total number of donors (deceased and living) is still insufficient to meet the increasing need for transplant candidates. According to the Organ Procurement & Transplanatation Network (OPTN) database, there are currently more than 123,000 transplant candidates on the waiting list in the United States, as opposed to an annual number of transplants ranging from 24,000 to 29,000 over the past 10 years (OPTN data as of January 30, 2015).

The widening gap between supply and demand not only drives the surgical development toward maximal utilization [14] and an expansion of the donor pool [15, 16] but also assumes the use of strategies to enable transplantation between individuals with high-immunological risk profiles. These strategies include treatment protocols to increase the likelihood that sensitized patients can successfully receive "human leukocyte antigen" (HLA-) or ABO-mismatched live- or deceased-donor organs.

For many years, the adaptive immunity, and particularly T-cells has been considered the central regulatory and effector hub in alloreative immune response [17]; this view has strongly shaped the development of immunopharamacologic interventions. However, critical events in organ transplantation, including brain death, ischemic-reperfusion injury, and acute and chronic immune responses still represent a clinical burden with high risk for graft failure. A recent 20-years survey from more than 250,000 renal transplant patients in the United States has shown that the progress in transplant outcome has mainly been made in improving short-term results, whereas the attrition rate of graft loss in the long term has remained essentially constant [18]. Undoubtedly, these data indicate that the traditional immunosuppressive strategies used so far have been successful in reversing the early acute immune response, but they have failed to prevent sustained graft damage. Thus, the former paradigm of T-cell dominance in allogeneic immunity has increasingly been questioned as to not entirely representing the complex interaction of events from donation to transplantation and beyond. So, what is missing?

Innate immunity is a non-adaptive and evolutionarily preserved arm of the immune system that distinguishes self from non-self structures in the body. It represents the first line of defense in the presence of noxious stimuli such as invading pathogens or physical or metabolic insults [19]. For many years, innate immunity was considered less relevant to transplant biology, and interest in developing pharmaceutical therapies targeting this ancient part of the immune system was restricted. An increasing appreciation for the innate immunity, however, has emerged from several important clinical observations. Kidney transplant candidates have the option to receive organs from either live or deceased donors. Transplantation involving living donors is highly appreciated not only because of the elective nature of the procedure, but also because organs from live donors are not exposed to critical events such as brain-death or prolonged cold preservation which are invariably the case in deceased-donor transplantation. When outcomes are compared between living-unrelated HLA-mismatched transplants and HLA-matched deceased-donor kidney transplants, it is intriguing to find that "unmatched" live-donor organs have the same or even better survival than do their "matched" deceased-donor counterparts [20]. A recent meta-analysis further revealed that the emergence of delayed kidney-graft function, mostly as a consequence of ischemic-reperfusion injury, is associated with a 41 % increased risk of graft loss in the long-term [21]. These observations have for many years suggested the existence of another non-antigen specific immune reaction, which appears to have a devastating effect on transplant outcome.

Today, mounting evidence has revealed an integral role for innate immunity in the non-specific and specific inflammatory reactions that are generated during critical transplant-related events, such as brain death, surgical trauma, organ preservation, ischemic-reperfusion injury, and humoral and cellular immune responses [19, 22–25]. In this review we will focus on antibody-mediated rejection (AMR) and role that complement plays in this adverse reaction.

13.2 The Complement System in Transplant Biology

13.2.1 Activation Mechanisms

The proteins, approximately 50 in number, that form the complement system are found either in body fluids in the form of zymogen molecules or on cellular surfaces as receptors and/or regulators. Activation of the system occurs via three pathways utilizing different recognition molecules that bind to structures on target surfaces, thereby generating two different proteolytic enzyme complexes, the classical/lectin pathway (C4b2a) and the alternative pathway (C3bBb) C3 convertases, which cleaves C3 into C3b, and the anaphylatoxin C3a. The classical pathway (CP) is triggered by antibodies, primarily IgM, IgG_1, and IgG_3, that are either present in antigen-antibody complexes or bound to a target surface. Such antibodies bind the C1 complex (C1q, $C1r_2$, and $C1s_2$) via its C1q moiety. Recent data show that IgG bound to a cellular surface forms hexamers by noncovalent interactions between their Fc segments, resulting in an avidity for C1q that is high enough to accomplish subsequent complement activation [26].

Similarly, the lectin pathway (LP) is initiated by the binding of mannan-binding lectin (MBL) or ficolins, thereby activating three different MBL-associated serine proteases (MASPs). These events all result in the assembly of the CP/LP C3-convertase. The alternative pathway (AP) may be triggered by various surfaces of biological or non-biological origin, e.g., microorganisms or biomaterials, which do not fully regulate the AP C3-convertase. Furthermore, regardless of which pathway is the initial trigger of the activation, the AP provides a powerful amplification loop, since C3b is a subunit, and therefore every deposited C3b molecule is a potential nucleus of a novel AP C3-convertase complex.

At this point, the pathways converge into a common terminal pathway (TP) in which the first step is the proteolytic activation of C5 into the anaphylatoxin C5a and the C5b fragment, which in turn represents the first component of the membrane attack complex (MAC). In its complete form, the MAC consists of C5b, C6, C7, C8, and multiple copies of C9, all of which are inserted into the cell membrane. A current summary of our understanding of the complement system is given in [27].

13.2.2 Effector Mechanisms

The generated anaphylatoxins C3a and C5a mediate the recruitment (via chemotaxis) and activation of polymorphnuclear leukocytes (PMNs) and monocytes by binding to C3aR and C5aR1 on these cells. The binding of the cells to the target is mediated by various receptors (primarily CR1 (CD35), CR3 (CD11b/CD18) and CR4 (CD11c/CD18)) via the target-bound C3 fragments C3b and iC3b, thereby mediating phagocytosis and cytotoxicity. In addition, the formation of the MAC can cause cell lysis, but cell damage as well as inflammasome activation may also occur already at sublytic concentrations of MAC [28, 29]. It should be noted that complement activation of the TP not only generates cell-associated MAC but also soluble, incomplete complexes containing only one copy of C9 (sC5b-9), which remain in the in the blood plasma.

13.2.3 Regulation of Complement Activation

Complement activation on autologous cells is controlled by regulators of complement activation (RCA), which are operative mainly at the C3 convertase level. These proteins include complement receptor 1 (CR1 [CD35]), membrane cofactor protein (MCP [CD46]), and decay acceleration factor (DAF [CD55]), which all act by disrupting the convertase and/or functioning as co-factors for the plasma protease factor I. The action of these cell-bound inhibitors is further topped up by factor H and C4 binding protein (C4BP), which regulate the convertases of the AP and CP/LP, respectively. These proteins are pulled down from the plasma as a result of their inherent affinity for the proteoglycans present on cellular surfaces. In addition, cell-bound CD59 and fluid-phase clusterin both inhibit the assembly of the MAC. C1 inhibitor (C1-INH) controls the activity of C1r and C1s of the C1 complex as well as the MASPs (-1, -2, and -3) of the LP [30].

13.2.4 Therapeutic Complement Inhibition

The clinical development of complement inhibitors has proved challenging [31, 32], and at present, there are only two drugs available in the clinic. Preparations of the regulator C1-INH (e.g., Cinryze, ViroPharma) block both the CP and the LP, but they are not complement-specific and also inhibit serine proteases of the coagulation and contact activation/kinin systems. The humanized anti-C5 mAb eculizumab (Soliris, Alexion), in contrast, blocks the TP by preventing the cleavage of C5 into the anaphylatoxin C5a and C5b, which initiates the formation of the MAC.

13.2.5 Assays to Monitor Complement Activation

The subject of complement diagnostics is reviewed in detail in [33, 34]. Quantification of individual complement proteins (e.g., C1q, C1-INH, MBL, C4, C3, factor B) is in general performed using specific immunoassays. During humoral rejection of a transplanted organ induced by the presence of DSAs, activation of the CP leads to the generation of complement activation products in the fluid phase and a concomitant consumption of the corresponding precursor molecules. This sequence of events can be monitored by measuring the levels of C3a by enzyme immunoassay

(EIA) or C3dg by nefelometry.. In order to compensate for varying concentrations of native C3, determination of the total amount of C3 should also be performed, and the ratio of C3a/total C3 or C3dg/total C3 should be calculated as a true measure of C3 activation.

C5a and sC5b-9 complexes can be readily detected in the fluid phase by EIA. However, the interpretation of the data for C5a (as compared to C3a) is difficult, since a substantial amount of the generated C5a will inevitably bind to leukocytes because these cells express much higher amounts of C5aR1 than C3aR [35].

Consumption of complement components within the CP and TP leads to a depression in hemolytic activity. This loss of function can be quantified, either by hemolytic assays specific for the CP using sheep erythrocytes sensitized with IgM antibodies (e.g. conveniently in a single tube CH-50 like analysis) [36], or by an EIA (Wielisa, Wieslab) that allows simultaneous determination of all three activation pathways [37].

Finally, histological evaluation of biopsies using antibodies against the C4 split product C4d is routinely used as a measure of complement activation [38]. However, caution should be observed when interpreting these data, since it is always possible that the detected C4 or C4 fragments are generated by the transplanted organ itself, in response to cytokine generation, and not deposited as a consequence of complement activation [39].

13.2.6 Local Production of Complement in the Kidney

Humoral allo-response, as observed in ABO-incompatible (ABOi) and HLA-sensitized transplant patients, are particularly of concern because they result in serious hyperacute/acute AMR and a high rate of graft loss [40]. Complement has long been implicated in the outcome of AMR, and the activation product C4d is currently being used as a biomarker of this condition [41].

An important study has shown that the C3 synthesis in the transplanted kidney is increased and is dependent on the duration of the cold ischemia [42]. C3 production by the donor kidney reached a peak after reperfusion, indicating the potential for using complement inhibition during organ collection as means of decreasing tissue damage and improving the transplant outcome.

13.2.7 Anti-ABO and Anti-HLA Antibodies

Major immunological obstacles in allogeneic transplantation are the ABO and the HLA barriers. Humans have so-called "natural antibodies" against the A and/or B carbohydrate antigens, antigens that they do not express. These antibodies are produced in a T-cell independent manner, without previous immunization, and are thus predominately of the IgM type; in addition, natural antibodies of the IgA and IgG_3 isotypes have been reported [43]. In contrast, individuals who are exposed to allogeneic HLA may become immunized, leading to the production, under the influence of T-cells, of IgG antibodies directed against the alien HLA. Such immunization can occur as a response to HLA-mismatched transplants, in multiparous women, or in patients who have received multiple leukocyte-containing blood transfusions [44].

13.3 The Clinical Burden of Pre-sensitization

The risk of hyperacute rejection is high in the presence of naturally occurring blood group antibodies when the ABO blood group barrier is being crossed [45]. Consequently, pre-transplant blood group typing of the donor and recipient is imperative to prevent accidental ABO-antibody mediated hyperacute rejection.

The term “sensitization” refers to the formation of anti-HLA antibodies prior to transplantation. When they receive a transplant, these sensitized patients have an increased risk of AMR and graft loss. In 1969, the concept of pre-transplant complement-dependent cytotoxic (CDC) cross-matching was introduced by Patel and Terasaki [46]. For the first time, the introduction of CDC testing enabled the identification of pre-formed cytotoxic antibodies that could be linked to hyperacute rejection and immediate graft failure [47, 48]. Despite the implementation of more advanced assays, such as Luminex for flow cytometric crossmatching (FCXM), CDC is still the golden standard that has dramatically decreased the risk of hyperacute rejection as a consequence of pre-formed HLA-linked cytotoxic antibodies.

In an ideal clinical situation, avoidance of humoral incompatibility can provide excellent short- and long-term results [49]. However, the current clinical landscape is far away from ideal, since the sensitization phenomenon has become a public health obstacle. In the United States, 30 % of patients awaiting kidney transplantation are sensitized (OPTN data as of May, 2014). Within the same population, the annual transplant rate among patients with high anti-HLA antibody burden (PRA > 80 %) has been reported to be as low as 6.8 % [50], indicating the low probability for “matched” transplantation and ultimately longer time on the waiting list which are directly associated with increased risk for mortality [51, 52]. The same picture is evident for the Scandinavian and other European populations. According to the Scandia transplant database, 1,362 patients are at present on the transplant waiting list. Of these, 28 % have anti-HLA antibodies, and 44 % of them are highly immunized (PRA > 80 %) (http://www.scandiatransplant.org). An analysis of the UK’s national kidney transplant waiting list in March 2009 revealed that 41 % of adult patients and 58 % of pediatric patients were sensitized (www.bts.org.uk).

Naturally occurring anti-ABO antibodies create similar disparities, and the time to transplant varies substantially between different blood groups [53]. Patients with willing living-donors share the same fate when either ABO- or HLA incompatibility is encountered.

13.4 Desensitization for ABO-Incompatible Transplantation

13.4.1 ABO-Incompatibility

Within the general population, there is a 35 % probability for two random individuals to be ABO-incompatible (ABOi). This statistic also implies that transplant candidates have the same likelihood of having willing but ABOi living donors. Given the current severe shortage in organ availability, treatment strategies to overcome the ABO barrier have become attractive options for maximizing the donor pool. Strategies enabling ABOi transplantation are based on the following fundamental principles: (1) removal of anti-A and/or anti-B antibodies by plasma exchange or immunoadsorption, (2) maintenance immunosuppression, (3) immunomodulation with intravenous immunoglobulins (IVIG), and (4) reduction of the B-cell reservoir by splenectomy or more recently, with anti-CD20 monoclonal antibody rituximab.

Traditionally, ABO-incompatible transplantations were effectively possible with a combination of intense immunosuppression, frequent plasma exchange, and splenectomy. However, excessive immunosuppression, combined with increased surgical trauma, were deterrents to the broader acceptance of this procedure. In 2003, Gunnar Tydén presented the Swedish protocol that replaced splenectomy with a single dose of rituximab combined with standard CNI-based immunosuppression. Furthermore, the protocol implemented antigen-specific immunoadsorption (GlycosorbABO; Glycorex, Lund, Sweden), rather than plasma exchange, using specific blood group antigens A (GalNAcα1–3(Fucα1–2)Gal and B (Galα1–3(Fucα1–2)Gal covalently bound to a Sepharose matrix [54, 55]. Combining efficacy with a reduction in immunosuppression and surgical trauma, this protocol dramatically increased acceptance of a broader implementation of ABOi transplantations [56, 57]. Today, ABOi transplantations are successfully performed on a global scale, with outcomes comparable to those of ABO-compatible transplantations [58, 59]. One key point in this success is a phenomenon known as accommodation.

13.4.2 The Concept of Accommodation

The initial target in acute vascular rejection is the vascular endothelium, ultimately followed by microvascular inflammation and thrombosis, then ischemia, apoptosis, or necrosis, and finally graft failure. Accommodation is defined as a condition in which the transplant elicits no rejection while maintaining normal function, despite the presence of anti-donor antibodies and fully functional effector complement components in the plasma [60].

In clinical setting, accommodation was first described in renal transplantation across the ABO barrier, when ABO-antibodies were temporarily eliminated prior to transplant [61, 62]. In order for accommodation to establish, either low anti-graft antibody level or lowered complement function seems to be required at the time of graft implantation. Accommodation is primarily induced in transplantations when the antigen elicits non-T cell dependent antibody responses. Typically, these responses are elicited by carbohydrate antigens such as the ABO and the Gal antigens. However, accommodation has also been reported for anti-HLA antibodies *in vitro* and in a presensitized allogeneic non-human primate transplantation model [63, 64]. The mechanism by which this non-responsive state is achieved is not fully understood. Several investigators [65] have reported an overexpression of the genes HO-1, A20, Bcl-2, and Bcl-x, which inhibit activation of the transcription factor NF-κB and thereby promote an anti-inflammatory state and downregulating proinflammatory cyto/chemokines. It has been reported that many of the processes involved in accommodation, including apoptosis triggered by TNF, and necrotic cell death caused by the MAC of complement, are prevented by pretreatment of the endothelial cells with the TH2 cytokines IL-4 or IL-13. This requires activation of phospholipid synthesis, in association with preservation of mitochondrial structure and function [66]. Also, up-regulation of complement regulators such as CD59 has been reported as an additional mechanism of accommodation [41, 67].

13.4.3 Complement Modulation in ABO-Incompatible Transplantation

In a survey of 19 consecutive patients undergoing desensitization for ABOi kidney transplantation, we previously demonstrated a 100 % patient and graft survival for a median follow-up of 40 months [68]. The treatment algorithm for this protocol is illustrated in Fig. 13.1.

For the same cohort, the median glomerular filtration rate at last follow-up was 82 mL/min, also indicating excellent long-term renal function. Given the ability of anti-ABO antibodies to induce a vicious humoral response, these results are somewhat surprising, even in the context of accommodation, which is presumably established within the first 2 weeks post-transplantation [69]. We previously assumed the presence of an earlier mechanism complementary to accommodation that could be important for promoting good outcomes after ABOi transplantation. During humoral rejection, the complement cascade is activated (through the CP) via the cleavage of C3 into C3b and C3a, finally resulting in the formation of the membrane attack complex C5b-9. This mechanism could be responsible for the speed and severity of the humoral rejection process in renal transplantation.

In order to evaluate the possibility of interference between desensitization and complement activation, we measured the serum levels of the complement products C1q, C3, C3a, and sC5b-9 from the start of preconditioning on day −30 to 1 month post-transplant in the aforementioned cohort. We found that antigen-specific immunoadsorption had an inhibitory effect on complement activation by depleting complement compounds. These results were further confirmed by our detection of C1q,

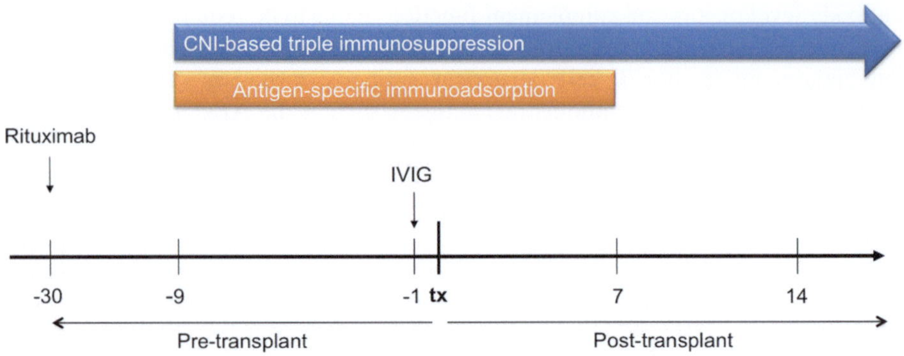

Fig. 13.1 Swedish desensitization protocol for ABOi living-donor kidney transplantation with center-specific adjustments. In the pre-transplant phase, patients are treated with a single dose of rituximab (day −30), triple-based maintenance immunosuppression (starting day −9), antigen-specific immunoadsorption (starting day −9, guided by isoagglutinin levels) and intravenous immunoglobulin (IVIG, day −1). Anti-IL-2 receptor antibody infusions are given on days 0 and 4. In the post-transplant period, immunoadsorption treatment is guided by the level of isoagglutinin titers. Immunoadsorption treatments are usually discontinued at 1 week post-transplantation

C3, and C3a in the eluates from used absorption columns. These findings were further confirmed by detection of C1q, C3 and C3a in the eluates from used immunoadsorbents [68].

A recent randomized clinical trial on 14 ABOi transplant recipients, confirmed our result by showing the same depleting effect of desensitization, including antigen-non-specific immunoadsorption on complement factors C3, C3a, and C1q [70]. In the line with these findings, it is conceivable that the complementary effects of complement depletion and antibody elimination are an important link in the success of transplantation across the humoral incompatibility.

A relevant question, however, is whether novel treatment strategies to specifically target complement activation could minimize the need for excessive antibody elimination and still provide comparable results. If so, what benefits would this treatment modification bring to clinical practice?

As mentioned above, desensitization protocols include serial apheresis sessions (immunoadsorption or plasma exchange) for elimination of anti-ABO antibodies in the pre-transplant period. The rationale for this time-consuming treatment is to create a short interval of time in which ABOi transplant can safely be performed with virtually no risk of severe humoral rejection. Hence, the feasibility of the protocol requires the availability of a living donor for elective planning of the transplant within this short "window of operability." It is therefore conceivable that patients with no access to living donors (such as those listed on transplant waiting lists) are considered ineligible for this successful treatment option. For these patients, this limitation entails disadvantages in terms of both donor availability and organ allocation, particularly for waitlisted patients with blood type O. It is generally known that blood type O individuals lack both A and B antigens. Therefore, blood type O donors are considered "universal donors," since their organs can be transplanted to A-, B-, and AB-recipients as well as O recipients. Most organ allocation policies prioritize optimal HLA matching, in particular for HLA-sensitized patients (e.g., the acceptable mismatch program in the Eurotransplant and Scandiatransplant areas) in order to maximize transplantation outcomes.

For this purpose, blood type O organs are being allocated to HLA-matched non-O transplant candidates (www.eurotransplant.nl). In a recent 12-year survey of more than 1,000 waitlisted transplant candidates within the Eurotransplant area, it was calculated that 14 % of all O organs have been allocated to non-O recipients. This significant "drainage" of O organs to non-O recipients was associated with increased accumulation of O patients on the waiting list, which further resulted in longer waiting time compared to non-O waitlisted transplant candidates. As a consequence, O patients not only had a higher waiting list mortality but also a concomitant inferior outcome when given a transplant [71].

Apart from the allocation issues with O recipients in general, the strict separation of the waiting list by blood type also causes local issues. At the Uppsala University Hospital in Sweden, efficient utilization of organs has also been hampered by the need for blood type compatibility between donor and recipient. This situation has been particularly apparent for pancreatic allografts, given the limited number of diabetic patients on the waiting list. In this respect, pancreatic allografts have been discarded in some cases because of a positive donor-specific cross-match in blood group-compatible recipients. On the other hand, crossmatch-negative recipients with an incompatible blood group have not been considered suitable candidates. As a consequence, pancreatic allografts have been underutilized.

In an attempt to find a solution to this clinical obstacle, we have assessed transplantation across the ABO barrier as a logical solution for optimal organ utilization. In deceased-donor transplantation, logistics during organ allocation and the legitimate demand for short ischemia time restrain the possibilities for time-consuming preconditioning, including antibody elimination. However, a simple abundance of excessive antibody elimination gives rise to serious concerns because the elevated level of anti-ABO antibody titer at the time of transplantation could induce severe AMR and immediate graft failure. Previously, we discussed the importance of antibody elimination, together with complement depletion, for success in ABOi transplantation. Given the pivotal role of complement in the pathophysiology of the AMR, novel strategies to specifically target complement activation could minimize the significance of preformed antibodies. Principally, this means that complement inhibition can increase the safe acceptance of higher antibody levels at the time of the ABOi transplantation.

Previously, we have introduced an "overnight" protocol including eculizumab and complement monitoring for patients undergoing ABOi deceased-donor transplantation [72] (Fig. 13.2).

To increase the safety margin in terms of preventing an early humoral response, we considered the following aspects to be important in the planning and design of the protocol: (1) Pre-transplant antibody elimination by a single plasma exchange combined with rituximab, (2) induction with thymoglobulin and maintenance of

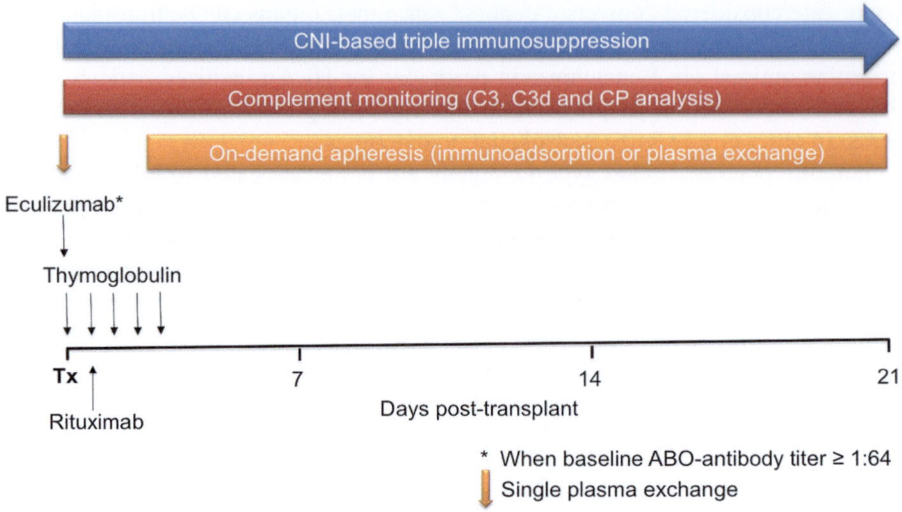

Fig. 13.2 Overnight protocol for ABOi deceased-donor transplantation. Treatment starts at the day of transplant (Tx) with single plasma exchange prior to surgery. In patients with high ABO-antibody titers (≥1:64), eculizumab is given just before the circulation of the graft is re-established. Immunosuppression consists of thymoglobulin induction followed by rituximab and CNI-based triple maintenance. Short term post-transplant aphereis with either antigen-specific immunoadsorption or plasma exchange is guided by the level of ABO antibody titers

triple immunosuppression, and (3) eculizumab induction for patients with high baseline anti-ABO antibody titers (≥1:64) or on-demand treatment basis in the case of humoral rejection. Furthermore, daily measurements of complement parameters C3 and C3d (for calculation of the C3dg/C3 ratio), as well as estimation of CP function, were implemented as a supplementary tool for the early diagnosis of complement-driven humoral rejection after the transplantation. The systematic analysis of complement activation and extensive surveillance revealed new insights into the potential and pitfalls of complement modulation using C5 blockade in "overnight" ABO-incompatible transplantation.

The first case involved a blood group type B donor who was allocated to a kidney-pancreas transplant recipient with blood type A1 and low isoagglutinin baseline titer. Despite all the measures taken to minimize the risk of humoral complications, a distinct biopsy-verified AMR was encountered on day 9, which was preceded by an increase in isoagglutinin. At the same time, monitoring of complement activation revealed a drop in CP function, and an increase in the C3d/C3 ratio, indicating significant complement consumption. The occurrence of humoral rejection during the rebound of isoagglutinins and the absence of anti-HLA antibodies prompted us to assume a rejection mediated by ABO reactivity. This rejection episode was accompanied by a rapid decline in kidney function and severe abdominal pain. The pivotal role of complement as the key inflammatory mediator during this particular type of humoral response was demonstrated by the observation that both the kidney function and the severe inflammatory response were rapidly normalized after a single administration of eculizumab. Considering our empirical knowledge of accommodation, short term protection from humoral damage by further suppression of complement function was favored. Here, CP function, was used as an analytical tool to monitor the therapeutic level of C5 blockade and guide the second pre-emptive administration of eculizumab, which was given after an interval of 4 days. Currently, after more than 1,200 days, the patient has both excellent kidney and pancreas graft function, with no further incidence of any type of rejection.

Encouraged by this successful case, we further introduced eculizumab and complement monitoring into the traditional Swedish desensitization protocol for a subgroup of high-risk patients with inferior response to antibody elimination and increased antibody presence at the time of living-donor ABOi transplantation. In this patient population, modification of the protocol to include eculizumab induction was considered important for prevention of early humoral complication (Fig. 13.3).

Overall, 18 patients were included in the ABO desensitization protocol for deceased-donor (n=8) and living-donor transplants (n=10). At a median follow-up of 538 days, the patient survival was 100 %. Three patients lost their grafts: In one ABOi kidney-pancreas transplant recipient (AB to O), the kidney was removed on day 270 because of a severe local fungal infection causing fungal septicemia. In this patient, the pancreas graft function was consistently normal after more than 1,000 days with no need for exogenous insulin. Two patiens developed de-novo anti-HLA DSAs and biopsy proven rejection episodes resulting in pancreas graft failure at days 240 and 940, respectively. At a median follow-up of more than 500 days, the median creatinine and HbA1c was normal for the remaining patients, indicating excellent intermediate long-term graft function.

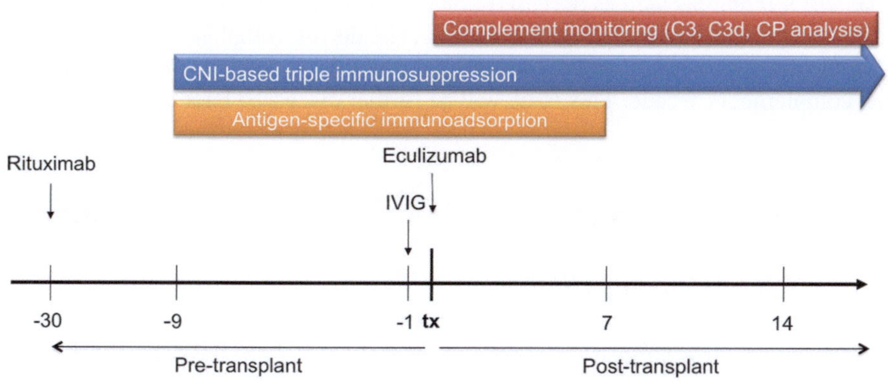

Fig. 13.3 Modified Swedish protocol for ABOi living-donor transplantation. Modification consists of inclusion of eculizumab induction and complement monitoring for high-risk ABOi patients

The data thus far have shown that the current desensitization protocols make high-risk ABOi transplants feasible for both deceased- and living-donor recipients. However, there were also obstacles that could potentially restrict the broader implementation of ABO desensitization protocols, including C5 blockade. One major challenge frequently observed was an early post-transplant increase in isoagglutinin titers. Indeed, elevated ABO antibody levels were observed in eight of the 18 cases, and most cases occurred in the deceased-donor transplant population (six of eight cases).

As a consequence of exposure to an increased level of post-transplant ABO antibodies, AMR was frequently observed, with an overall incidence of 44 %. However, it was intriguing to find that AMR was also evident in patients treated with eculizumab when CP function measurement showed complete blockade of complement in hemolytic assay. In one case, this phenomenon was extremely clear: In the high-risk pancreas-kidney transplant recipient with a double blood-type mismatch (AB to O), consistent C5 blockade (indicated by 0 % CP function) was achieved by eculizumab during the first 2 weeks after the transplantation. However, despite complete terminal complement inhibition, the patient presented with AMR on day 9, with ongoing complement consumption (increased C3 turnover) as indicated by an elevated C3d/C3 ratio, paralleled by an increase in isoagglutinin levels. Given the persistent C3 consumption during C5 blockade, the rejection treatment was successfully completed with C1-INH (Berinert), which further normalized the C3d/C3 ratio, indicating disruption of the upstream complement activity [73].

Persistent upstream complement activity and cleavage of C3 during complete C5 blockade is a novel finding with strong clinical relevance, as further indicated by our data. Among the peptide fragments generated by C3 consumption, C3a is one of the most potent anaphylatoxins, not only stimulating T-cell proliferation and longevity but also promoting T-cell responses directed against alloantigens, illustrating the bridge between complement-driven humoral damage and cellular rejection [74–76]. Implementation of C5 blockade in the ABO-desensitization protocol enables high-risk ABOi living- as well as deceased-donor transplantation. However, the upstream complement activity and the occurrence of AMR during complete C5 blockade indicate the need for a more specific complement inhibition tackling the early stages of the complement cascade. This issue will be further discussed later in this chapter.

13.5 Desensitization for HLA-Incompatible Transplantation

13.5.1 HLA-Incompatibility

For a subgroup of highly sensitized patients, desensitization still remains the treatment of choice to enable transplantation. Currently, there are a variety of different treatment modifications to achieve desensitization, all sharing the same fundamental principles: (1) antibody elimination by plasma exchange or immunoadsorption; (2) maintenance immunosuppression, including T-cell-depletive induction (with alemtuzumab, thymoglobulin); (3) immunomodulation and antibody clearance with high-dose intravenous IVIG; and (4) reduction of the B-cell reservoir (with rituximab) and/or antibody production (with bortezomid) [77]. The overall purpose of combining these treatment modalities is to lower the level of anti-HLA donor-specific antibodies to a safe level at the time of the transplantation, a strategy that was previously discussed for ABO desensitization. Indeed, desensitization for HLA has recently been useful in providing survival benefit when compared to sensitized patients remaining on the dialysis [78]. However, following desensitization, prolonged exposure to DSAs still remains a challenge. Moreover, long-term antibody elimination (e.g., by plasma exchange or immunoadsorption) is not a realistic treatment option. As a consequence, desensitized patients are often faced with an increased risk of acute humoral, subclinical, or even chronic rejection, as well as a high incidence of transplant glomerulopathy at 1-year post-transplantation [40, 79].

13.5.2 Complement Modulation in HLA-Incompatible Transplantation

Given the importance of the complement system in antibody-mediated injury, the introduction of eculizumab into the conventional HLA desensitization protocol was a reasonable attempt to address the sustained humoral damage apparent in desensitized transplant patients. Initial experiences with the use of eculizumab were reported from the Mayo Clinic in the United States, concerning 26 highly sensitized patients with a positive pre-transplant cross-match, who underwent living-donor kidney transplantation using a protocol that included eculizumab induction with subsequent prolonged administration for 1 month post-transplant. At 3 months, the authors were able to demonstrate a 7.7 % incidence of biopsy-verified AMR, as compared to 41 % in the historical control, despite equal DSA exposure in both groups. Moreover, protocol biopsies at 1-year revealed a lower rate of transplant glomerulopathy in the eculizumab group than in the historical control [80].

Although these initial data are promising, the efficacy of eculizumab in treating AMR, in particular in the presence of prolonged DSA exposure, is controversial. Here, experiences on therapeutic intervention with exulizumab have mainly been accumulated from the Mayo Clinic again. In a series of 16 sensitized kidney transplant recipients with a positive cross-match, 38 % of the patients showed evidence of chronic humoral rejection after eculizumab treatment [81]. In a recent report on 30 sensitized transplant recipients, the incidence of microvascular injury, including glomerular and capillary inflammation (an indicator of humoral rejection), was 28 % at 3 months for eculizumab-treated patients [82]. Indeed, the incidence of microvascular injury in eculizumab-treated patients did not differ from that of patients given plasma exchange only, which indicates an inferior effect of terminal complement inhibition on sustained humoral damage. In line with these findings, the same group also demonstrated the occurrence of AMR despite complete therapeutic effect of eculizumab (as indicated by CP function) [83].

One possible explanation for the inconsistency of terminal complement inhibition in preventing humoral allo-responses could be the involvement of the upstream complement pathways. This hypothesis is supported by our results on the use of eculizumab in high-risk ABOi transplantations. As previously discussed, increased ABO antibody exposure after transplantation was highly associated with AMR and increased upstream complement activity, as indicated by an increase in C3 turnover. This upstream activity was also evident in patients presenting with AMR, despite having a full therapeutic level of eculizumab as indicated by CP function level. Indeed, in one patient, C1-INH was successfully added to target the upstream complement activity (as indicated by an elevated C3d/C3 ratio) when an AMR occurred during eculizumab treatment.

Given these accumulated data, the concept of upstream complement inhibition is gaining increasing interest. C1-INH, either in its recombinant form or as a concentrated extract from human serum, has been successfully used in pre-clinical models to prevent both allogeneic and xenogeneic humoral immune responses [84–87]. Currently, eculizumab and C1-INH Berinert (a concentrated extract from human serum) are the only approved drugs with indications for clinical complement inhibition. In a recent randomized, placebo-controlled clinical trial, C1-INH Berinert was introduced into a desensitization protocol including IVIG and rituximab for sensitized patients with pre-formed donor-specific antibodies and positive cross-reactivity (CDC−, FCMX+) [88]. Here, C1-INH or placebo was given as an induction treatment that was continued twice weekly for 21 days. At the 6-month follow-up, there was no difference in the overall incidence of AMR between the recipients of C1-INH and placebo. However, in contrast to the placebo group, AMR was absent from the CI-INH-treated patients during the treatment period. Moreover, the author observed a reduction in complement-binding anti-HLA antibodies (by C1q analysis) in the CI-INH patients when compared to the control patients; however, data on the incidence of microvascular injury were not presented. Overall, these initial data are promising, but further larger studies are warranted to validate the potential advantages of C1-INH over C5 blockade.

13.6 Current and Future Perspectives

Although eculizumab has been useful for the prevention and/or treatment of AMR, it is not a universal remedy. In view of potential importance of upstream complement activity in transplantation across humoral incompatibility, exploration of alternative complement-specific therapeutics has become of significant clinical value. Currently, the armamentarium of complement therapeutics is limited to eculizumab and C1-INH (Berinert); C1-INH is being evaluated for its safety and efficacy in sensitized transplant recipients. Notably, there is a growing interest in the development of new strategies aiming to block complement activation at the level of C3 [31]. Indeed, there are preclinical data indicating that therapeutic intervention at the level of C3 strongly prevents AMR and promotes accommodation.

The Yunnan-cobra venom factor (Y-CVF) is a cobra venom and a structural and functional analog of complement factor C3 [89]. Y-CVF has the ability to bind factor B and to form the complex Y-CVF/Bb. Y-CVF/Bb acts as a convertase, cleaving both C3 and C5. Since Y-CVF/Bb is not easily deactivated by endogenous regulators such as protein H and I, sustained consumption of C3 results to its total depletion and an overall complement inhibition [90]. In a recent study, Y-CVF was combined with standard CNI-based triple immunosuppression in skin-presensitized rhesus monkeys undergoing kidney transplantation [63]. Y-CVF was given daily beginning on day −2 to achieve complete C3 depletion by the time of the transplantation. Post-transplantation, the treatment was repeated every other day for a period of 2 weeks. While, non Y-CVF animals lost their grafts at a median of 3 days, long term survival was achieved in the Y-CVF group. Indeed, three of five animals had still normal serum creatinine after more than 715 and 1,000 days, respectively. The reasons for the loss of the remaining grafts were hemorrhagic complication post-biopsy (day 41) and euthanasia (day 140) because of a poor general condition. Moreover, AMR was completely absent in the Y-CVF group, despite the persistence of donor-specific antibodies. Taken together, these encouraging data indicate that short tem systemic interception of C3 might be beneficial for the prevention of AMR and the promotion of accommodation in allogeneic transplantation.

Compstatin peptides are a family of cyclic synthetic peptides that bind to C3 and prevents its activation by convertases [91]. Therefore, they inhibit he initiation, amplification, and the terminal sequence of all three activation pathways. The current lead analog (Cp40) features a nanomolar $IC_{50,}$, and it binds C3b with sub-nanomolar affinity [92]. In *in vitro* and non-human primate studies, compstatin has been successfully used for a wide range of clinical indications that feature uncontrolled C3 activation, such as sepsis [93], experimental hemodialysis [94], age-related macular degeneration [95], paroxysmal nocturnal hemoglobinuria [96], and periodontal disease [97]. A thorough survey of the molecular development of compstatin and its potential clinical applicability is available [98].

In the context of desensitization across humoral incompatibility, future protocols including C3 blockade may constitute an attractive treatment option that could potentially fill the therapeutic gaps in the current desensitization strategies, including terminal complement inhibition. Considering the accumulating data on C3 interception thus far, future strategies might not only be more efficient in preventing AMR but also in promoting accommodation and long-term survival. The future may not be far away, since the Cp40-based therapeutic AMY-101 (Amyndas Pharmaceuticals) is currently being evaluated for clinical trials in uremic patients undergoing ABOi kidney transplantation.

Clearly, there is a current surge in novel treatment strategies as well as therapeutic interventions tailored to enriching the treatment alternatives for sensitized patients awaiting transplantation. However, successful implementation of these novel treatment strategies requires the knowledge and professional competence of the medical community to ensure sound and reliable clinical judgments concerning the adoption and appropriate of these therapies. It is conceivable that close interdisciplinary collaborations between physicians, transplant surgeons, immunologists and immunogenetic laboratories will remain important for the provision of adequate medical care to high-risk sensitized patients with terminal end-organ disease.

Acknowledgments We thank Dr. Deborah McClellan for excellent editorial assistance. This work was supported by grant 2013-65X-05647-34-4 from the Swedish Research Council (VR), the European Community's Seventh Framework Programme under the grant agreement n°602699 (DIREKT), the Kidney Foundation (Swedish Association of Kidney Patients), the Eriksson and Bergholm foundations, and faculty grants from the Linnaeus University.

References

1. Druml W, Druml C. Emerich Ullmann (1861–1937): not only a pioneer of kidney transplantation. J Nephrol. 2004;17(3):461–6.
2. Legendre C, Kreis H. A tribute to Jean Hamburger's contribution to organ transplantation. Am J Transplant. 2010;10(11):2392–5.
3. Merrill JP, Murray JE, HARRISON JH, GUILD WR. Successful homotransplantation of the human kidney between identical twins. J Am Med Assoc. 1956;160(4):277–82.
4. Hamburger J, Vaysse J, Crosnier J, Auvert J, Lalanne CM, Hopper J. Renal homotransplantation in man after radiation of the recipient. Experience with six patients since 1959. Am J Med. 1962;32:854–71.
5. Merrill JP, Murray JE, Takacs FJ, Hager EB, Wilson RE, Dammin GJ. Successful transplantation of kidney from a human cadaver. JAMA. 1963;185(5):347–53.
6. Murray JE, Merrill JP, Harrison JH, Wilson RE, Dammin GJ. Prolonged survival of human-kidney homografts by immunosuppressive drug therapy. N Engl J Med. 1963;268(24): 1315–23.
7. Kahan BD. Cyclosporine. N Engl J Med. 1989;321(25):1725–38.
8. Borel JF, Feurer C, Gubler HU, Stähelin H. Biological effects of cyclosporin A: a new antilymphocytic agent. Agents Actions. 1994;43(3-4):179–86.
9. Kino T, Hatanaka H, Hashimoto M, Nishiyama M, Goto T, Okuhara M, et al. FK-506, a novel immunosuppressant isolated from a Streptomyces. I. Fermentation, isolation, and physico-chemical and biological characteristics. J Antibiot. 1987;40(9):1249–55.
10. Ekberg H, Tedesco-Silva H, Demirbas A, Vitko S, Nashan B, Gurkan A, et al. Symphony - comparing standard immunosuppression to low-dose cyclosporine, tacrolimus or sirolimus in combination with MMF, daclizumab and corticosteroids in renal transplantation. Transplantation. 2006;82(1):83.
11. Rana A, Gruessner A, Agopian VG, Khalpey Z, Riaz IB, Kaplan B, et al. Survival benefit of solid-organ transplant in the United States. JAMA Surg. 2015;150(3):252–9.
12. Courtney AE, Maxwell AP. The challenge of doing what is right in renal transplantation: balancing equity and utility. Nephron Clin Pract. 2009;111(1):c62–7. discussion c68. Karger Publishers.
13. Hsu C-Y, Vittinghoff E, Lin F, Shlipak MG. The incidence of end-stage renal disease is increasing faster than the prevalence of chronic renal insufficiency. Ann Intern Med. 2004;141(2):95–101.
14. Biglarnia A-R, Bennet W, Nilsson T, Larsson E, Magnusson A, Yamamoto S, et al. Utilization of small pediatric donors including infants for pancreas and kidney transplantation: exemplification of the surgical technique and the surveillance. Ann Surg. 2014;260(2):e5–7.
15. Maggiore U, Oberbauer R, Pascual J, Viklický O, Dudley C, Budde K, et al. Strategies to increase the donor pool and access to kidney transplantation: an international perspective. Nephrol Dial Transplant. 2015;30(2):217–22.

16. Biglarnia A-R Minimizing risks and morbidity in live kidney donors. Acta Universitatis Upsaliensis, Uppsala; 2010.
17. Colvin RB, Smith RN. Antibody-mediated organ-allograft rejection. Nat Rev Immunol. 2005;5(10):807–17.
18. Lamb KE, Lodhi S, Meier-Kriesche HU. Long-term renal allograft survival in the United States: a critical reappraisal. Am J Transplant. 2010;11(3):450–62.
19. Farrar CA, Kupiec-Weglinski JW, Sacks SH. The innate immune system and transplantation. Cold Spring Harb Perspect Med. 2013;3(10):a015479-9.
20. D'Alessandro AM, Pirsch JD, Knechtle SJ, Odorico JS, Van der Werf WJ, Collins BH, et al. Living unrelated renal donation: the University of Wisconsin experience. Surgery. 1998;124(4):604–11.
21. Yarlagadda SG, Coca SG, Formica RN, Poggio ED, Parikh CR. Association between delayed graft function and allograft and patient survival: a systematic review and meta-analysis. Nephrol Dial Transplant. 2009;24(3):1039–47.
22. Baldwin WM, Ota H, Rodriguez ER. Complement in transplant rejection: diagnostic and mechanistic considerations. Springer Semin Immunopathol. 2003;25(2):181–97.
23. Damman J, Hoeger S, Boneschansker L, Theruvath A, Waldherr R, Leuvenink HG, et al. Targeting complement activation in brain-dead donors improves renal function after transplantation. Transpl Immunol. 2011;24(4):233–7. Elsevier B.V.
24. Lin T, Zhou W, Sacks SH. The role of complement and Toll-like receptors in organ transplantation. Transpl Int. 2007;20(6):481–9.
25. Chen G, Chen S, Chen X. Role of complement and perspectives for intervention in transplantation. Immunobiology. 2013;218(5):817–27.
26. Diebolder CA, Beurskens FJ, de Jong RN, Koning RI, Strumane K, Lindorfer MA, et al. Complement is activated by IgG hexamers assembled at the cell surface. Science. 2014;343(6176):1260–3. American Association for the Advancement of Science.
27. Ricklin D, Hajishengallis G, Yang K, Lambris JD. Complement: a key system for immune surveillance and homeostasis. Nat Immunol. 2010;11(9):785–97.
28. Nguyen HX, Galvan MD, Anderson AJ. Characterization of early and terminal complement proteins associated with polymorphonuclear leukocytes in vitro and in vivo after spinal cord injury. J Neuroinflammation. 2008;5(1):26.
29. Triantafilou K, Hughes TR, Triantafilou M, Morgan BP. The complement membrane attack complex triggers intracellular Ca2+ fluxes leading to NLRP3 inflammasome activation. J Cell Sci. 2013;126(Pt 13):2903–13. The Company of Biologists Ltd.
30. Zipfel PF, Skerka C. Complement regulators and inhibitory proteins. Nat Rev Immunol. 2009;9(10):729–40.
31. Ricklin D, Lambris JD. Complement in immune and inflammatory disorders: therapeutic interventions. J Immunol. 2013;190(8):3839–47.
32. Emlen W, Li W, Kirschfink M. Therapeutic complement inhibition: new developments. Semin Thromb Hemost. 2010;36(6):660–8. © Thieme Medical Publishers.
33. Mollnes TE, Jokiranta TS, Truedsson L, Nilsson B, Rodríguez de Cordoba S, Kirschfink M. Complement analysis in the 21st century. Mol Immunol. 2007;44(16):3838–49.
34. Nilsson B, Ekdahl KN. Complement diagnostics: concepts, indications, and practical guidelines. Clin Dev Immunol. 2012;2012(4):962702–11.
35. Zwirner J, Götze O, Begemann G, Kapp A, Kirchhoff K, Werfel T. Evaluation of C3a receptor expression on human leucocytes by the use of novel monoclonal antibodies. Immunology. 1999;97(1):166–72.
36. Nilsson UR, Nilsson B. Simplified assays of hemolytic activity of the classical and alternative complement pathways. J Immunol Methods. 1984;72(1):49–59.
37. Seelen MA, Roos A, Wieslander J, Mollnes TE, Sjöholm AG, Wurzner R, et al. Functional analysis of the classical, alternative, and MBL pathways of the complement system: standardization and validation of a simple ELISA. J Immunol Methods. 2005;296(1-2):187–98.

38. Sapir-Pichhadze R, Curran SP, John R, Tricco AC, Uleryk E, Laupacis A, et al. A systematic review of the role of C4d in the diagnosis of acute antibody-mediated rejection. Kidney Int. 2015;87(1):182–94.
39. Hamer R, Molostvov G, Lowe D, Satchell S, Mathieson P, Ilyas R, et al. Human leukocyte antigen-specific antibodies and gamma-interferon stimulate human microvascular and glomerular endothelial cells to produce complement factor C4. Transplantation. 2012;93(9):867–73.
40. Stegall MD, Chedid MF, Cornell LD. The role of complement in antibody-mediated rejection in kidney transplantation. Nat Rev Nephrol. 2012;8(11):670–8.
41. Sacks SH, Zhou W. The role of complement in the early immune response to transplantation. Nat Rev Immunol. 2012;12(6):431–42.
42. Farrar CA, Zhou W, Lin T, Sacks SH. Local extravascular pool of C3 is a determinant of postischemic acute renal failure. FASEB J. 2006;20(2):217–26.
43. Panda S, Ding JL. Natural antibodies bridge innate and adaptive immunity. J Immunol. 2015;194(1):13–20.
44. De Clippel D, Baeten M, Torfs A, Emonds M-P, Feys HB, Compernolle V, et al. Screening for HLA antibodies in plateletpheresis donors with a history of transfusion or pregnancy. Transfusion. 2014;54(12):3036–42.
45. Sheil AG, Stewart JH, Tiller DJ, May J. ABO blood group incompatibility in renal transplantation. Transplantation. 1969;8(3):299–300.
46. Patel R, Terasaki PI. Significance of the positive crossmatch test in kidney transplantation. N Engl J Med. 1969;280(14):735–9.
47. Williams GM, Hume DM, Hudson Jr RP, Morris PJ, Kano K, Milgrom F. Hyperacute renal-homograft rejection in man. N Engl J Med. 1968;279(12):611–8.
48. Kissmeyer-Nielsen F, Olsen S, Petersen VP, Fjeldborg O. Hyperacute rejection of kidney allografts, associated with pre-existing humoral antibodies against donor cells. Lancet. 1966;2(7465):662–5.
49. Opelz G, Döhler B. Effect of human leukocyte antigen compatibility on kidney graft survival: comparative analysis of two decades. Transplantation. 2007;84(2):137–43.
50. Vo AA, Lukovsky M, Toyoda M, Wang J, Reinsmoen NL, Lai C-H, et al. Rituximab and intravenous immune globulin for desensitization during renal transplantation. N Engl J Med. 2008;359(3):242–51.
51. Wolfe RA, Ashby VB, Milford EL, Ojo AO, Ettenger RE, Agodoa LY, et al. Comparison of mortality in all patients on dialysis, patients on dialysis awaiting transplantation, and recipients of a first cadaveric transplant. N Engl J Med. 1999;341(23):1725–30.
52. Meier-Kriesche H-U, Kaplan B. Waiting time on dialysis as the strongest modifiable risk factor for renal transplant outcomes: a paired donor kidney analysis. Transplantation. 2002;74(10):1377–81.
53. Leichtman AB, Cohen D, Keith D, O'Connor K, Goldstein M, McBride V, et al. Kidney and pancreas transplantation in the United States, 1997–2006: the HRSA breakthrough collaboratives and the 58 DSA challenge. Am J Transplant. 2008;8(4 Pt 2):946–57. Blackwell Publishing Ltd.
54. Tydén G, Kumlien G, Fehrman I. Successful ABO-incompatible kidney transplantations without splenectomy using antigen-specific immunoadsorption and rituximab. Transplantation. 2003;76(4):730–1.
55. Rydberg L, Bengtsson A, Samuelsson O, Nilsson K, Breimer ME. In vitro assessment of a new ABO immunosorbent with synthetic carbohydrates attached to sepharose. Transpl Int. 2005;17(11):666–72.
56. Becker LE, Süsal C, Morath C. Kidney transplantation across HLA and ABO antibody barriers. Curr Opin Organ Transplant. 2013;18(4):445–54.
57. Mamode N, Marks SD. Maximising living donation with paediatric blood-group-incompatible renal transplantation. Pediatr Nephrol. 2012;28(7):1037–40.

58. Opelz G, Morath C, Süsal C, Tran TH, Zeier M, Döhler B. Three-year outcomes following 1420 ABO-incompatible living-donor kidney transplants performed after ABO antibody reduction: results from 101 centers. Transplantation. 2015;99(2):400–4.
59. Takahashi K, Saito K. ABO-incompatible kidney transplantation. Transplant Rev. 2013;27(1):1–8.
60. Park WD, Grande JP, Ninova D, Nath KA, Platt JL, Gloor JM, et al. Accommodation in ABO-incompatible kidney allografts, a novel mechanism of self-protection against antibody-mediated injury. Am J Transplant. 2003;3(8):952–60.
61. Bannett AD, McAlack RF, Morris M, Chopek MW, Platt JL. ABO incompatible renal transplantation: a qualitative analysis of native endothelial tissue ABO antigens after transplantation. Transplant Proc. 1989;21(1 Pt 1):783–5.
62. Chopek MW, Simmons RL, Platt JL. ABO-incompatible kidney transplantation: initial immunopathologic evaluation. Transplant Proc. 1987;19(6):4553–7.
63. Chen Song S, Zhong S, Xiang Y, Li J-H, Guo H, Wang W-Y, et al. Complement inhibition enables renal allograft accommodation and long-term engraftment in presensitized nonhuman primates. Am J Transplant. 2011;11(10):2057–66.
64. Narayanan K, Jendrisak MD, Phelan DL, Mohanakumar T. HLA class I antibody mediated accommodation of endothelial cells via the activation of PI3K/cAMP dependent PKA pathway. Transpl Immunol. 2006;15(3):187–97.
65. Dehoux J-P, Gianello P. Accommodation and antibodies. Transpl Immunol. 2009;21(2):106–10.
66. Suhr BD, Black SM, Guzman-Paz M, Matas AJ, Dalmasso AP. Inhibition of the membrane attack complex of complement for induction of accommodation in the hamster-to-rat heart transplant model. Xenotransplantation. 2007;14(6):572–9.
67. Tan CD, Sokos GG, Pidwell DJ, Smedira NG, Gonzalez-Stawinski GV, Taylor DO, Starling RC, Rodriguez ER. Correlation of donor-specific antibodies, complement and its regulators with graft dysfunction in cardiac antibody-mediated rejection. Am J Transplant. 2009;9(9):2075–84.
68. Biglarnia A-R, Nilsson B, Nilsson Ekdahl K, Tufveson G, Nilsson T, Larsson E, et al. Desensitization with antigen-specific immunoadsorption interferes with complement in ABO-incompatible kidney transplantation. Transplantation. 2012;93(1):87–92.
69. Takahashi K. A new concept of accommodation in ABO-incompatible kidney transplantation. Clin Transplant. 2005;19 Suppl 14:76–85.
70. Eskandary F, Biesenbach P, Sandurkov C, König F, Schwaiger E, Perkmann T, et al. ABO antibody and complement depletion by immunoadsorption combined with membrane filtration – a randomized, controlled, cross-over trial. Nephrol Dial Transplant. 2014;29(3):706–14.
71. Glander P, Budde K, Schmidt D, Fuller TF, Giessing M, Neumayer H-H, et al. The "blood group O problem" in kidney transplantation – time to change? Nephrol Dial Transplant. 2010;25(6):1998–2004.
72. Biglarnia A-R, Nilsson B, Nilsson T, von Zur-Mühlen B, Wagner M, Berne C, et al. Prompt reversal of a severe complement activation by eculizumab in a patient undergoing intentional ABO-incompatible pancreas and kidney transplantation. Transpl Int. 2011;24(8):e61–6.
73. Biglarnia A-R, Tufveson G, Nilsson T, Karlsson-Parra A, Nilsson Ekdahl K, Nilsson B. The use of eculizumab in ABO-incompatible transplantation. Aegean Conf Proc. 2013;73:33.
74. Li K, Anderson KJ, Peng Q, Noble A, Lu B, Kelly AP, et al. Cyclic AMP plays a critical role in C3a-receptor-mediated regulation of dendritic cells in antigen uptake and T-cell stimulation. Blood. 2008;112(13):5084–94.
75. Strainic MG, Liu J, Huang D, An F, Lalli PN, Muqim N, et al. Locally produced complement fragments C5a and C3a provide both costimulatory and survival signals to naive CD4+ T cells. Immunity. 2008;28(3):425–35.
76. Sacks SH. Complement fragments C3a and C5a: the salt and pepper of the immune response. Eur J Immunol. 2010;40(3):668–70.

77. Iyer HS, Jackson AM, Zachary AA, Montgomery RA. Transplanting the highly sensitized patient. Curr Opin Nephrol Hypertens. 2013;22(6):681–8.
78. Montgomery RA, Lonze BE, King KE, Kraus ES, Kucirka LM, Locke JE, et al. Desensitization in HLA-incompatible kidney recipients and survival. N Engl J Med. 2011;365(4):318–26.
79. Lefaucheur C, Loupy A, Hill GS, Andrade J, Nochy D, Antoine C, et al. Preexisting donor-specific HLA antibodies predict outcome in kidney transplantation. J Am Soc Nephrol. 2010;21(8):1398–406.
80. Stegall MD, Diwan T, Raghavaiah S, Cornell LD, Burns J, Dean PG, et al. Terminal complement inhibition decreases antibody-mediated rejection in sensitized renal transplant recipients. Am J Transplant. 2011;11(11):2405–13.
81. Cornell LD, Gloor JD, Nasr SH. Chronic humoral rejection despite C5 inhibition after positive-crossmatch kidney transplantation (plus XMKTx). Am J Transplant. 2010;10:125.
82. Dean P, Park W, Cornell L, Schinstock C, Stegall M, William J. Early subclincial alloantibody associated injury in positive crossmatch kidney transplantation (+XMKTx) despite terminal complement blockade. Transplantation. 2014;98(supp 1):79.
83. Bentall A, Tyan DB, Sequeira F, Everly MJ, Gandhi MJ, Cornell LD, et al. Antibody-mediated rejection despite inhibition of terminal complement. Transpl Int. 2014;27(12):1235–43.
84. Hecker JM, Loss M, Klempnauer J, Winkler M. C1-inhibitor for therapy and prophylaxis of acute vascular rejection of porcine kidneys in cynomolgus monkeys. Transplant Proc. 2002;34(6):2382.
85. Kirschfink M. C1-inhibitor and transplantation. Immunobiology. 2002;205(4-5):534–41.
86. Tillou X, Poirier N, Le Bas-Bernardet SEP, Hervouet J, Minault D, Renaudin K, et al. Recombinant human C1-inhibitor prevents acute antibody-mediated rejection in alloimmunized baboons. Kidney Int. 2010;78(2):152–9.
87. Poirier N, Blancho G. Recombinant human C1-inhibitor inhibits cytotoxicity induced by allo- and xenoantibodies. Transplant Proc. 2008;40(2):581–3.
88. Vo AA, Zeevi A, Choi J, Cisneros K, Toyoda M, Kahwaji J, et al. A phase I/II placebo-controlled trial of C1-inhibitor for prevention of antibody-mediated rejection in HLA sensitized patients. Transplantation. 2015;99(2):299–308.
89. Kock MA, Hew BE, Bammert H, Fritzinger DC, Vogel C-W. Structure and function of recombinant cobra venom factor. J Biol Chem. 2004;279(29):30836–43. American Society for Biochemistry and Molecular Biology.
90. Vogel C-W, Fritzinger DC. Cobra venom factor: Structure, function, and humanization for therapeutic complement depletion. Toxicon. 2010;56(7):1198–222.
91. Sahu A, Kay BK, Lambris JD. Inhibition of human complement by a C3-binding peptide isolated from a phage-displayed random peptide library. J Immunol. 1996;157(2):884–91.
92. Qu H, Ricklin D, Bai H, Chen H, Reis ES, Maciejewski M, et al. New analogs of the clinical complement inhibitor compstatin with subnanomolar affinity and enhanced pharmacokinetic properties. Immunobiology. 2013;218(4):496–505.
93. Silasi-Mansat R, Zhu H, Popescu NI, Peer G, Sfyroera G, Magotti P, et al. Complement inhibition decreases the procoagulant response and confers organ protection in a baboon model of Escherichia coli sepsis. Blood. 2010;116(6):1002–10.
94. Reis ES, Deangelis RA, Chen H, Resuello RRG, Ricklin D, Lambris JD. Therapeutic C3 inhibitor Cp40 abrogates complement activation induced by modern hemodialysis filters. Immunobiology. 2014;3.
95. Chi Z-L, Yoshida T, Lambris JD, Iwata T. Suppression of drusen formation by compstatin, a peptide inhibitor of complement C3 activation, on cynomolgus monkey with early-onset macular degeneration. Adv Exp Med Biol. 2010;703(9):127–35. New York, NY: Springer New York.
96. Risitano AM, Ricklin D, Huang Y, Reis ES, Chen H, Ricci P, et al. Peptide inhibitors of C3 activation as a novel strategy of complement inhibition for the treatment of paroxysmal nocturnal hemoglobinuria. Blood. 2014;123(13):2094–101.

97. Maekawa T, Abe T, Hajishengallis E, Hosur KB, Deangelis RA, Ricklin D, et al. Genetic and intervention studies implicating complement C3 as a major target for the treatment of periodontitis. J Immunol. 2014;192(12):6020–7.
98. Mastellos DC, Yancopoulou D, Kokkinos P, Huber-Lang M, Hajishengallis G, Biglarnia A-R, et al. Compstatin: a C3-targeted complement inhibitor reaching its prime for bedside intervention. Eur J Clin Invest. 2015;45(4):423–40.

Index

J.D. Lambris et al. (eds.), *Immune Responses to Biosurfaces*,
Advances in Experimental Medicine and Biology 865,
DOI 10.1007/978-3-319-18603-0

MIX
Papier aus verantwortungsvollen Quellen
Paper from responsible sources
FSC® C105338

If you have any concerns about our products,
you can contact us on
ProductSafety@springernature.com

In case Publisher is established outside the EU,
the EU authorized representative is:
Springer Nature Customer Service Center GmbH
Europaplatz 3, 69115 Heidelberg, Germany

Printed by Libri Plureos GmbH
in Hamburg, Germany